MAP MEN

TRANSNATIONAL LIVES AND DEATHS OF GEOGRAPHERS
IN THE MAKING OF EAST CENTRAL EUROPE

MAP MEN

Steven Seegel

THE UNIVERSITY OF CHICAGO PRESS

CHICAGO AND LONDON

The University of Chicago Press, Chicago 60637
The University of Chicago Press, Ltd., London
© 2018 by The University of Chicago
All rights reserved. No part of this book may be used or
reproduced in any manner whatsoever without written
permission, except in the case of brief quotations in critical
articles and reviews. For more information, contact the
University of Chicago Press, 1427 E. 60th St., Chicago,
IL 60637.
Published 2018
Printed in the United States of America

27 26 25 24 23 22 21 20 19 18   1 2 3 4 5

ISBN-13: 978-0-226-43849-8 (cloth)
ISBN-13: 978-0-226-43852-8 (e-book)
DOI: 10.7208/chicago/9780226438528.001.0001

Library of Congress Cataloging-in-Publication Data

Names: Seegel, Steven, author.
Title: Map men : transnational lives and deaths of geogra-
phers in the making of East Central Europe / Steven Seegel.
Description: Chicago ; London : The University of Chicago
Press, 2018. | Includes bibliographical references and index.
Identifiers: LCCN 2017034931 | ISBN 9780226438498 (cloth :
alk. paper) | ISBN 9780226438528 (e-book)
Subjects: LCSH: Geographers—Europe, Central—
Biography. | Geographers—Europe, Eastern—Biography. |
Geographers—United States—Biography. | Penck, Albrecht,
1858–1945. | Romer, Eugeniusz, 1871–1954. | Rudnyts'kyi,
Stepan, 1877–1937. | Bowman, Isaiah, 1878–1950. | Teleki, Pál,
gróf, 1879–1941.
Classification: LCC G67 .S44 2018 | DDC 910.92/2437—dc23
LC record available at https://lccn.loc.gov/2017034931

♾ This paper meets the requirements of ANSI/NISO
Z39.48-1992 (Permanence of Paper).

Biographies are like seashells; not much can be learned from them about the mollusk that once lived inside them.

Czesław Miłosz, *Miłosz's ABCs* (2001)

I like maps, because they lie.
Because they give no access to the vicious truth.
Because great-heartedly, good-naturedly
they spread before me a world
not of this world.

Wisława Szymborska, "Map" (2011)

.

CONTENTS

ACKNOWLEDGMENTS

What started as inquiring chats about geography between Sweden, England, Germany, East Central Europe, and the Americas has now become a book. I am grateful to many people for these conversations, and to those who invited me to present the research in progress at various institutions. Venues include Uppsala University in Sweden, the Polish Academy of Sciences in Berlin, Harvard's Radcliffe Institute for Advanced Study, the Harvard Ukrainian Research Institute, Indiana University, Columbia University, Pomona College (CA), the University of Illinois at Chicago, University of California, San Diego, University of Münster, the Herder Institute for East Central European Research in Marburg, Center for Urban History of East Central Europe in Lviv, and the Prague University of Economics. Additional travel and research was made possible by grants from the American Geographical Society Library (AGSL), the Herder Institute, the Center for Urban History, and, on multiple occasions, the Faculty Research and Publications Board (FRPB) at the University of Northern Colorado.

For making use of archives, libraries, and special collections, my deepest debts are to James Stimpert, Kelly Spring, and Jackie O'Regan at Johns

Hopkins University; John Hessler at the Library of Congress; Jovanka Ristic, Kay Guildner, and Susan Peschel at the AGSL of the University of Wisconsin-Milwaukee; Ksenya Kiebuzinski at the University of Toronto; Chloe Raub and Richard P. Tollo at George Washington University; Barbara Bulat and Anna Graff at Jagiellonian University Library; Árpád Magyar and Gabriella Petz at the Archive of the Hungarian Geographical Society; László Csiszár at the Hungarian Academy of Sciences; György Danku at the National Széchényi Library; János Kubassek at the Hungarian Geographical Museum; Heinz-Peter Brogiato and Bruno Schelhaas at the Leibniz-Institut für Landeskunde in Leipzig; and Christian Lotz at the Herder-Institut in Marburg. Colleen Stewart hunted down many items through Interlibrary Loan. I owe a special thanks to Laura Connolly in Humanities and Social Sciences, and Joan Clinefelter, Nick Syrett, Fritz Fischer, and Diana Kelly in History for their continued support. I am grateful to all of my immediate colleagues for conversation, in particular Mike Welsh (and his early morning aroma of blueberry bagels), Nick Syrett, T.J. Tomlin, Robbie Weis, Corinne Wieben, and Aaron Haberman who suggested things to read. My graduate students Amber Nickell, Chandra Powers-Wersch, Nicole Farina, Brian Zielenski, and John Všetečka opened new avenues for me to explore.

In addition, I am lucky to have knowledgeable friends and colleagues who took time to offer critical feedback, support, and advice. Among them are Nathan Wood, Tim Snyder, Sue Schulten, Jeremy Crampton, Mark Bassin, Mike Heffernan, Geoffrey J. Martin, Ute Wardenga, Norman Henniges, Kelly O'Neill, Serhii Plokhy, Alfred Rieber, Kate Brown, Li Bennich-Björkman, Gunnar Olsson, Christian Abrahamsson, Elżbieta Święcicki, Debbie Cohen, Tarik Cyril Amar, Per Rudling, Susan Pedersen, Pamela H. Smith, Pey-Yi Chu, Padraic Kenney, Jeff Veidlinger, Hiroaki Kuromiya, Katie Hiatt Matilla, Patrick Michelson, Marina Mogilner, Keely Stauter-Halsted, Małgorzata Fidelis, Ilya Gerasimov, Laura Hostetler, Michael Müller, Kai Struve, Claudia Kraft, Guido Hausmann, Olga Bertelsen, Peter Haslinger, Agnes Laba, Matt Pauly, Kathryn Ciancia, Liliya Berezhnaya, Heidi Hein-Kircher, Amelia Mukhamel Glaser, Martha Lampland, Patrick Patterson, Masha Rybakova, Ed Beasley, Lynn Lubamersky, Karin Friedrichs, Ilya Vinkovetsky, Igor Barinov, Jan Surman, Ina Alber, Darek Gierczak, Oleh Shablii, Oleksandra Vis'tak-Posivnych, Barbara Romer, Iryna Matsevko, Andrij Bojarov, Sofia Dyak, and Karl Schlögel. My former Brown-Harvard mentors Pat Herlihy and the late Abbott (Tom) Gleason, Mary Gluck, Dominique Arel, Roman Szporluk, and Omer Bartov appear in one form or another on the pages.

Perhaps the book's most daunting challenge was to understand where Hungary fits in a global context, in and beyond East Central Europe. Back in 2010, I decided to take the plunge as a Slavist and learn Hungarian from scratch—from a classically trained punk musician. It turned out to be one of the most rewarding choices I have ever made. Köszönöm szépen to my patient language teachers Menyus Böröcz, Szilvia Kristofori-Bertényi, and Valéria Varga. Historians Balázs Ablonczy, István Deák, Holly Case, Monika Baár, Madalina Veres, Alexander Vári, Bálint Varga, Steve Jobbitt, and Iván Bertényi inspired me to keep plugging away. I am grateful to the two anonymous readers at Chicago, who graciously put aside busy schedules and pushed me to design a stronger book. I have been very fortunate to work with Mary Laur and Christie Henry at the University of Chicago Press, and to benefit from their wide knowledge, timeliness, and confidence in the project. Thanks to Morika Whaley for the indexing, to Michael Koplow for his excellent work in copyediting, and to Rachel Kelly and Miranda Martin. Not least, there is Annika Frieberg, my co-shuffler who relived these lives and deaths together with me. This is a story for you.

AUTHOR'S NOTE

On the inevitable issues posed by variable toponymy in East Central Europe, there are cases in which a place has multiple names or has been contested. In general, I try to adhere to the views of the person I am discussing, while also retaining this variability. I employ parentheses or slashes (Szatmár/Satu Mare, Kiev/Kyiv, or Lemberg/Lwów/Lviv/Lvov) in reference to languages as they appear in the context of the story, or where there is a fair amount of ambiguity.

All translations in the text from French, German, Hungarian, Polish, Russian, and Ukrainian are my own, and I remain accountable for all errors of selection, fact, and interpretation.

INTRODUCTION

In 1902, his first year as an undergraduate student at Harvard, the brilliant geographer Isaiah Bowman had just one suit, and because of the school's dress code, he had to wear it all the time. Bowman, a naturalized U.S. citizen, came from rural poverty in southern Ontario's farmlands. At Harvard, he earned just $2,500, a modest stipend.[1] His mentor, William Morris Davis—himself a cofounder of the Association of American Geographers (AAG) and the charismatic force behind the professionalization of geography in the United States—recognized Bowman's limited means and even ventured to have breakfasts arranged for him. In fall 1904, Davis introduced him to the Vienna-based Albrecht Penck, a pioneering geomorphologist, his visiting colleague. The Harvardite placed great trust in his ambitious protégé, whom he handpicked to be Penck's personal assistant. Penck and Bowman together labored through heavy German to English translations for the Lowell Memorial Lectures, each day for at least two hours. In the end, Albrecht was deeply grateful. Once the famed professor took leave, he gave Isaiah a generous personal gift in an envelope. Inside of it was money the young geographer needed badly. He used it immediately to buy a second suit.[2]

Bowman's maps and clothes defined him at least as much as his ideas, research jaunts, pedigrees, or these homespun tales. The American Bowman—who eventually succeeded Davis as director of the American Geographical Society (AGS) in 1915 and became the U.S. chief territorial specialist in 1918 in advance of the Paris Peace Conference and later the president of Johns Hopkins University—went eastward to be free, a frontiersman eager to accomplish great things in an uncertain global world. This book takes a fresh, skeptical look at such stories, his life and work, by contacts he had with other like-minded geographers of East Central Europe, scientists in search of a deeper purpose. They gained knowledge of the earth by skill and luck, dedication and chance. Beyond their countries, Bowman, Davis, and Penck were groomed into an emergent middle class and its establishment values. They held fast to reason and progress, family and place, yet neither their maps nor their selves were ever completely rational or modern. The geographers were what I will call "map men," an educated class sharing in the projects and goals of forging a place called home, crafting narratives of belonging in life and in death, by the dissemination of maps as graphic tools and anxiously performed markers of multigenerational status and self-hood. Those who are featured were a bevy of contradictions. They combined defenses of Western civilization with biological racism, anticommunism with anti-Semitism, discomfort in cities or academe with heightened provincialism, global applications of rights with sexism, liberal imperialism with frustrated revisionism.[3] When young Bowman arrived at Harvard at the cusp of a new century, he was a backtracked Thoreau with a tie, part Alger myth and part Fitzgerald's Jimmy Gatz. All our tale spinners were educated within a network of eager service-oriented geographers ("map men" into the 1950s were mostly men, though not limited here to a biological category) who sent their works, maps, and letters across oceans and upwards to the powerful.

ARGUMENT: A TRANSNATIONAL LOVE STORY

At the heart of the spaces of map men's lives in each other's company was a giant, even utopian cooperative enterprise. Let us call it a love story, for dramatic effect. The deepest bonds they set up were confraternal, socially coded in the natural world's outdoor spaces. Map men were drawn together first by interactions in the name of scientific progress, such as the Penck-Davis exchanges and the AGS's Transcontinental Excursion of 1912. After World War I came the Paris Peace Conference of 1919 and congresses of the International Geographers' Union (IGU), which allowed interwar geographers (excluding those from German and revisionist powers) and facilitated long-distance exchanges of maps and personal letters. Through world war and revolution, private/public and personal/professional bonds became more complex, and in new ways. An aspiring class held on to sentimental attachments after 1918–19, the sure result of an upward spike in Europe's literacy from the nineteenth century, what Richard J. Evans has finely termed "the golden age of letter writing."[4] Maps were sent with letters to bridge gaps within a guild of white-collar experts. In cooperation and in conflict, maps encoded ideals of artificial civility and objectivity, all of which proved hard to sustain in an age of propaganda, empire and nationalism, party politics and public relations. The book's core argument is that interest in maps was often pathological, a sign of frustration and unfulfilled personal ambition along with a host of other emotions—fear, petty jealousy, and resentment—that nestled inside provincial, contradictory, and closed professional worlds of privilege, learning, and authority.

Sorting out this complicated love story from German-speaking East Central Europe, our protagonists learned to "speak map." They became world-class exhibitors of map-related tools. Throughout the book, I will look at what happens once we pull back the curtains of civility and dedication to shared scientific principles. The collective biography of five geographers in Germany, Poland, Hungary, Ukraine, and the U.S. rests on four additional supporting points about map men as a category of analysis—who they were, how they lived, where they traveled, what they loved, and how they died. These are, in turn: (1) the place-based *homo geographicus* who "spoke map"; (2) the basic, illiberal tasks of geographers and geography as a science; (3) stateside geography as revisionist; and (4) geography as affective, in and through letters.

First, the geographers in becoming professional represented mobile yet place-sensitive men, the *homo geographicus* in socially constructed spaces guided by institutional norms. Modern academics repackaged the ancient discipline of Herodotus, Ptolemy, and Strabo. Maps were symbols, texts, and artifacts. They functioned as markers of civilization, special graphic tools of power for replicating patterns of patronage and nationalist difference. Second, barely beneath geography as a science was an astounding amount of unsavory detail. Map men were anxious about their status as illiberal, provincial pre-1914 hyphenated Anglophile Germans (they all spoke German). Captivated by projects of soul-widening travel and mobility, they envisaged geography as a new mega-science. Third, their grasp of maps and geography was largely antimodern, anti-urban, and, in some cases, anti-Semitic as a defense of privilege and Europe's grand explorer tradition in East Central Europe, on top of colonial fantasies on frontiers. They reworked love of travel into twentieth-century "revision," getting into or out of territorial cages of a liberal international order and the post-1918 modern nation-state. Fourth, the maps they sent to each other were *affective*, not just rational tools. Contra Benedict Anderson, maps were less like national nineteenth-century censuses, museums, or planned-out grids than moody, messy Rorschach blots, trace psychological evidence left behind of transnationally based emotional worlds.[5] To "speak map" was to belong and not belong, perform in a surreal, visually charged language, and participate in the cross-border sciences of geography and cartography, as a spatial medium for intensely personal politics.

A FIVE-HEADED CAST: DEFINING MAP MEN

Map men may be defined as a coterie of professionals, aspirants (both men and women) who longed in old-fashioned ways for power and personal bonds, all while trying to fulfill generational dreams of security and adventure so characteristic of a roving intelligentsia of experts. In the book's dramatis personae, our main cast looks like this:

1. *Albrecht Penck* (1858–1945) of Germany was a famed geomorphologist who proposed the 1:1 million map of the world in Bern in 1891; chair of geography in Berlin starting in 1906; longtime friend of Isaiah Bowman and visiting professor in 1908–9 at Columbia University; colonial geographer and former rector of Berlin (now Humboldt) University; supported the kaiser's war aims in World War I; alerted the authorities in Vienna and Warsaw to the Polish atlas by his former student Eugeniusz Romer; advocated *völkisch* revision of the Treaty of Versailles and anti-Polish *Ostforschung* in the 1920s-30s; involved in Leipziger Stiftung in 1926–33; fled from the Allied bombing of Berlin in 1944; died in Prague in March 1945.

2. *Eugeniusz Romer* (1871–1954) of Poland had origins in an old Polish aristocratic family dating back to the fifteenth century; was one of Penck's students and his greatest rival; was chair of geography in late Austria-Hungary in Lwów starting in 1911; wrote the *Geographical-Statistical Atlas of Poland* (1916), which was smuggled to the West through neutral Sweden and used by President Wilson's U.S. Inquiry and at Paris in 1919; mapped the east at Riga in 1921; founded a Cartographical Institute in the 1920s for maps in Poland's Second Republic; was a lifelong friend of Bowman and coorganizer of the International Geographical Congress (IGC) in Warsaw in 1934; survived Nazi and Soviet occupations; hid in a monastery in Lwów from 1941–44; died in Kraków in January 1954.

3. *Stepan Rudnyts'kyi* (1877–1937) came from Ukraine, and was the country's most famous geographer; competed in pre-1914 Galicia with Romer; was also a student of Penck; was a disciple of Mykhailo Hrushevs'kyi and an advocate of Ukrainian independence in World War I and during the Polish-Ukrainian War of 1918–19; lost his academic position in interwar Poland and fled to Prague and Vienna; relocated to Soviet Kharkov/Kharkiv in 1925, where he worked in institutions for Ukrainian geography and cartography; was arrested in 1933, charged with "bourgeois" fascism during Stalin's purges; was executed on the Solovki Islands in November 1937.

4. *Isaiah Bowman* (1878–1950) was born in poverty in Canada, and became a U.S. citizen in 1899; studied geography at Harvard and Yale; was a lifelong friend of Penck, Romer, and Teleki; was an expert on settlement issues in South America in the 1900s; turned during World War I to political geography in East Central Europe; became director of the American Geographical Society in 1915; was appointed chief territorial specialist of Wilson's U.S. Inquiry in preparation for Paris; wrote *The New World* in 1921; was the leading light of the International Geographers' Union (IGU) in 1920s-30s; was president of Johns Hopkins University from 1935–48; was an advocate for the internationalization of geographic science and the United Nations in 1945; died in Baltimore in January 1950.

5. *Count Pál Teleki* (1879–1941) of Hungary was born in Budapest; was from an aristocratic political family in Transylvania dating back to the fourteenth century; rose to prominence in late 1900s as a traveler to the Sudan; researched Europe's early modern cartography of Japan; devised the "Carte Rouge" of 1918–19, based on population in Hungary; conservative advocate of anticommunist, anti-Semitic policies under Admiral Horthy; supported territorial revision for post-Trianon Hungary in the 1920s and 1930s; was prime minister of Hungary twice, the last time in 1939–41; committed suicide in Budapest in April 1941.

Our men were transnational Germans. At least this was how they came of age. In response to cataclysmic events, experts adhered to a common habitus and set of prescribed manners, within the civilizing confraternity of a scientific community.[6] By the mid to late nineteenth-century, however, colonial explorers became ever more domesticated as academic professionalization and geography's institutionalization took over.[7] When men actively sought out violence in their adult emotional lives, they embraced colonial fantasies of power and conquest, as Adam Hochschild has illustrated in his Conradian biography of King Leopold II (r. 1865–1909) of Belgium.[8] Psychohistory can take us only so far, however.[9] Love among our men did not always lead to modern violence, nor was it necessarily a desire for sex or power, or same-sex love. Homosocial or queer male friendship on frontiers, which usually tended to exclude women, was never confined to the erotics of "modern" city space or (sub)cultural life.[10] Yet bonds and contacts were quite intense. Maps, like history in general, became stories of a gendered place and time, dressed up aesthetically in scientific garb.[11]

If confidence in civilization and progress was shaken first by the political earthquakes of World War I, geographers limped along after 1918–19 as frustrated border crossers and treaty arbiters in motion, part of a past century's transformative contact in new sciences. In an age of fast-developing technology, maps were made accessible by optimized space-time cultures, not to mention postal services.[12] Explorers turned into academics. They belonged to a club of coded colonials, expert-centric internationals in an era of antagonism, prior even to the League of Nations or United Nations.[13] Once frontiers closed or were limited by political borders, holdover myths and dreams of adventure persisted.[14] As ominous prejudices emerged such as the "judeobolshevik" in the exclusionary national maps of East Central Europe of 1917–21, territory met geography and geography intersected with the geopolitics of nation-states, a "spatial turn" into the 1920s and 1930s.[15] Yet bodily metaphors against perceived threats of "penetration" persisted, especially when revisionist anger directed at the Paris settlements and borders of interwar nation-states was hitched to geopolitics.[16] Under such circumstances, our men took refuge in their maps, their letters, and each other—for instance, many made their wives and children into centers (and centrisms) of emotionally mapped worlds. Until their deaths in the late 1930s to early 1950s, our map men still held stubbornly to the idea that geography was fully rational, a science unaffected by desire or the octaves of human emotions.

EPISTOLARY GEOGRAPHY

The book's source base relies strongly on letters, memoirs, archival documents, and reviews and writings as they appeared in geographical journals in English, French, German, Hungarian, Polish, Russian, and Ukrainian. I read back from scientific literature to see what has slipped between the cracks. Geographers expressed worlds in unpublished diaries and letters as they became interested in, and obsessed with, various map technologies. Geographers tried, and most of them failed, to assimilate and narrate myths of origin by place into twentieth-century nation-states, out of dynastic Europe's pre-1914 empires from which they emerged.[17] As persons aspiring to join the ranks of experts, they defy easy categorization by nationality, or as nationalists, conservatives, (proto)fascists, or geopoliticians.[18] Reading back from their articles, speeches, and reviews shows the instability of the performed myths they managed, and the everyday pressures they faced, for leading publicly respectable lives. Letters once thought lost or irrelevant to résumés and careers reveal quotidian truths about "identity" as a limited category of analysis.[19]

As the book's method of choice, I define epistolary geography as a spatial strategy for charting out the biorhythms of mobile professionals' lives, a place-sensitive, transnationally source-based means of historical study. By delving back into this kind of subjugated knowledge, the intimacy of geography is shown in a new way. Letters show uncertain historical paths to assimilation, how insecure map men hid behind science in networks of power. Some manicured or destroyed confidential items, as in the cases of Teleki, Penck, and Bowman. What they left behind as "legacies" do as much to conceal their personalities as to define them. Men and maps offer a history of science as emotion, not just tales of heroic experts, founding fathers, or objective Doktorvater academics with résumés to be lauded by acolytes jockeying for position. In their attachments, map men were acutely aware of their status, if not always of broader politics. They defended privilege for their "sons" in a biological and academic sense, in emergent subdisciplines of twentieth-century geography and cartography as information science, and later digital technology.

In transnational East Central Europe, I draw from a number of studies to bring out how the map men communicated. Guntram Henrik Herb's erudite analysis of German cartography from 1918 to 1945 is a fine place to start.[20] Peter Haslinger's comprehensive study of late imperial Habsburg and Czech geography and geographers treats frames for geopolitics smartly in intersubjective frames, for maps were part of mediated struggles that figured heavily in cross-border territorial ideologies and claims.[21] Pieter Judson's research on language activists is one of my keys to understanding frontier diversity, and, following Tara Zahra and James Bjork, the layers of national indifference in Central European lands.[22] Deborah R. Coen has innovatively challenged the private/public dichotomy of lives in her history of the Exner family of scientists in late imperial Vienna.[23] Guido Hausmann has looked at how and why the racial/colonial practices of geographers in Europe's East became integral to the politics of World War I.[24] Willard Sunderland focuses on problems posed by life stories in war and revolution across Eurasia, the issue of how much we can speculate about any one person's inner world.[25] Kristin Kopp offers a convincing feminist postcolonial critique of German "othering" of Poland in the literary, cultural, and geopolitical discourses of map-related fantasies, particularly in the 1920s.[26] Amir Weiner and Peter Holquist focus on modern tools of intelligence and population politics, and on how Europe's techniques of landscape management defined difference after two world wars and during the Cold War.[27] Short of having a clear identity, what we learn about geographers is that they all had flaws and vices. Vices made them vulnerable men. To turn Robert Musil on his head, map men were *men with qualities*.

It's a truism nowadays to say that maps are socially constructed—if nothing else, a way for academics to find solidarity, escape traps of nations, or just congratulate each other with their shared conventional wisdom. This book takes things a step further, dealing with what Matthew Edney calls "processual map history."[28] Maps give us spatial snapshots of life stories and risky human choices.[29] My study shows map men in motion, not of a place but moving through places. I call for a new spatial history of maps less as censuses or grids than as scarcely effable affect and fantasy, an epistolary geography of coded images.[30] I follow the lead of insights by feminist geographers such as Doreen Massey, Pamela Moss, and Lynn Staeheli, to rework canons of knowledge into cross-disciplinary (auto)biographies of place and space.[31] This is what I mean by "transnational," a word that does not equal modern or twentieth-century history, as Jürgen Osterhammel's geographically sophisticated nineteenth-century history of an interconnected world illustrates.[32] I combine Zygmunt Bauman's notion of liquid modernity with Erving Goffman's dramaturgy of self and audience and Judith Butler's thoughts on how identities are subverted when performed.[33] Seeing geography as a sensory history of encounter, I move away from modernity toward "geographies of loss" and the tribulations of groundlessness, the all-too-human feeling of never finding comfort in one's place or one's own skin.[34] The book thereby offers a *transnational* perspective as Sven Beckert outlines the term, of lives that "transcend . . . any one nation-state, empire, or politically defined territory."[35]

TRIPTYCH

By "Ostmitteleuropa," I take leave of Milan Kundera's chimerical space and focus mainly on Germany, Poland, Hungary, and Ukraine in the book. Among the experts, Rudnyts'kyi fell victim to Stalin's purges in 1937. Teleki committed suicide in 1941. Of the five men, three had papers that were partially or fully destroyed—Teleki (likely by his own hands) in 1941, Penck (probably by his acolytes, though they kept his relics in a cabinet or "Penckschrank") in

1945, and Bowman (partly by his family) after 1950, before they were donated to Johns Hopkins University. After Rudnyts'kyi's death in 1937 and Romer's in 1954, communist authorities censored their works and maps. Today, Romer's archive is at the Jagiellonian University Library in Kraków, a city where he did not live until after the Second World War, displaced there in the last nine years of his life. Penck's salvaged papers, lost in war and transit from Berlin after his death, today are at the Leibniz-Institut für Landeskunde in Leipzig, brought there by the Berlin scholar of geography, Norman Henniges. To date, only Rudnyts'kyi's full correspondence, due in part to venerations of him as a national scientist in independent Ukraine, has been published in full, in German and Ukrainian.[36]

I use "East Central Europe" interchangeably with "Ostmitteleuropa." I only wish that I could cover every major geographer there, but my focus is on *character* in life and death, done with an aim to disentangle lives from modernity and identity discourse, and remove reduction of groups by "national" language.[37] The fact that a place or country has or once had many ethnicities does not make it diverse, much less tolerant of diversity. Frankly, I conceive of diversity as a range of social perspectives on justice, not just mapped by language; in any case, most of the book's geographers are bi- or multilingual. It's true nonetheless that I could include more geographers writing in any number of European languages, some not represented by a state of their own. A lot more can be done. A fine Polish monograph on multilingual professors' wars by Maciej Górny, a postcolonial study of ethnocentrism among geographers, anthropologists, psychologists, and race specialists between 1912 and 1923 in Central and Southeastern Europe, is particularly impressive in this regard.[38] As I introduce my supporting cast, I hope to make the reader more skeptical of national-heroic and literalist readings of lives and maps. Especially in Europe's east, one faces the difficulty of storytelling in the letters, maps, and other documents that sometimes do not survive.

The triptych scans out as follows. Chapter 1 begins with Professor Penck's pupils and demonstrates how their dreams came to be anchored in provincial lives, friendships, and travel plans before 1914. Chapter 2 shows how World War I and revolution crystallized geographers' prejudices. Chapter 3 looks at the map men's changed confraternity in its trans-Atlantic reach, through letter exchanges at the heart of the Bowman-Romer friendship after 1918–19. Chapter 4 studies how revisionism became institutionalized in Germany and Hungary as a protest against the treaties of Versailles and Trianon. Chapter 5 looks at efforts to bring back the pre-1914 confraternity of geographers via the International Geographers' Union (IGU) from 1928 to 1934, culminating in Bowman's efforts to heal the rift between Penck and Romer. Rudnyts'kyi's transnational life came to an abrupt end. In chapter 6, geographers experienced disappointment, exile, and death where they tried to chart a future. Chapter 7 reexamines the map men generationally through World War II, in the cases of Penck (d. 1945), Bowman (d. 1950), and Romer (d. 1954). All of our protagonists collapsed politics into hidden sagas of the *homo geographicus*, a reflection of their age, their maps, and the tense professional lives they led.

CHAPTER ONE

PROFESSOR PENCK'S PUPILS

Many of Europe's greatest nineteenth-century adventurers, men like Alexander von Humboldt, Cecil Rhodes, and David Livingstone, found their bearings as geographers on the frontiers of America, Africa, and Asia before 1914. Their achievements are noteworthy, but the formative encounters of such explorers are often neglected.[1] A key one came on 24 August 1912, when the Transcontinental Excursion of the American Geographical Society (AGS) began. Sponsored by the organization founded in New York City in 1851 after similar societies were established in Paris (1821), Berlin (1828), London (1830), and St. Petersburg (1845), the trip was formally kicked off by the U.S. government's incorporation and recognition of Alaska in 1912, purchased for $7.2 million from Russia in 1867. The gentlemen of the AGS celebrated over sixty years of existence by inaugurating a new building at Broadway and 156th Street, eventually the headquarters for President Wilson's U.S. Inquiry of 1917, a team of experts responsible for remapping a postdynastic world. Steeped in colonial geography, the American organizers scheduled the two-month trip with the goal of studying environmental landscapes, also bringing into personal contact forty-three European and seventy U.S. men within specialized fields of knowledge. The Harvard professor William Morris Davis handpicked a young Bowman, just back from his second research stint in South America, to serve as one of the three lead marshals. Scores of British, Italian, German, Austrian, and Russian geographers arrived with their expertise and fantasies of America, to discover landscapes pleasing to the eye, and apply for the first time the emerging subdisciplines of geography.[2]

Those involved in the AGS 1912 excursion were pupils of each other, lovers of nature, and products of a century's firm belief in progress and the advancement of geography. The fellowship of geographers was a confraternity of scientists across borders. Many would become intimate friends. Like Penck in German-speaking Central Europe, Count Teleki was eager to learn things abroad. He traveled with his friend and Hungarian compatriot, the geographer Jenő Cholnoky (1870–1950). Eugeniusz Romer arrived in New York from Poland, also representing Austria-Hungary in the register. Theirs was a long, exciting, wending voyage. When Teleki and Cholnoky left Budapest, they embarked on a journey by land and sea that would cover around thirteen thousand miles.[3] In the United States, the Hungarians were most impressed by the cities of Seattle, Tacoma, Portland, San Francisco, Denver, and Santa Fe. Teleki especially loved the Grand Canyon. Isaiah Bowman was in awe of the leading lights whose works he had read as a graduate student under Davis at Harvard. They were kindred experts and diviners of the world's great outdoors: Teleki and Romer, Eduard Brückner (1862–1927) of Austria, Emmanuel de Martonne (1873–1955) and Lucien Gallois (1857–1941) of France, the Germans Joseph Partsch (1851–1925) and Harry Waldbaur (1888–1961).[4] These men belonged to Europe's grand explorer tradition as it evolved into a new multidisciplinary profession in the twentieth century. As an elated Davis and Bowman secured their contacts to develop American geography from the U.S. East Coast westward, the men expanded the enterprise quickly and with as much alacrity as their predecessors Humboldt, Karl Ritter, and Friedrich Ratzel had done in East Central Europe's recent nineteenth-century past.

SAXONY, 1858

Whenever Isaiah Bowman looked for inspiration to Ostmitteleuropa, he had a heroic German geographer of choice. That man's name was Friedrich Karl Albrecht Penck, born 25 September 1858 in Saxony, in the Reudnitz district of eastern Leipzig (figure 1.1). The Penck family was steeped in the region's history. Ludwig Emil, Albrecht's father, was born in Dresden in 1829 and moved

FIGURE I.I. Albrecht Penck (1858–1945) and his rock collection in Berlin. [No date, likely before 1914.]

to Leipzig as a young man, where he became a successful local book merchant. Elisabeth Starke, Albrecht's mother, came from a small town named Pillnitz. Both of Penck's parents were devout Christians, from Reformed evangelical Lutheran families. We learn the history primarily from Albrecht's memoirs in Berlin during World War II, in which he began with a *Heimat* saga of eighteenth-century provincial Saxony. Penck told sentimental tales of kinship similar to Bowman's, of a moral ascendancy into German academe and his social rise into a professional bourgeoisie. Penck's paternal grandfather, Ludwig Friedrich, had been a papermaker living in the village of Ilfeld (today Nordhausen), also from Saxony (Thuringia after 1946). Such genealogies are tantalizing and often uplifting, yet incomplete. In search of *Heimat*, Albrecht actually did not know the family tree, or elected not to unearth it, for any "premodern" part or member of his family. Boundaries therefore are not clear. This stylized memory was a search for the comforts of provincial place, which in Europe's age of industrial modernity and war was often imagined, through cultural geography, by strained romance and mythic continuity.[5]

What was known was that the Pencks from the 1850s to the 1870s lived in Leipzig-Reudnitz and attended Lutheran services there. After Albrecht, Ludwig and Elizabeth had two daughters, Johanna, who was called Hanni (1862–1948), and Elsbeth (1868–1930). Protestant Saxony was the locus of Penck's identity, integral to the opportunities and privileges he enjoyed as a young, white European man.[6] An affluent Protestant donor to Scandinavia, Auguste de Wilde of Leipzig, financed his early schooling away in Munich and his first trip abroad. She made confessional identity a precondition of eligibility.[7] Young Albrecht was precocious, willful, and goal-oriented. Inspired by Humboldt, Ritter, and Ratzel, he soon found a calling in the study of geology and geography. He entered the University of Leipzig in 1875, rather close to home, at age seventeen where he came into a German ethos of specialization in the natural sciences.[8] He published his first paper in 1877 on glacial deposits, then went to

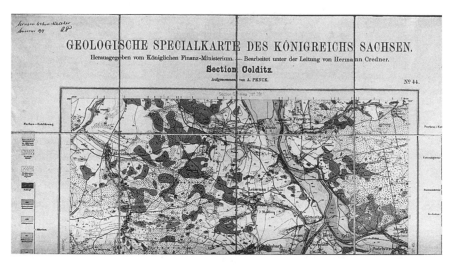

FIGURE 1.2. Albrecht Penck, *Geological Survey Map of the Kingdom of Saxony, Colditz Section* (no. 44), published by the Royal Ministry of Finance in 1879. Prepared under the direction of Hermann Credner, lithographed and printed by the firm of Giesecke & Devrient in Leipzig. Penck also prepared a similar 1:25,000 map for the Grimma Section (no. 28). Courtesy of the Archiv für Geographie, Leibniz-Institut für Landeskunde, Leipzig, Germany.

work for the geological survey of Saxony in 1878, for which he prepared some of his earliest maps (figure 1.2). In 1879, Penck published a major research paper on the formation of boulder clay in the German lowlands. Twenty years after Charles Darwin's *Origin of Species* and midway between science and religion, Penck articulated geosophical and biogenetic explanations for the origins of Nordic and Alpine landscapes. Discoveries of deposits led him to suggest an ice sheet's threefold movement into northern Germany.[9] When he found more boulder clay deposits near his family's home in Leipzig, he argued that their origins were not local, but in Scandinavia. Thus began Penck's lifelong global-to-local Nordophilia. Decades later, in 1905, he was honored by the Royal Swedish Academy of Sciences as one of their own. Penck in his life would become a close friend and ally of the pro-German Swedish geographer Sven Hedin (1865–1952), the Eurasian explorer and anti-Semitic advocate (Hedin was actually part Jewish) for German militarism during the two world wars.[10]

Moving into the confraternity, Penck made his transnational career as a German geographer out of looking wider—and, when possible, being elsewhere. Leaving home in Saxony, the Protestant man became a creature of Prussian and Habsburg German-speaking Ostmitteleuropa.[11] A geographer for Penck was a highly educated man, not power-seeking but an objective scientist, an academic who explained terrains. German geographers as such were wise experts, not mandarins in a closed caste but plein-air explorers of nature, Europeans open to outdoor laboratories. Appointed to positions of privilege, they had a civic duty to serve their governments. Penck's ascent into modern geography followed Otto von Bismarck's three wars of the 1860s and early 1870s, against Denmark, Habsburg Austria, and France. Historians' quarrels about post-1848 Germany's Sonderweg (special path) notwithstanding, the two Lutherans had something in common. Penck owed his authority to maps, states, and censuses, the mid to late nineteenth-century projects for grouping populations by confession and nationality in order to fix boundaries, control subjects, describe people statistically, and develop a common economic space.[12] In Prussian, Saxon, and Habsburg lands, colonial map men like Penck joined in the ventures of geographical societies. This took place just as departments were formed and even more chairs of geography in Europe were created, the first one being for Ritter in Berlin in 1828.

In the life spans of individuals and countries, such modern or illiberal parallels after the 1848–49 revolutions beg for inspection. The geographer Penck did not serve in the military like Ritter. Nor was he of noble birth, or a poet-scientist in the mold of Humboldt. Rather, he secured an ivory tower habitus after Bismarck's project of German unification and the Franco-Prussian War of 1870–71. Disciplinary knowledge made him portable. He was also no amateur. He became a Privatdozent, equivalent to an assistant professor, in 1883 at the University of Munich at the age of twenty-seven. In 1885–86, Penck was appointed chair of physical geography at the University of Vienna, after the retirement of the renowned alpinist Friedrich Simony (1813–96). Touted for his research in geomorphology, he focused on the Ice Age in German valleys of the Alps and broadened himself into geology, climatology, and glaciology. Penck was not alone in his Nordic theories of glacial deposits; other geologists of this school in the 1880s were his colleague Eduard Brückner (1858–1945) and the Scandinavian academics Otto Torrell (1828–1900) and Gerard de Geer (1858–1943).[13] In 1887, he coauthored *Das Deutsche Reich* with his friend Alfred Kirchhoff (1838–1907), the chair at the University of Halle since 1873, a major study of the Second Empire's geography. The survey framed Bismarck's unified *kleindeutsche* Germany as a sum of its regions, in effect Europe's newest empire. In Penck's Prussian-Saxon harmony of man and nature, German geography was an aspirational world science.[14] The German tongue, the poetic language of high learning and of Goethe, Schiller, and Humboldt, coordinated everything into an organic unity, a political cosmology that was also common globally to late imperial cultures.[15]

Propelled by these new opportunities and ideas in the 1880s, Penck then made a great modern discovery of something else, the U.S. West from afar.

From boyhood, he had loved *Heimat* literature and the adventure stories of Karl May (1842–1912), devouring the tales of Old Shatterhand and Winnetou. He read May's cinematic fictions of cowboys and Indians, stock characters on the frontier. May's pulp storytelling combined racial escapism with European ethnocentrism for a middle-class readership of his era.[16] Nature for Penck was geo-coded in German space by ethnicity. Public patriarchal norms seeped into private life, where distant lines and local places of his parents' family romance blurred. In 1886, a year after he had been appointed chair of geography in Vienna, he married Ida von Ganghofer, sister of the successful *Heimat* novelist Ludwig von Ganghofer (1855–1920). Ida and Ludwig were children of August von Ganghofer, the powerful ministerial councilor of Bavaria. Ludwig endowed the mountainous peoples in Alpine and Tyrolean climes with virtue, evoking a fertile German south and east. Penck adored this literature in a dark age of empire and industry. His frontier space blurred into pastoral idyll. Bavaria, Saxony, and rural America were the stuff of Penck's home, a colonial explorer's open world with borders yet to be defined. Lands and oceans could be traversed transculturally by the expert's yearning for travel and curious gaze.[17]

In Penck's passion for geosciences, networks of knowledge transfer (*Wissenstransfer*) and modern science occurred transnationally.[18] Like many who came before and after, Penck saw himself as open to new knowledge everywhere. In any middle-class academic's life, he surmised, travel broadened the mind and satisfied the soul. So he took globe-trotting expeditions abroad. In the 1890s and 1900s, the professor's fame as a geomorphologist peaked, resulting in offers to lecture around the world. In 1891, Penck boldly proposed the first 1:1,000,000 map of the world at the International Geographical Congress (IGC) in Bern, Switzerland, in an attempt to standardize scale.[19] The conservative was ahead of his time. The proposal, his brainchild, was taken up as the International Map of the World, or Millionth Map, later on in 1913.

In 1894, he published his masterwork, the two-volume *Morphologie der Erdoberfläche*. He wrote thousands of pages on fluctuations of the Ice Age and deposits in the valleys of the Alps. In 1897, the British Association for the Advancement of Science (founded in 1831 in opposition to the Royal Geographical Society, it was modeled popularly on the Gesellschaft Deutscher Naturforscher und Ärzte, which started in Leipzig in 1822) invited him to visit England and North America for the first time.[20] From 1899 to 1906, Penck engaged with the Association for the Dissemination of Scientific Knowledge in Vienna, to advance literacy and bring geography to an educated public. When the geographers Ferdinand von Richthofen (1833–1905) and Erich von Drygalski (1865–1949) founded the Berlin Oceanographic Institute in 1900, he had an integral role. Penck coauthored *Die Alpen in Eiszeitalter*, a vast three-volume study of Alpine formations in the Ice Age, between 1901 and 1909 with his Vienna colleague Eduard Brückner (1862–1927).[21]

By 1906, Penck was at the apex of his career, a scientist appointed to Richthofen's chair of geography at the University of Berlin. When the museum of the Oceanographic Institute opened, he spoke at the kaiser's inauguration of it. As his résumé grew impressive, the filler, pedigree, and titles of academics in Germany mattered more. In the 1908–9 year, he visited Columbia University in New York City on exchange. More than half a century before the cultural diplomacy of Fulbright scholarships and the German Academic Exchange Service (DAAD), Albrecht had friendly contacts with William Morris Davis, Bowman's Harvard mentor and the leading geomorphologist in the United States. Penck's geography before 1914 was indeed transnational and transformational.[22] He was synonymous with geography as a border-hopping pursuit, the result of global empirical research, the essence of higher specialized education. In short, he was part of a nineteenth-century civilizing endeavor.

WEST GALICIA, 1871

Another of Bowman's heroes and Penck's own pupils was Eugeniusz Mikołaj Romer (1871–1954), Poland's most esteemed geographer after Copernicus (figure 1.3). Born in Lemberg/Lwów/Lviv (now part of Ukraine) on 3 February 1871, Romer was a subject of the Habsburg emperor Francis Joseph I (r. 1848–1916) and the Austro-Hungarian Dual Monarchy.[23] Though later nationalized as Polish, Romer spoke fluent German and French. He also knew Russian,

FIGURE 1.3. Eugeniusz Romer (1871–1954), taken in Paris in 1919. Romer was fond of sending not only maps and atlases, but also caricatures and autographed images of himself. Sent in December 1919 from Romer in Lwów to Bowman in New York City. Courtesy of Special Collections, The Milton S. Eisenhower Library, The Johns Hopkins University, Baltimore, Maryland.

Ukrainian, and English. In autonomous Galicia, Eugeniusz was of mixed aristocratic origins, his family having the alternate spellings "von Römer," "Rommer," "Remer," "Rejmer," and "Roemer." It is mostly forgotten that the family had Saxon origins, for Romers settled in Poland and produced many scholars, diplomats, scientists, and artists. The large clan's national identity was ambiguous, but it was symbolized by the retention of estates from Livonia in the north and today's Lithuanian border with Latvia to the southern foot of the Western Carpathians, bordering Hungary, the Slovak Republic, and Ukraine.[24] Eugeniusz came from a branch that settled around Jasło, in southern Poland, during the fifteenth century.[25]

Where Penck's linear, Bismarckian, modern German bourgeois story has traceable early modern gaps, the Romers' trajectory is also unclear in a different way. With the first partition of Poland-Lithuania by imperial Russia, Prussia, and Austria in 1772, the "lawful" acquisition of the Kingdom of Galicia and Lodomeria by Empress Maria Theresa (r. 1740–80) was a tragic event that affected the family's transformational belonging to Poland's early modern noble nation in Habsburg Galicia.[26] Acculturation to high German, at least on the provincial level, seemed for the first time necessary and unavoidable. In Vienna in 1784, Emperor Joseph II (r. 1780–90) conferred rights to Eugeniusz's direct line, and in 1818, Emperor Francis I (r. 1804–35) made Romer's grandfather Henryk (Heinrich) a count. Count Henryk identified with Poland, which after its last partition in 1795 no longer existed as a state. By the Congress of Vienna of 1815, the Kingdom of Poland was placed under Tsar Alexander I. Henryk joined in Poland's November Uprising of 1830–31 against Tsar Nicholas I (r. 1825–55), who suppressed the Decembrist revolt of 1825. Henryk had two sons, Edmund and Władysław. The latter was a medical student who joined in Poland's January Uprising of 1863–64. He perished. Eugeniusz's father Edmund, who also took part, was fortunate to survive. Chastened by the event, Edmund pledged loyalty to Habsburg rule in the wake of the Austro-Prussian War of 1866. Under the crown, Edmund worked as a lawyer by profession and a civil servant in Rzeszów and Lwów. He married Irena Körtvelyessy de Asguth, an affluent Hungarian noblewoman whose family owned lands in what became parts of Romania and Yugoslavia (after the Treaty of Trianon in 1920).

Yet Eugeniusz boldly rejected his parents' Austro-Hungarian compromise. Coming of age in Galicia, he transferred dreams of Polish independence to his elder brother, Jan Edward Romer (1869–1934). Jan was an officer in the Austro-Hungarian army, who later fought with the paramilitary forces of Marshal Józef Piłsudski (1867–1935).[27] By the 1890s, Eugeniusz's "anti-imperial choice," as the historian Yohanan Petrovsky-Shtern has aptly termed the dilemma for

Ukraine's Jewry, was really a calculated gamble against the family's Saxon past and decades-long liberalism in Habsburg lands.[28] He nationally departed from Mitteleuropa, a hegemonic idea developed by Friedrich Naumann (1860–1919), a Protestant and liberal politician who, like Penck, was born near Leipzig. In German geopolitics and economic and political geography, as Michael Heffernan has shown, it was a framing device.[29] This layered "German" spatial microworld of the family's "Polish" history remained taboo in Europe's post-1848–49 ethnocentric era of modern exclusionary politics.[30]

Since Romer rejected the Mitteleuropa of the 1870s and Bismarck's Second Reich of 1870–71, one might presuppose a natural German-Polish antagonism, or Catholic-Protestant faultlines of *Kulturkampf*. Such tensions were real, but clashes of civilization and myths of ethnic hatred even from family lore were only partly true. More essential was Romer's knowledge, acquired from Penck and others, of Europe's emerging sciences. In 1889, Eugeniusz enrolled at Jagiellonian University and took classes with Franciszek Czerny-Schwarzenberg (1847–1917), a professor from a Habsburg triloyal family of Bohemian landowners and another chair of geography in East Central Europe (since 1877).[31] Czerny-Schwarzenberg inspired Romer to undertake field research near Kraków, much as Penck had done near Leipzig. In 1891–92, Romer's third year there, he traveled abroad to Halle-an-Saale to study geography with Kirchhoff, coauthor with Penck of *Das Deutsche Reich* in 1887. In Halle and then in Berlin and Vienna, Romer took hundreds of pages of notes, in fluent German and Polish, on giants such as Ritter, Humboldt, Ratzel, and Penck.[32] In the 1893–94 year, he studied under Antoni Rehman (1840–1917) in Lwów, a professor of geobotany and founding member of the geography department at the University of Lwów (since 1882). The magnanimous Rehman became Romer's father-mentor, what Davis and Penck were for Bowman. He looked out for Eugeniusz, who taught geography briefly in fall 1893 at the Franz Joseph I Gymnasium No. 3 in Lwów, where he was accused of taking part in "anti-Habsburg disturbances" in the school's reading hall. Inspired (and perhaps talked into it) by Rehman, Romer ultimately decided against a teaching career.[33]

Young Eugeniusz learned to keep his head down—and here is the key—until he earned his scientific credentials out of Ostmitteleuropa. In spring 1894, he finished his thesis and got his geography degree with a specialization in climatology, a science in increasing demand.[34] At age twenty-four, Eugeniusz received a stipend to study abroad and went to learn geomorphology in the fall 1895 term under Penck in Vienna, followed by the spring 1896 term with Richthofen in Berlin. When Edmund lost his position (then in Brześć) after a dispute with Count Kazimierz Badeni (1846–1909), the governor of East Galicia, his parents'

Habsburg triloyalism nearly unraveled. Disillusioned by the failures of liberalism, Eugeniusz looked alternatively to Penck and Richthofen in pursuit of climatology and other subdisciplines of glaciology, geology, tectonics, and meteorology. By the time Romer reenrolled at the University of Lwów in 1896–97, he was well-versed in all these main currents of German geographic science.

Then in 1899, Romer got married to Jadwiga Rossknecht, the daughter of the director and coowner of the Okocim brewery, one of the largest in Poland. Named for the fourteenth-century queen of Poland, Jadwiga also came from a German-speaking family. Yet her parents raised her Catholic and Polish in Kraków with her four siblings. Three years younger than Eugeniusz, their families knew each other. They had met in their teens. She was capable and smart, educated to become a teacher of German and Polish. The marriage by all accounts was of two people in love. It was also a strategic partnership, in a different way than Penck's. In the patriarchal confines of Catholic Galicia, the aristocrat's choice of a "Polish" partner was a provincial protection of privilege in Habsburg lands. He had to reckon with the realities of diminished noble power for the nineteenth-century Polish szlachta in the nation. The couple got married in the usual Catholic ceremony, though Romer's "German" (actually, Austro-Hungarian) parents' household was mostly secular. Eugeniusz and Jadwiga eventually had two sons, Witold (b. 1900) and Edmund (b. 1904), and the family of four settled in Lwów in a brick house on ul. Ujejskiego 6, near the Lwów Polytechnical School. By the end of a liberal-positivist era in the 1890s and 1900s, the professional Romers joined into a nationalizing, Galician, and increasingly urban Polish technical intelligentsia.

Like Bowman and Penck, Romer was a precocious talent. In 1899, he defended his doctoral habilitation on the asymmetry of river valleys, when he was just twenty-eight years old. In a mixed city of mostly Poles, Jews, and Ukrainians, Romer became part of a Polonophone scientific community. He taught geography at the Lwów Academy of Commerce. With the help of Rehman, he secured his first position in Habsburg lands, as an assistant professor in the university's geography department. Throughout the 1900s, Romer edited German-to-Polish adaptations of maps and atlases. As early as 1903, following the model of chairs of academic geography in Berlin and Vienna, educated Poles lobbied for their own in Lwów, against the counter-aims of Ukrainian activists. Romer came out in full support of the Polonization of Galician schools. He opposed the legislative effort in 1907 of Ukrainians and Polish Ukrainophiles to institutionalize Polish-Ukrainian bilingualism in schools and at the university (the so-called *utrakwizacja*, in Polish). In the city's changing illiberal spaces, he drifted farther from his father's triloyalism. In Galicia, however, no choice of identity was simple. His patriotic brother Jan Edward married a Ukrainian (Uniate) Catholic from East Galicia, Stefania Lityńska, who assimilated as Polish by marriage, language, and religion after she converted to Roman Catholicism.[35]

If Penck confronted modernity by staying a Saxon and a man of Ostmitteleuropa in global search of premodern places, Romer became an assimilationist Galician, promoted in 1908–9 to the position of professor (full docent) at the University of Lwów. He conducted student excursions around Lwów, in Podolia and Pokutia, to the Tatras and Carpathians. The trips were of the same kind that he took in his study abroad, in 1895–96. Then in 1909, his fate was sealed when Rehman handpicked him as his successor for the geography chair. The continuity of place for the two Habsburg men ran deep—so deep, in fact, that Eugeniusz and Jadwiga purchased Rehman's house near the university on ul. Długosza 25. They moved there straight from ul. Ujejskiego 6 in 1909. Living quarters of the new house were decorated with antique Galician furniture, of which the parents were especially proud. The abode featured what the sons Witold and Edmund called a "mystery room," their father's office for research and the act of designing his magical maps. The multilingual Romer sons, both travelers and technically adroit, would follow the family's long path of *Wissenschaft* in their twentieth-century lives.[36]

Mannered academics like to disguise grievances in high-minded language of standards and scientific debate, but Romer's prewar professionalism was noblesse oblige, a form of polite learning much like Penck's. In 1909, the Swiss geologist Maurice Lugeon (1870–1953) invited him to present his findings on glaciers, much as Penck had done decades earlier. Romer was elated. His work on Alpine geomorphology rejected the concept of Mitteleuropa and challenged Penck's Nordic theory of origins for landforms. In 1910, Romer also traveled with a team of scientists on an imperial Russian geological and mining expedition to the Far East. In 1911, after his benefactor Rehman retired in Lwów, Romer was promoted to chair of the geography department. When politics came up, Romer soon stressed the precedence of Polish over German geographic knowledge. In a major work in 1912, *The Natural Basis of Historical Poland*, Romer rejected the notion of the liberal Warsaw geographer Wacław Nałkowski that Poland was "transitional." Instead, he offered the poignant cultural symbol in Europe of a greater Polish "bridge" between frontiers and stretching from the Baltic to the Black Sea and Carpathian Mountains in the south, and from the Oder in the west to the Dnieper and Dniester river basins in the east.[37]

Romer's Polish Carpathians, not the Urals, were now imagined as Europe's spiritual bulwark. Instead of relying on anthropocentric evidence, as Nałkowski had done in his adaptation of Ratzel, Romer knew what he had learned from Penck, the cultural application of the physical sciences of climatology, geomorphology, and glaciology. Romer spatially depicted the Vistula as Poland's main wellspring, for it flowed north to south and organically joined the "two important Ruthenian rivers, the Dnieper and Dniester, separated otherwise by wilderness."[38] He utilized the German Eurocentric map of civilization he had imbibed from Humboldt and Ritter and in Halle, Vienna, Berlin, Kraków, and Lwów, now to criticize Penck in his Berlin chair, as he mapped and claimed Silesia at the next IGC in Paris in 1913. This battle between Prussia and Galicia and over German-Polish frontiers became a kind of sublimated family quarrel. Stakes were raised fast: Romer's prewar contacts with American, British, and French geographers would prove invaluable in his search for support of Poland's independence after August 1914.

EAST GALICIA, 1877

With aims to forge a country called Ukraine from empires before World War I, Stepan L'vovych Rudnyts'kyi was a rival of Romer and a student of Penck (figure 1.4). That he originated from East Galicia and was a Ruthenian (*Ruthene*, in the German or Austrian designation) not of noble origin mattered significantly. His branch of the Rudnyts'kyi family came from the village of Avhustov (Augustów), not far from Ternopil, and his grandfather Denys was a Greek Catholic priest. His father Lev, born in 1851, completed gymnasium in Berezhany and studied in Vienna and Lemberg/Lviv. The learned Lev, who earned a degree in German and in history and geography, became the director of classical gymnasia. He moved jobs with the family to Peremyshl in 1873, Lviv in 1879, Ternopil in 1887, and back to Lviv in 1891. Lev married Emilia Tabors'ka in 1874; she was of Armenian extraction and also came from

FIGURE 1.4. Stepan Rudnyts'kyi (1877–1937). [No date, likely around 1919.]

a Ruthenian family of priests. Stepan, born in Peremyshl in 1877, was closest to his mother, who managed the household and encouraged all four of her children to learn German and Ukrainian and pursue higher intellectual paths.[39]

Language activists on frontiers embodied Habsburg diversity, and Rudnyts'kyi was no exception. His eldest brother Levko was a jurist and Ukrainian activist for the Lemko population, while his younger brother Iurii became a writer. Iurii and his sister Sofia, who graduated from the prestigious Vienna Musical Academy and married the legal scholar Stanislav (Stash) Dnistrians'kyi (1870–1935), Rudnyts'kyi's trusted friend, each translated school textbooks from German into Ukrainian. Unfortunately, the Rudnyts'kyi family suffered tragedy twice over, first in 1896 when Emilia died of tuberculosis, and again in 1898 when Lev had a fatal heart attack. At the start of a new century, the two eldest sons, Levko and Stepan, had to care for their younger siblings. In their busy and intellectually ambitious household, ethnic identity and labels for Ruthenians or Ukrainians was less a factor than performed social and intrafamily roles.

Other intangibles of character were in play. Symbolic places and mobility in East Galicia in the late Habsburg Empire offer clues to Rudnyts'kyi's outlook as a geographer, given that we lack the luxury of a written memoir. First was transportation. Modernity was accelerated in the provinces by new railroad lines between Peremyshl, Lviv, and Ternopil. Second, education was vital. The Rudnyts'kyi clan strove to gain access in Galicia as nationalizing intelligentsias after the revolutions of 1848–49 sought to build identity through shared history and geography, high culture, and language. Third was intolerance. While Vienna grew less welcoming to minorities in the mid-1890s, the University of Lemberg/Lwów was dominated by a Polish administration. (The anti-Semitic populist Karl Lueger [1844–1910], founder of Austria's Christian Social Party, was elected mayor of Vienna in 1897.) Due partly to Habsburg leniency on diverse frontiers and partly to Ukrainian pressure, discrimination did not extend to student admissions.

Rudnyts'kyi took advantage by earning his *matura* in 1895. He was accepted into the philosophy department in Lviv for the 1895–96 year. To finance his studies and follow his father's path, he gave lessons in history, geography, and German to children of comfortable Polish families like the Romers. Charting his own course, he devoured history, geology, geography, and paleontology and took new offerings in Ukrainian language and literature. He was nurtured by patronage and patriotism when the historian Mykhailo Hrushevs'kyi (1866–1934), the professor and president of the Shevchenko Scientific Society (Naukove Tovarystvo im. Shevchenka, or NTSh, est. 1873), opened doors for the geographer's pursuits. Hrushevs'kyi, the charismatic force behind the NTSh on the model of Europe's academies of science, sought to gather professionals into Ukrainian studies. He broadened geography in Europe into jurisprudence, medicine, and the physical and natural sciences. Many such enlighteners were Galicians, dabblers and polymaths in learning, part of a dispersed intelligentsia across the two sides of Zbruch/Zbrucz River, the imperial Habsburg-Russian border.[40]

Then in 1899, Rehman in Lemberg/Lwów/Lviv trained Rudnyts'kyi in oceanography and orography (regional studies). He also supervised his thesis in climatology, which was published in Polish in the journal *Kosmos*, and in Ukrainian in the NTSh journal. Rehman, who made a virtue of open pluralism toward the empire's nationalities, found a stipend for his Germanophone student Rudnyts'kyi to study abroad, just as Romer had done in 1891–92 and 1895–96. Rehman sent Rudnyts'kyi to study with Penck in Vienna, then under the meteorologist Karl Uhlig (1872–1938) in Tübingen, and Penck's colleague Brückner in Berlin.[41] Rudnyts'kyi took on advanced work in geomorphology, physical geography, anthropogeography, climatology, and astrogeography. He avidly read scholars in German, French, Polish, Ukrainian, and Russian. In 1901, he published the first study of astrogeography in Ukrainian, a major breakthrough. It earned him full membership in the NTSh. Given Hrushevs'kyi's wide accomplishments in history, an inspired Rudnyts'kyi dreamt of a national Ukrainian school of geography to include all the cutting-edge fields of Ostmitteleuropa—in climatology, meteorology, glaciology, oceanography, earth magnetism, geophysics, volcanism, and hydrography.[42]

When Stepan married in 1902, it could be interpreted as a kind of frontiersmanship and social calculus. Yet categories of the "intimate" or constructed erotic spaces are hard to evaluate, especially in provincial places and gendered contexts of Europe's colonial world.[43] Unlike in the cases of Penck or Romer, Rudnyts'kyi's marriage did not result immediately in upward mobility. He got married relatively young for a professional, in Lviv at the age of twenty-five, to Sybilla Schenker, just twenty-one. She was the daughter of a retired Austro-Hungarian officer from Stryi/Stryj. By all accounts, he and his wife deeply loved one another. The first of their three children was born in Lviv in 1902, and they named her Emilia after Stepan's late mother. Levko, the second child, was born in 1908 and called Lev, after Stepan's father. Iryna, the third child, whom the parents called Orysia, was born in 1912. It seems that Stepan and Sybilla combined family and work very seriously. They were civic-minded teachers who shared the drive in Galicia for Ukrainian education and statehood in Europe.

On the Habsburg margins, the couple experienced serious financial difficulties. For one thing, Sybilla was a professional who absolutely had to work for a living (it is unclear whether that is what she wanted). She earned a salary as a teacher in the village of Kormanychi, near Peremyshl. She took an active part in women's social welfare organizations for the care of Ruthenian children.[44] While culturally progressive in economically "backward" Galicia at least in terms of gender equality, these obligations took a harsh toll. Sybilla had a congenital heart condition for which she required treatment in European sanatoria. She fell constantly ill. In Galicia, the nonnoble family could not afford the expenses of specialized care. During summer breaks, the teachers and their three children retreated to the area of Hutsul'shchyna in the Carpathians, a mecca for the Ukrainian intelligentsia, to restore their health and let nature take its course.[45]

Stepan thus had many obstacles in his path in the 1900s, but his transnational Galician bearings were significant throughout his life.[46] His professionalization in Ukrainophone lands ran up invariably against competing Polish aspirations. When the NTSh in 1902 petitioned Vienna to set up a Ukrainian university, Poles blocked the initiative. Rudnyts'kyi had to pursue his habil while teaching at a secondary school job in Ternopil. In summers between 1902 and 1905, he went on expeditions to the basin of the Dniester River. He presented the results of his first geomorphological travels at the University of Vienna's Geographical Institute in June 1904, a moment of Austrian (rather than Polish) recognition and an early highlight.[47] At the urging of Hrushevs'kyi, Rudnyts'kyi collected materials in order to write Ukraine's first geography, in a national and popularly readable form. In 1907, Rudnyts'kyi began drafting a new ethnographic map of Ukrainophone dispersals for the NTSh, an activity that aroused Romer's suspicion. Romer had a pointed exchange with his Ruthenian counterpart, in which he denied Rudnyts'kyi's academic merits. The confrontation foreshadowed World War I and foretold of the enduring conflations, well after Paris in 1919, of *Wissenschaft* with East Central Europe's violence on demographically mixed frontiers.[48]

Rudnyts'kyi and Romer thus shared parallel Galician tracks into Europe's transnationally mobile confraternity of pre-1914 scientists and geographers. Stepan too had studied Humboldt, Ritter, Penck, Richthofen, and Ratzel. His dossier included original work on the Subcarpathian basins, based significantly (if derivatively) on Richthofen's work. He submitted his habil in Lviv in 1907 with supporting documents and letters of reference in German, Polish, and Ukrainian. Because the Ministry of Culture and Education in Vienna had to approve all applicants, not until July 1908 was he deemed qualified to defend the work. Although Rehman and the geologist Rudolf Zuber (1858–1920) supported his candidacy, nine months passed before he heard a response. As with so many other ceremonial rituals in Austria-Hungary, hurdles were real but not impossible to overcome. He had to give a lecture in the university's auditorium, half in Ukrainian and half in Polish in accordance with university protocol. His lecture topic was "The Black Sea."

His efforts finally rewarded, Rudnyts'kyi was appointed to Privatdozent in geography in August 1908, the stepping-stone position Romer had back in 1899. Habsburg leniency allowed Rudnyts'kyi to teach courses in Ruthenian (so designated) in the 1908–9 year, and he prepared by assembling a German-Ukrainian geographical lexicon, the first of its kind for modern Ukraine, which was published by the NTSh in Lviv.[49] Perhaps justifiably, Stepan viewed terminology in German as more precise and rich than English or French. He considered Mitteleuropa geography, without the "Ost-," farther advanced than that of anywhere else in the world.

There were darker sides to progress, and it did not equal modernity either. As our map men gazed in and out from Europe's margins, they assigned grand nineteenth-century purposes to their work. Young Stepan belonged to an aspiring middle-class intelligentsia, susceptible to Habsburg political fantasy cultures since the Enlightenment.[50] He missionized geography in Europe, imperially in this Galician sense, just as Penck from Saxony aimed to get to the centers of Vienna and Berlin, and Romer toward a Polonophile trans-Atlantic West. Stepan worked to train cadres, develop new field studies, and learn research methods. He finished the draft for a national geography of Ukraine in 1909. The first volume on Ukraine's physical geography was printed by the Lan publisher in Kiev/Kyiv at the end of 1910. Unfortunately, the publisher lost the second manuscript, his Ukrainian syncretism of Ratzel's human (or cultural) geography in *Anthropogeographie*. Rudnyts'kyi, explicitly having copied Ratzel's neologism, was impelled to rewrite it in full. Not until summer 1914 was it printed in Lviv, right before the Russian imperial army marched into Galicia.

By 1914, Rudnyts'kyi saw for Ukraine in geography what Hrushevs'kyi located in history in the last prewar years, a nation- and state-building means to unite groups of like-minded professionals and kin-state people, across borders into newly settled spaces for a territorial homeland. He produced the first Ukrainian school atlas in 1912 (it was reprinted in 1917, 1919, and 1928), just as Romer had done for Poland in 1908. This was a world atlas with maps of Europe's great powers, and a demographic survey of Ukraine based on recent census statistics. When war broke out, he fled Lviv for Vienna. The end of Europe's dynastic empires afforded him not just tragedy but a new opportunity, from East Galicia to Vienna in 1914 to produce maps, pamphlets, and books for independence. Almost overnight, out of his Ostmitteleuropa of friends, father-mentors, and rivals, Stepan Rudnyts'kyi became the world's leading geographic expert on Ukraine.

All of our geographers may be called transnational Germans. While Penck recounted his plucky rise from provincial Saxony and Bavaria, Isaiah Bowman manifestly lauded his family's American rural and Anglophile origins. Actually, he was born a Canadian, a subject of Queen Victoria, on 26 December 1878 in Berlin, Ontario. The son of Samuel Cressman and Emily Shantz Bowman, Isaiah was of Swiss German, English, and Mennonite background. Bowman's grandfather Moses was a teacher turned preacher. Both sets of his Canadian grandparents seem to have been financially stable well into the 1870s. They had large families, which diffused their wealth. The hardest time, evidently, was when Samuel had to change his occupation from teaching to farming in order to make a better living for the family. If one believes Bowman's story (we'll see shortly why this is problematic), his daddy emerges as a kind of ne'er-do-well. When Isaiah was just eight weeks old, Samuel and Emily moved with him and his two sisters, ages two and four, just across the border to a 140-acre farm in Brown City, Michigan, about sixty miles north of Detroit. Isaiah was the third of their eight children. In the financial crisis of 1873, the family nearly went broke. Isaiah's pro-American spin is the one that has survived, in praise of their down-home, patriotic virtues, which persisted above all else. In his folkloric family tale, he insisted that the Bowmans were "strongly endowed with religious feeling and a sense of duty and responsibility."[51]

As many in America were, Bowman was noticeably silent about his German roots and language knowledge in the course of the two world wars. He described his father's station in life as "a farmer in the middle-western sense," although by level of literacy a "school teacher" who owned just a few books. Isaiah singled out his father's few favorites given to him: an English dictionary, Captain Cook's *Voyages*, and Henry Morton Stanley's *Darkest Africa*. As a young boy, he loved English adventure tales. He read them along with a biography of Alexander the Great and the "Indian stories" (as he called them) starring Daniel Boone.[52] He regarded his mother, Emily, as a nurturing type and cast her in the role of a warm, engaged parent who encouraged him to read and develop his talents. The classic American race-, class-, and gender-based myths of success—the prosperity within reach for the hardworking few—pervaded Bowman's family romance. The rags-to-riches man, the self-made man, and the pioneer settler were these performed melodramas in a Gilded Age, despite the fact that his family had not lacked means before the 1873 crash.[53] Bowman appealed to the Puritan colonial past as a foundation myth, since "records indicate that many of the Bowmans were of extremely ancient and honorable lineage . . . Among the first of the name to emigrate to America was Nathaniel Bowman, who came from England in the fleet with Governor Winthrop in 1630 and settled at Watertown, Mass."[54] He recalled how some Bowmans fought as American officers in the Revolution across the Canadian border. These mappings enabled him to incorporate Ontario as a place into English-speaking America, with the Midwestern frontier space of his youth in heroic terms, a formative site for how he as a boy became a man.

Bowman's world was oriented spatially into a civilization, an American harmonious self, provincially gendered as pride of place in the nineteenth century. It was peppered with folksy stories that prescribed a great destiny. He narrated the pride as a coming to literacy; geography was a form of literacy in the early modern U.S.—knowing where you dwelled, what your limits were, who your family knew, the behavior of others as they treated you, what you could come to expect.[55] He captured the feeling when he showed his mother the "A" he earned for drawing his first map in school. It was powerful—and manly, too. In 1896, while a young man in remote Brown City, Bowman studied with the owner of a local grocery store, a sea captain who taught him geometry and navigation. In Michigan, he passed the teachers' exam and began his career in nearby St. Clair County, just a few miles to the south. On his twenty-first birthday in 1899, the year Joseph Conrad published *Heart of Darkness*, Bowman became a naturalized citizen of the United States of America. His path in becoming a teacher was not a foregone conclusion. He was small in stature, 5'3" tall and 125 pounds, and he compensated with a stern demeanor. At the cusp of a novel century of educational advancement (not to mention mass warehousing), he loathed being confined indoors.

In 1900, young Bowman registered for geography courses at the Ferris Institute in Bog Rapids, Michigan. When in 1901, he entered the Michigan State Normal School, now called Eastern Michigan University, he began a period of formal tutelage in geography under Mark Jefferson (1863–1949), Davis's master's student. Bowman embraced geography as a profession, with a neophyte's zeal. Jefferson, who had spent six years in Argentina, specialized in cartography and did empirical research on land settlement issues. He later became the chief cartographer of the American Delegation to Paris in 1919. It was Jefferson, the former Harvard man, who joined Bowman to the confraternity. Ypsilanti was a kind of nursery for Cambridge. He fed Bowman's dreams of civilization, just as Penck did for his pupils from Germany and Austria-Hungary. Bowman quickly fell in love with his new pursuit.

Then in 1902, Bowman got his big break. Jefferson, fifteen years his senior, recommended him to Davis. For a young man without means, the trip from

Michigan to Cambridge was like embarking to a foreign country.[56] Like Romer, Bowman had tried, and failed, in his job as a schoolteacher. After a summer working for the U.S. Geological Survey (USGS) in Michigan, in September he enrolled as a scholarship student at Harvard. He studied under Davis and the geologist Nathaniel Southgate Shaler (1841–1906), both steeped in German and European currents. Socially awkward and class-conscious, Bowman struggled to fit into the unforgiving, status-conscious Ivy League world of networking. He skipped class meetings and refused invitations. The dynamic of ruling-class elitism and its occasional déclassé flirtations with socialism turned him off. Lost in an alien milieu, he turned to serious moral growth through *Bildung* and *Wissenschaft* in geography, dreaming of exploration.

How Bowman found his bearings in a nation and profession is inseparable from how he learned from Davis and Shaler to appreciate Europe's explorers. Davis and Shaler were men passionately devoted to finding patterns in the natural world, divining the unlearned secrets of the earth's forms and evolutionary processes. Trans-Atlantic crossings were made by Ratzel's American protégé, Professor Ellen Churchill Semple (1863–1932), who studied in Leipzig. Shaler, Davis, and Semple were tuned into Progressive Era values, champions of reason and service who advanced education to develop a coterie of stateside experts. Ratzel, trained as a zoologist, conceived of an organic national vision of government in which Germany, like the U.S. West, was largely agricultural and a laboratory for development where populations settled and geographers supported tasks of land surveying and mapping. By developing his idea of *Lebensraum* in his 1898 *Political Geography*, Ratzel imagined that modern Germany was a kind of America, characterized by settlement and resource management. In this model, which traced back to Johann Gottfried von Herder (1744–1803) and the German Enlightenment and romanticism, in which diverse cultures and civilized nations grew like plants in a garden, peoples settled down in order to survive and occasionally mingle with the rest of humanity. Scientists used their expertise for exploiting natural resources. Families of nations and states were pioneers who preserved the species, outcompeted rivals, and survived. Bowman fused together German cultural nationalism and Ratzel's political geography with a white American Puritan ethic, settler colonialism, the authority of geographic science, and frontier exceptionalism.[57]

Davis, Bowman's mentor, was deeply inspired by the nineteenth-century German scholarship of Humboldt, Ritter, Ratzel, Richthofen, and Penck. He applied European works of physiography, geology, and human geography. Davis invited Penck to Harvard in 1903 to present his influential book, *Die Morphologie der Erdoberfläche* (1894), a founding text in geomorphology. In

1904, Davis was among the founders of the Association of American Geographers (AAG), the organization that (then as now) viewed geography as a giant discipline. The AAG was a gathering site for the country's learned, specialized, traveling experts. He authored in his long career over five hundred articles and books, the main contribution being his "universal" theory for the cycle of erosion. His goal in America was for its geographers to build upon and surpass in science what the Europeans, especially the Germans, had already accomplished. By personality alone, Davis at Harvard was an iconic figure and Bowman's sponsor into academe. Yet the Harvard professor's contradictions of privilege and meritocracy went unresolved. For many years, his ongoing tensions in academe significantly shaped and defined Bowman's life.

Above all, Bowman in the 1900s inherited the insecurity of Davis's social habitus, in which an American credentialization through modern geography was a civilizer's rite of passage. He got to Europe first through books, then by meeting "great men." When the geology department gave a dinner at the Colonial Club with Penck and the Scottish oceanographer Sir John Murray (1841–1914) as guests of honor, Davis invited his protégé to join other Harvard academics. In fall 1904, Bowman wrote to his Michigan teacher Jefferson enthusiastically, "One of my ambitions has been to see and hear Penck—the real *Penck*! I can scarcely wait to see him. I suppose you know of the monograph he's getting out, in sections, on the glacial features of the Alps? I'm reading Ratzel on 'Die Erde und das Leben' and find him even more interesting here than in 'Anthropogeographie.' These old German boys make me realize that not all the geography is west of the Atlantic."[58] Bowman showed off his specialized mastery. In a second letter to Jefferson, he again spoke of the opportunity: "Through Professor Davis' kindness I am having the great opportunity of becoming thoroughly acquainted with Penck. He finds the English phrase somewhat elusive and on his having expressed a desire for someone who knew German and could help him in writing out his lectures Davis spoke of me. I meet him for two hours each day."[59] Bowman's letters illustrated Davis's abiding influence and the outsized ambitions for American map men and geography in spaces they shared.

At Harvard in 1904, Bowman's encounter with Penck happened in the halls of German-American knowledge transfer. Davis and Shaler doubtless saw their Harvard students as junior versions of themselves—curious, entrepreneurial, and disciplined. Bowman attended the very first meeting of the Association of American Geographers (AAG), where he presented his first paper on "The Deflection of the Mississippi." In spring 1905, Jefferson urged Bowman to take the American civil service exam, while he was completing his Bachelor of Science from Harvard. Davis warned Bowman before graduation that few jobs were available for geographers in the inchoate profession. Bowman earned money in summer 1905 by working once again for the U.S. Geological Survey (USGS). Jefferson at the Michigan Normal School advised Bowman to take a more permanent teaching position at Ypsilanti Normal College in Michigan. Just as Romer rejected becoming a provincial teacher, Bowman pursued the research path of famed world travelers like Humboldt. Yale offered him a summer internship, followed by a full-time instructorship in physical geography in its new geography department. Bowman in New Haven set out to complete a doctorate in geography. He happily wrote to his hero, Professor Albrecht Penck. In winter 1905, Penck was on a scientific expedition to South Africa while also in his final year of instruction and at the University of Vienna. Penck was delighted to hear the news. The supportive professor sent a friendly postcard back to young Bowman.[60]

In his next ten extraordinarily prolific years at Yale, Bowman finished his dissertation and published four books in physical geography, geomorphology, and resource management. In 1906, he was elected to the AAG. In 1907, he published his *Water Resources of the East St. Louis District*, using his knowledge of the U.S. Midwest and experience of working for the USGS. He undertook fieldwork expeditions in 1907 to Peru and Bolivia, as part of his dissertation research. That was his first research stint out of the country. He defended his dissertation, "The Geography of the Central Andes," in 1909 when he was only thirty years old. Its analysis was pretty derivative from his Harvard sponsors, Davis and Shaler, but it was important nonetheless. Bowman's biological racism came out in sketches of indigenous populations in Peru and Ecuador, while he stressed as superior the discourses of white colonial settlement.[61]

By the time Bowman next encountered Penck, in December 1908, Penck held the chair of the Kaiser Wilhelm II Professor of Geography at the University of Berlin. He was invited to Yale to deliver the Sillman Memorial Lectures. Penck's third scientific trip to the United States in 1908–9 was a family affair of status and bourgeois mobility. His son Walther followed in his footsteps and became immersed at Yale in advanced geography and geology. Ida Penck also accompanied her husband from Berlin. In New Haven, Penck was elated to learn of Bowman's travels in South America, his dissertation defense, and his marriage. Bowman's peers elected him in 1909 as associate editor of the *Bulletin of the American Geographical Society*.

Among the map men, Bowman's Ivy League makeover into East Central Europe still was not complete without an upwardly mobile marriage. On 28 June 1909, he wed Cora Olive Goldthwait (1874–1952) at an Episcopalian church in Lynn, Massachusetts. They first met in 1902, after Bowman came to Harvard. Cora came from a well-off upper-middle class Protestant family (figure 1.5).

FIGURE 1.5. Cora Olive Goldthwait (1874–1952). [No date, c. 1909.] Taken likely just after her marriage to Isaiah Bowman in Lynn, Massachusetts. Courtesy of Special Collections, The Milton S. Eisenhower Library, The Johns Hopkins University, Baltimore, Maryland. Isaiah Bowman Papers (IB-P Ms. 58), Series I, Box 3, Item 9.

She played the violin and cello. She earned a degree in English literature from Radcliffe College. (So-called joint instruction was not allowed for Harvard men and Radcliffe women until 1943.) Cora introduced him even more to the East Coast ruling class; clearly, she was born and bred to marry a well-to-do New Englander. After a six-year courtship, Bowman proposed in March 1908, but the couple waited to marry until Bowman earned his Yale doctorate. Penck was so pleased by Bowman's news that he handwrote a letter to him,

> My dear Bowman, Many congratulations upon your wedding, which was in future when I left you, and which may last for many happy years! I came home in the middle of May, and since that time I am nearly drowned by the work stored up in the time of my absence. There are only moments when I can enjoy the lasting impressions of my stay in North America, and have the pleasure of remembering Yale with its good people. Give all my heartiest compliments, to the geologists and to the members of the faculty club. But above all: remember me to your wife, and tell her, how happy I shall be to see

her with her husband on this side of the Atlantic—but do not show her my english [*sic*] mistakes! Very sincerely yours, Albrecht Penck.[62]

Bowman's marriage was a final component to his East Coast progress and success. It became a key aspect of his vision of America's newfound role in the world. He and Cora had three children, Walter (b. 1910), Robert (b. 1912), and Olive (b. 1915). In Bowman's last years at Yale before World War I, he became more involved with the burgeoning AAG (Davis was president from 1904–6, Jefferson in 1916) and its reaches of geographic power. He grew in concern with securing U.S. business interests and resource exploitation south of the border. The Yale geographer published *Forest Physiography and Physicography of the United States and Principles of Soils in Relation to Forestry* in 1911, *Well-Drilling Methods* also in 1911, and *South America: A Geography Reader* in 1915. He continued to study Ratzel, in full vogue since Semple published *The Influences of Geographic Environment* in 1911. Of Bowman's twenty-four articles in this period, twenty were on South America. He traveled on his second expedition in 1911 and his third in 1913 (figure 1.6). He continued to review scholarly works in U.S. and

FIGURE 1.6. Photo(s) of Isaiah Bowman (1878–1950), before 1914, on one of his three expeditions to South America. Courtesy of Special Collections, The Milton S. Eisenhower Library, The Johns Hopkins University, Baltimore, Maryland. Isaiah Bowman Papers (IB-P Ms. 58), Series I, Box 3, Item 10a.

European geography.[63] Like his mentor Davis, he looked to South America as a kind of playground and continent of the explorer, to be opened to global commerce and academic study. His turn was into the guild of Europe's men into which he was granted entry—and a new suit—in the 1900s and 1910s by Davis, Jefferson, and Penck. Bowman came to join Ostmitteleuropa by his lofty American aspirations, but without ever setting foot there before 1914.

BUDAPEST-TRANSYLVANIA, 1879

Just as no map is ever 100 percent original, no map man was a blank slate. Of those who came from East Central Europe's aristocratic milieu, few carried more baggage than Count Pál János Ede Teleki de Szék, born in Budapest on 1 November 1879 (figure 1.7). Teleki, a Transylvanian subject of the Austro-Hungarian Dual Monarchy, was shaped by his family's remarkable past, belonging to a multigenerational clan of landholders in Habsburg-Ottoman borderlands.[64] The Telekis were a mix of Calvinists and Roman Catholics of varying piety. No consensus yet exists on their genealogy. One version details the family's escape from the Ottoman Turks when Muslims conquered the Balkans, in which the Telekis in the fourteenth century obtained estates in southeastern Hungary. A second set of claims traces back to the merging of the Garázda and Teleki families in mid-sixteenth-century Transylvania. Mihály II Teleki (1630–90), the clan's most famous ruler, supported Poland and cooperated with Jan Sobieski (r. 1674–96). Mihály II, who negotiated for Transylvania's autonomy, had thirteen children. In 1685, he received the title of count from the Habsburgs, after Sobieski's 1683 defeat of Ottoman forces in the battle of Vienna. Count Sámuel Teleki (1739–1822) established the Bibliotheca Telekiana in 1802, as a public library in Marosvásárhely. It was set up the same year as Hungary's grandest library, the Széchényi Library in Budapest. József

FIGURE 1.7. Photo of Count Pál Teleki (1879–1941). [No date, likely before 1919.]

(1790–1855), Pál's paternal uncle, became a prominent jurist, a historian of the Renaissance and the Hunyadi ruling family, the first president of the Hungarian Academy of Sciences, and governor of Transylvania. Pál's cousin Sámuel (1845–1916), an explorer of Africa, was credited with "discovering" Lake Rudolph in northern Kenya and Lake Stefanie in southern Ethiopia. He inspired Pál's expeditions in the nineteenth-century explorer tradition of Humboldt, Livingstone, and Rhodes.[65]

Stirring even more controversy, the Telekis of Hungary were deeply affected by the revolutions of 1848–49. Pál's father Géza (1843–1931) was only six years old in 1849, but he vividly remembered in Paszmos/Posmuș when Romanians and Germans looted, burned, and partially demolished the family's estate owned by Pál's grandfather Ede (1813–75), who was in the Hungarian parliament. The Piarists educated Count Géza, like many nobles in Catholic lands. As with Edmund Romer in Austria-Hungary, he was reared to become, predictably, a lawyer or civil servant. Due to revolution, the Telekis were forced to relocate to a smaller estate in Bagybánya/Baia Mare, Kővárvidék County. Géza, who spoke French and German and considered himself European by high culture, returned to Transylvania and became honorary sheriff of Kővárvidék County. After the *Ausgleich* in 1867, he became the recorder in the county, a high if boring and generally secure position in Habsburg lands.

Son Pál inherited his father's Habsburg bureaucratic localism and involvement in politics. His mother was Irén Muráti, a Greek Orthodox woman and the daughter of a prosperous merchant family in Pest. Though many powerful families were Calvinist, Pál, born in 1879, was baptized Roman Catholic according to his father's wishes. Géza in 1875 began representing the Nagynyires district in the parliament. When the influential Kálmán Tisza (1830–1902), the liberal supporter of Magyarization, formed a new government in 1889, he asked Géza to come to Budapest from his seat in Transylvania and become the new minister of the interior. Géza agreed, but after Tisza was forced to resign in 1890, that Teleki ordeal ended fast. Overall, modern party politics in Hungary hardly offered a stable life. The noble family spent summers on their Pribékfalva estate and winters in Budapest.

Gifted at languages, Pál easily learned German. He picked up Magyar and Greek; he learned to write French very well; he had an English private tutor; he could speak some Dutch, Italian (he loved Italian opera), and Romanian. Bred to be a European gentleman, he read avidly through many of the 3,200 volumes of books, including scientific works, in the library of the family's estate. He had tutors from early on and took final exams each May to meet standards of the Ministry of Education. Pál attended Piarist schools for noblemen in Budapest, and in 1897, his father sent him to study in the faculties of law and political sciences at the University of Budapest, a common track. While the Teleki family owned nearly 30,000 acres of land, Géza's holdings were actually quite modest. By the late part of the century, he maintained an estate in Pribékfalva, Szatmár county (today Satu Mare, Romania), with "only" around 2,500 acres of land. The future of their lives of relative privilege was mobile, and appeared to lie outside Transylvania.

While politics and jurisprudence were Géza's preferences, Pál was drawn by his bookish habits toward travel and the disciplines of geography. It was not Humboldt but Karl May's tales of the U.S. West that became a favorite comfort read, a guilty pleasure well into his adult years.[66] Once at the University of Budapest, Pál took formal courses by the Hungarian geographer Lajos Lóczy (1849–1920), a natural scientist, professor of geology, and chair of the geography department. Educated in the German tradition, Lóczy in Hungary was the successor to Janos Hunfalvy (1820–88), who had studied under Ritter in Berlin. Lóczy, a Transylvanian aristocrat as well, transmitted Penck's work and became a kind of father figure for Teleki. His Central Europe was impressively expansive, for he had also studied in Zürich. He became a leading scholar, in the Richthofen and Penck tradition, of the geomorphology of land formations in the Southern Alps and Tyrol.

In 1880, he participated in a famed expedition to China and the Trans-Himalayas led by Count Béla Széchenyi (1837–1908), of one of the most influential families in Hungary. (Széchenyi visited the U.S. in 1862 and published a well-received travelogue the following year.) Lóczy extended his research locally to the Banat Mountains in the West Carpathians, and in 1891, to geological and geomorphological expeditions to Lake Balaton, the Transylvanian Basin, and the Carpathian Basin. He earned memberships in geographical societies in Berlin, Leipzig, Vienna, and Bern. Elected by his peers in Budapest, he became the president of the Hungarian Geographical Society (Magyar Földrajzi Társaság, or MFT) in 1891–92 and from 1905–13.[67]

Inspired by Lóczy's success, Teleki found a way both to embrace European science and to parlay the explorer's vigorous life into a career. The MFT, started in Budapest in 1872 under Lóczy's predecessor Hunfalvy, who served as its first president, was a civic association for academically minded gentlemen. It connected Budapest to Europe's older geographical societies. Its members elected a board and published a journal, the *Geographical Bulletin* (*Földrajzi Közlemények*), to which Teleki contributed often. The membership doubled in size by 1900. Patriotic researchers focused on geography, ethnography, statistical demography, cartography, geology, climatology, and many other disciplines.

The MFT complemented the Hungarian Academy of Sciences (Magyar Tudományos Akadémia, MTA), started by Count István Széchenyi (1791–1860) in Pest, on the bank of the Danube in 1825. Count József Teleki had served as the MTA's first president. Young Pál developed a leading role in streamlining these institutions. He heralded Lóczy's results from the Lake Balaton expeditions, some thirty-two volumes of data and analysis, as one of the greatest successes of the MFT. Under Lóczy's influence, Teleki turned to the organic theories of Ratzel, whose *Anthropogeographie*, translated into Hungarian in 1887, also inspired the "German" Rudnyts'kyi.[68]

Pál Teleki was no artist in the *fin-de-siècle*, but he was a collector of artifacts. His dilettantish interests extended to the gathering of maps. At a December 1898 meeting of the MFT, he presented his first major work, a study of early European cartography in East Asia. Teleki sought knowledge of geography in China and Japan, the United States, and sub-Saharan Africa. He did not know any Asian or African languages. Academe fit badly. He continued to dabble. In 1901, he enrolled for one semester at the design school in the University of Budapest's fine arts department, but having scarce artistic talent, he dropped out after one semester. Géza thought that his son Pál could serve the state properly by becoming an expert on agricultural reform. This dull pursuit also failed. When Pál matriculated briefly in 1901–2 at the Magyaróvár Agricultural School, a place for sons of large estate-owning families, he showed himself to be a poor student and unsuited for land management issues. In April 1902, the count failed the second major law exam in economics, financial law, and statistics. He flunked a second time in March 1903. When with all the political intrigue of Hungarian academe and the MFT, he pursued geography as a career in 1903–4, at what looked like a desperate moment. It was at this uncertain point when Professor Lóczy gave Pál a break, a modest equivalent of Bowman's new suit. The geographer kept Teleki on as his personal assistant. He strongly advised him to focus his attention on professional endeavors.

When the capricious Teleki listened to anyone, it seems, it was Lóczy. In a sense, he picked a father other than his biological one. In 1903, he completed his dissertation thesis, "The Question of Primary State Formation," under Lóczy's direction at Budapest University. The Transylvanian count argued that "the state was the result of national development," a faint and derivative echo of Ratzel, within the historical lands of the Crown of St. Stephen. Teleki averred that the Magyar nation, because it was "rational," had a tendency toward independent statehood. Casting out the Romanians and Ruthenians provincially from civilized nineteenth-century Europe, he wrote, "We don't even have to go so far afield. The Moldavian Romanian or the Máramaros Ruthenian would be equally incapable of . . . abstract reasoning."[69] No proof was needed. Teleki recycled some crude knowledge of Herbert Spencer's theories of race and evolution. Then, ignoring Lóczy's advice not to publish anything of scientific value for at least four years, he wrote in 1904 a lengthy endorsement of eugenics in the inaugural issue of the *Archiv für Rassen- und Gesellschaftsbiologie*. In effect, the aristocratic Teleki's modernity was the expansive enterprise of nationalizing geography into race science, cobbled together out of his Transylvanian father Géza's ethnocentric prejudices and some of his own. He placated his father in 1904 by working as a magistrate in Szatmár County. At age twenty-four, he ran for Hungary's parliament and was elected as a deputy in Transylvania for the same district his father had represented. Toward the end of 1905, he finally passed the law exam on the third try.

Like our other men, Count Teleki hated to have a day job. He did not want to be any son of empire, an ordinary teacher, lawyer, or bureaucratic servant, particularly not in the Habsburg Empire. In 1907, the map-man-in-training set off for Sudan, following in the footsteps of his cousin Sámuel Teleki, the explorer of Africa. When he returned, he got married. In November 1908, Pál wed Johanna von Bissingen-Nippenburg, the daughter of Count Rudolf von Bissingen-Nippenburg, a retired lieutenant in the Austro-Hungarian army. This, too, was a tryst of love, but Johanna also was well-connected. She was related to the Transylvanian family of Count István Bethlen (1874–1946), the future prime minister and opponent of Nazi Germany, through his wife Margit. The Bissingen family were military officers who had settled in Hungary. They owned around 2,900 acres in the counties of Temes and Krassó-Szörény. The Teleki-Bissingen marriage was a political event at St. Stephen's Basilica in Budapest. In the St. Stephen lands, Pál's marriage conferred privilege, and to preserve the family's legacy, he tried to join late Austria-Hungary's technical intelligentsia to a Magyar "historic" space and developing European professional class. The couple had one daughter, Maria (b. 1910), and one son, Géza (b. 1911).

For the count, maps of such guarded privilege went to the heart of many melodramatic pursuits in his life and work. On the way back to Budapest from Sudan in spring 1908, Teleki conducted research at the Royal Geographical Society and British Museum in London, the Bibliothéque Nationale in Paris, and the Rijksarchief in The Hague. He buried himself in early modern European maps of Japan and wrote up the project on the family's Transylvanian estate. Pál sent his scholarly manuscript to Cholnoky, his colleague and confidante in Koloszvár/Cluj, another of Lóczy's students, and his future traveling companion to the U.S.[70] Genuinely impressed, Cholnoky told Teleki to present his findings in Geneva in 1908, at the ninth International Geographic Congress (IGC). The result was a stunning achievement, the publication of a giant folio of maps in 1909. It launched Teleki's career. Almost overnight, the twice-failed lawyer and once-failed artist was remade into one of Hungary's best-known geographers. The 1909 *Atlas of the History of the Cartography of the Japanese Islands* was a historical atlas of 168 pages and 20 maps, based on representations by European travelers and cartographers.[71]

Remarkably, there is no evidence that Teleki traveled to Japan or held a conversation at any time with a Japanese scholar. The aristocrat knew no Japanese, just as he had no familiarity with African indigenous languages.[72] In the tradition of Abraham Ortelius and Gerardus Mercator, and of Enlightenment science, the count claimed a unique Eurocentric universal civilizer's natural model of the world.[73] Teleki's 1909 atlas of the historical cartography of the Japanese islands bulked up his résumé. It was published in Leipzig in German and Hungarian, with praise coming when Japanophilia was in full European vogue in the French visual arts, and in England with the works of H. G. Wells and Beatrice Webb. The Geneva International Geographical Society elected Teleki to its old map and chart committee. In 1911, the Paris Geographical Society, the oldest society of its kind in the world, awarded Teleki the prestigious Jomard Prize.

The French Orientalist Henri Cordier (1849–1925), the society's president and sponsor of the award, lauded Teleki and placed him in the European pantheon by comparing his work to the expeditionist Otto von Nordenskjöld (1869–1928) of Sweden.[74] The MFT president Lóczy got Teleki nominated for general secretary, at which point Teleki could reach out to a Hungarian public for old maps, charts, and manuscripts. He worked on a Magyar world atlas with this literacy project in mind.[75] At the tenth IGC (Rome, 27 March–3 April 1913), Teleki praised the line of Hungary's explorers and its evident progress in geography.[76] Teleki got his first position as a geographer at the Budapest College of Commercial Studies in the 1913–14 academic year. When his father Géza died on 27 September 1913, the Pribékfalva estate in Transylvania became his responsibility. Yet Pál could not go home, or not exactly. He hitched his fortunes to the Ostmitteleuropa of modern maps, geography, and boundary making, as mediated through Penck, Davis, Bowman, Ratzel, Lóczy, and others. Teleki hunted for a legacy. On the eve of World War I, the Transylvanian count envisioned a shrine to European geographical science in Budapest, modeled on the Museum for Comparative Regional Geography (Museum für Vergleichende Landeskunde) erected in Leipzig in 1892. Unable to secure funding from the Austro-Hungarian monarchy with its more obvious and pressing concerns in 1913, Teleki's fantasy museum dedicated to heroic exploration of the physical world could not come to be in the twentieth century—at least not yet.

———————

Returning from the AGS excursion in 1912, our aspiring map men belonged to a world of transnational frontiers. They loved nature and sought out new landscapes. They guarded their newfound status and selected out geographies that best suited their aims. In an era of colonial power, however, entire histories were lost or suppressed. Bowman, for instance, like his mentor Davis, never objected to explorers' conquests, or their involvement in the slave trade, or the merciless violence and dispossession of Native Americans due to white settlement.[77] Penck's pupils sought power and privilege, knowledge and science, by contacts near and far among professional like-minded men. Products of Europe's era of networks of knowledge across America, Asia, and Africa, our men became more vested nationally in their new middle-class status, grounded insecurely in their academic chairs—and among their "sons" in stateside institutions. As young boys became men and explorers became patriotic bearers of myths of progress, they grew bored. Given what privilege they had gained, then they grew worried.

Our geographers were dreamers. They sought to make great discoveries, to be the next Humboldt, Darwin, Rhodes, or Livingstone. They saw the world's frontiers as penetrable and mostly borderless. In scientific pursuits, the men saw little contradiction between nations and internationalism. They had notable patrons of an esteemed and chaired sort, charismatic proponents of intellectual cooperation. All were the new professional map men of East Central Europe, for in August 1914 Teleki searched for Hungary's fictive unity; Rudnyts'kyi advanced Ukraine's territorial integrity beyond mere ethnography; Bowman sought greater influence for the United States; Romer produced anti-Ruthenian maps for Poland; and Penck in Berlin supported the Second Reich. When the First World War broke out, our geographers finally would cloak in scientific garb the prepossessions of what they loved most of all—their nations and families, their lives and maps.

CHAPTER TWO

OBJECTIVITY

The lost transnational life of Albrecht Penck, the geomorphologist who proposed the first 1:1 million map of the world in 1891, reads like the history of geography itself. In spring 1914, the Royal Geographical Society in London bestowed upon him its prestigious Founder's Medal, and Oxford University gave him an honorary degree, putting his reputation at its height in the Anglo-American world.[1] In summer 1914, the famed professor was on an expedition to Australia, sponsored by the British Association for the Advancement of Science. Penck, the holder of the world's first academic chair in geography in Berlin after Ritter, was an immediate person of interest in England. He was the director of the Berlin Oceanographic Institute and a key figure of the Berlin Geographical Society, the world's second-oldest such society, set up in 1828. When the war broke out in August 1914, things unraveled quickly. When Penck returned to London, the British press stirred up a jingoistic populace. Scotland Yard detained him and confiscated his notes and maps. Though Penck surely supported the kaiser's war aims against British sea power, the scientist was not a spy. Penck was arrested and detained in London, between September 1914 and January 1915. Scotland Yard interrogators were interested in the maps of the man who was the acting head of the society. From the international expedition to Australia, he had in his possession, for instance, a German map of New Guinea. Trying to dispel the charge of espionage, Penck noted the cooperative spirit of hospitality and inquiry on board the steamliner. The Anglophile appealed to the standing of geography as an objective science.[2]

A creature of the nineteenth century, Penck sought to separate geography's sciences from national prejudice. He was of the opinion that the *Times* and *Guardian*, which vilified his Swedish friend Sven Hedin, whipped England's public into such hysteria that the mass media blew Germany's high seas rivalry with Great Britain way out of proportion.[3] Only in January 1915 could he return to his family in Berlin. In hindsight, Penck's apologetics for his country's war aims from 1914 to 1919 could be associated with any racist, protofascist, or neocolonial form of revision, *völkisch* geography or *Ostforschung*, German chauvinism, or even forcible removal of other populations. But if history is more than a twentieth-century search for culpability, Penck had a point. His prejudices were the codes of a confraternity. The accused spy's past was full of illusions shaped by interpersonal contact, his habits of learning, quests for institutional belonging, the hopes of a vanishing aristocracy, and the rise of a middling bourgeoisie. Coordinates of *Bildung*, *Wissenschaft*, *Kultur*, and *Objectivität* shaped his mobile sense of place. They were virtues and vices of a persistent, if ill-defined, Ostmitteleuropa in a global world.[4]

WWI COLLISIONS

In *Detained by England*, Penck's practically unreadable, overly dramatic account of his arrest in 1914 and early 1915, the layers of the past were evident. Geography may have acquired scientific status, but maps were still colored by power and privilege, anxiety and fear, biological notions of race, ethnicity, and sexual difference. Before he left for the expedition to Australia, Penck's family and German students bade him a warm farewell. Penck recalled lively exchanges with the Englishmen he met while on board. He invoked cordial relations with English scientists and their wives, sons, and daughters whom he met. He added tropes of *Wanderlust*, traveling the world like a Humboldt in search of geography's leading lights. He performed roles of a middle-class family man. He went to battle, in a false parallel of soldiers and athletes alike, with the special ability he possessed. The expert detailed Australia's geography, adding past memories of travel in South Africa from where, in fact, he and Bowman had corresponded in 1905. He held conversations with German-speaking Australians, impressed by their high German and knowledge of Goethe and Schiller. On board, the professor recalled a certain Pastor Krueger in London, a Protestant working in the London Home Office who was responsible for aiding German women and children. He spoke of his son, Walther, a geographer and doctor of geology who studied at Yale and followed his example. He regretted missing Walther's habil defense and the baptism of his granddaughter. Meanwhile, the British press worried Ida sick. To make the yearning for home complete, Albrecht described how his loyal spouse wrote to him while he was under arrest, concerned as she was for her husband's welfare.[5]

Already by early 1915, Penck was infuriated by everything he lost, and so quickly. World War I changed him and our other geographers, bringing out the emotions and fantasies of their fragile selves. Released from London and back in Berlin by February 1915, Penck put his expertise to use for the Kaiserreich's geography. Drawing from his friend's book, he rehashed Hedin's pro-German arguments as a justification for Germany's two-front war. He read and reviewed it once back in Berlin.[6] In a speech of 30 April 1915 at the University of Berlin, "What We Have Won and What We Have Lost in the War," Penck wrote,

> The Germany that I have encountered on my return from England was different from the one I left in the summer. The wave of enthusiasm at the outbreak of war . . . has raised our people to a higher plane, and made us come together in uniformity and solidarity. Certainly, we have been at a place in the sun. The war has given us the feeling of strength, unity, and will, without which a nation cannot rise to world power. Today, the war reveals for us the unity of will to victory, in which we sacrifice the best interests of our nation. This will to victory is our advantage, to develop the ideas of humanity that are now threatened by brutal violence.[7]

Penck published it among like-minded Germans in the journal of the Berlin Geographical Society. In "The War and the Study of Geography," he placed geography in the service of statecraft, framed into a national tradition of Humboldt, Ritter, Ratzel, and Richthofen. War, he argued, was a tremendous opportunity for growth. This was true not only for geography, but for the immersion of German pupils into a patriotic education, visually into wherever their kinfolk settled and accomplished great things on the globe.[8]

War nationalized Penck even further. He next applied Europe's colonially globalist ideas to Germany's wild west to the East, in Poland and Ukraine. In his 1916 "Ukraina" article, Penck mapped the steppes into a German sphere of influence.[9] Fixated on the wide-spanning green grasses and black soil for farming and settlement, Penck depicted Ukraine across Eurasia, in regions absent of cities, industry, modern politics, or social movements—in short, an ideal frontier space. He wholly ignored Habsburg multiethnic history and Ukraine's diverse linguistic and confessional populations. Penck spoke of "German folksongs from the forest . . . the wide steppe [and] areas of high grass." He described the black soil by the German term *Löß* (*Loess*, or "loess" in English), an archaic Alemaic word meaning "loose," a derivation supposedly from peasant settlers in the Rhine Valley, common as well to soils of the Great Plains and Mississippi River Valley in the United States. For Penck, the soil seemed to possess an erotic quality. He referred in detail to how Richthofen, his hero and mentor, experienced its sensuousness: "You can crush it between your fingers . . . it is fixed and loose at the same time . . . a dust that has fallen out of the wind, settled gradually, captured by the grasses of the prairie . . . thus appears the pattern in Ukraine."[10]

In this way, Penck's pre-1914 geomorphology collapsed into reverie about nature and emotionally charged stories. The explorer's outdoor life was set by frontiersmanship and the *Heimat* literature he loved. In Penck's relationship with Rudnyts'kyi, his former student, he clearly used him to play Ukraine's claims in Galicia off of Poland's for Germany's wartime aims. Singling out Stepan for acclaim, Penck elected to rely not on Romer's expert ethnographic knowledge ("falsifications," as he called them) for Ukraine, but rather on Germanophone intelligence and Rudnyts'kyi. When Romer's atlas of Poland was published in Vienna, Penck was outraged. The Berliner's appeals had already raised many eyebrows in 1914–15, drawing the suspicions of Polish geographers including Romer, who made use of sympathizers to the Polish cause and his contacts abroad.[11] Romer hustled to prepare his sizeable *Geographical-Statistical Atlas of Poland* in Vienna at the Military-Geographical Institute, to make persuasive arguments in graphic form. The giant folio was debuted before an audience of Polish academics in December 1915 in Kraków and published by the Viennese firm of Freytag & Berndt in early 1916. Polish activists sent his atlas via Sweden, for fear of confiscation, to London, Paris, and Washington. Disturbed that it was made in the capital of Germany's ally, the "friendly" Penck went out of his way to alert the proper authorities.

We shall return to the gist of the Penck-Romer conflict, but on 12 April 1916, the Berlin professor sent a letter from the museum of the Berlin Oceanographic Institute to Hans Hartwig von Beseler (1850–1921), a veteran of the Franco-Prussian War. At that time, von Beseler was the governor-general of Warsaw. Penck alerted von Beseler to Romer's anti-Mitteleuropa sentiments and played ethnic and conspiracy cards together. He noted, by a kind of *Kulturkampf*, that Poles were Catholics and had "powerful friends" in Austrian Vienna. Unsympathetically, Penck accused Poles in Galicia and Czechs in Bohemia and Moravia of "treason"; by contrast, he lauded Rudnyts'kyi's "excellent [and] rare knowledge of the country." He griped about German media at home for ignoring its own experts, just as he excoriated the British press in London. He stressed that it was hard "to penetrate deeper into Austrian conditions," by which he meant not Habsburg lands but a "German" Poland and Ukraine. Penck learned of Ukraine's geography "only thanks to my old friendly relations"—in other words, Rudnyts'kyi.[12]

Our old confraternity was collapsing fast. Romer, Penck's former student, was now his rival, the chair of geography in Galicia at the University of Lwów since 1911.[13] Significant here was the fact that despite creeping animosity, the geographers selectively retained their epistolary channels and codes of amicability. Penck saw Rudnyts'kyi as assimilated, like Hedin, by nationality a German or at least Germanophile. Complicated by his aristocratic origin, Romer posed an exception. A caesura between "German" Penck and "Polish" Romer was probably overdue in the era of modern nationalism, but it was not inevitable. The men exchanged postcards since at least 1902, when Penck in Vienna first wrote to his former student.[14] Once they became aware of their rift, in fact they tried *harder*, even in the tense year of 1916, by private correspondence to hold the relationship together. On 22 May 1916, Romer replied to his former professor "in anticipation of a friendly message." He signed it "your very devoted [Romer]."[15] Romer knew by then that Penck had received the atlas, and he defended his use of sources in both German and Polish, insisting they were not only "solid" but "first class." Romer stressed Polish unity of the lands, and he alluded to the "scientific lectures" that he had given previously in Vienna and Kraków. Romer brashly suggested that Penck should invite him to Berlin, to lecture at the Oceanographic Institute's museum on the topic of "The Poles and Their Relation to Central Europe." He flashed his credentials into German Mitteleuropa, though by 1912 he had ostensibly rejected the concept. Having researched Poland "long before the war," in Romer's words, he claimed to base his atlas solely on objective data for modern industry and agricultural progress. The West Galician formally concluded with "expressions of my highest consideration" to his esteemed former professor.[16]

Honor codes persisted when Penck aimed to prove Romer wrong about geography, even in the nationalizing climes of 1916–17. Gentlemen adhered to bourgeois rectitude, all the while reasoning that they could keep the quarrel beyond the nation, on a plateau of civility. Penck wrote to Romer from Berlin on 5 June 1916, "The first delivery of your atlas of Poland was waiting when I came back here from Vienna. . . . Only last week did I settle down again in Berlin, and was I able to devote myself to the manifold tasks that remain undone. . . . Above all, I want to thank you . . . for sending your atlas . . . [and] pursuing the Polish question." Penck noted Romer's technical deficiencies in presenting the maps. He broached its contents: "Frankly speaking, the atlas has not met expectations . . . [This] method for displaying political-geographical data hardly seems satisfactory to me . . . I would have had at least a representation of languages, including population density, etc." He complained about young Romer's misuse of population density, his binaristic reduction of peoples to Germans and Poles at the expense of other minorities, and his outlandish claim to the Polish "national purity of a people." Again the subtext was a fantasy—for it was Germans, not Poles, who in their civilization to the East were rooted in the natural fertility of soil and culture. Only "German" experts had the authority to arrange maps and statistics for aggrandizement and/or conflict resolution. Responding to the Galician's effort to invite himself to Berlin, Penck offered an old appeal to pre-1914 excursions, "to make materials familiar . . . objectively understand the extent of the Polish question . . . [in] visits to Poland."[17]

Using appeals to objectivity, the former nineteenth-century experts tried hard to persuade each other of nobler scientific standards, above prejudice.[18] Yet they were pushing national, even imperial, agendas in colonial language. In a letter in December 1916, Penck thanked Romer for the "second delivery" of his Polish atlas. He even complemented Romer for his "representation of the great German fertility of soil" (*Bodenerträgniss*).[19] Penck contained the Poles into the Congress Kingdom of 1815, his Prussocentric take on the Two Emperors' Manifesto of 5 November 1916. He then threw his weight behind von Beseler's commission in Warsaw. He wanted to obtain from the east handbooks, maps, atlases, and other intelligence. He was proud of how "our people [Germans] had worked hard from there [Warsaw] and have made vigorous efforts to close the gaps of knowledge that the Russian government has disrupted." He ended with his own authority, an allusion to the works on the geography of a "Slavic" east, without knowledge of or interest in their languages, for the Berlin Geographical Society. Penck wrote, "I am still fully engaged with my academic work and administrative activity." Colonially, the professor was out of place, but not out of line. We might parallel the Berlin transplant's statements here to Teleki's love of Japan and Africa in the 1900s, or Bowman's dreams of U.S. power in Latin America.

By the beginning of 1917, our geographers had become more frustrated and obsessed with moving wartime borders. Borders (*Grenzen*) doubled as open frontiers and sovereign limits around a claimed territory.[20] Men now busied themselves by disproving each other's fallacious arguments from authority. In a German letter from Lemberg on 31 December 1916, Romer resisted the idea of "natural boundary" determined by Vienna settlements of 1815. Romer now interpreted the 5 November 1916 declaration as a first step to a Greater Poland. The Two Emperors' Manifesto was in no way binding. At best it was a blueprint, at worst a false hope for the Poles. The Central Powers neither granted independence nor accounted for shrunken frontiers. Aware of this, Romer more deeply feared Penck's reputation in the West and the effects of German wartime disinformation. He worried that Penck would be regarded as the "primary authority in German geography . . . his fame looks out to the whole world, and at the same time he has friends in confidence and is counted among the top men in the state." Penck was a "highly placed and influential man" who had judged his atlas "very negatively" and "on German terms." The "political motives [of the] great Berlin professor," Romer declared, were no longer "rooted (*gewürzelt*) . . . in geographic and scientific interest."[21] Romer cited the *Politische Geographie* of Ratzel to the exact page in the 1903 edition, where the author referred to Poland and its fourteen million in population, how it would be a country four times as large as Belgium and Holland, in order to point out German inconsistencies. Even Ratzel claimed, "The state cut off from the sea, enclosed by Russia, Austria and Germany, would never achieve such a degree of identity, and thus political influence, above that of a small kingdom."[22] By using Ratzel to offset Penck, Romer played his pedigree out of Ostmitteleuropa. His knowledge of Ratzel encoded his status as a German-trained transnational expert. It was a clever way of ascending to objective norms. Avoiding arrest in Vienna, Romer's atlas of 1916 and quarrel with Penck recast Poland as nineteenth-century European civilization itself.[23]

PAN-AMERICAN CAREERIST

Meanwhile, back in the formally neutral U.S. of 1914, Isaiah Bowman put his notions of objectivity to use by finding a new global empire to serve. His visions of dominance began not in Europe or on an east/west gradient, but in the American north/south manner starting with the Monroe Doctrine of 1823 and the geography of a hemispheric power. In 1915, he completed his *South*

FIGURE 2.1. Eugeniusz Romer, title page of the *Geographical-Statistical Atlas of Poland* (1916).

America: A Geography Reader, which extended the Roosevelt Corollary (of 1904) to the Monroe Doctrine, a way to ensure U.S. expansionist commercial interests in its markets and spheres of interest. In 1916, he published *The Andes of Southern Peru*, another product of his Yale dissertation research. He moved into world geography, seeking alternatives to German "living space" and geographers such as Ratzel. For instance, he translated from French *La géographie humaine*, a seminal work of 1910 by Jean Brunhes (1869–1932), who had become the world's first chair of human geography in the Swiss city of Lausanne in 1907.[24]

Bowman's Anglophile Franco-German-Polish networking savvy paid off in 1915, when he was appointed the director of the AGS in New York City, succeeding his patron Davis. Immediately, he looked for friends in his black book.[25] At Yale, he invited his friend Romer from Vienna to teach geography in New Haven during the 1915–16 academic year, replacing him for the interim period. Although this proved impossible given the British-German rivalry on the high seas, the gesture was what mattered. It revealed their shared Anglo-Polish globalism that was at stake in World War I.[26] Bowman had known Romer ever since the AGS excursion of 1912. Both unmade their German-Saxon backgrounds while coming of age as geographers. Romer was keenly aware of the need to bend the ear of his new American, prospectively pro-Polish patron. Through a conduit, Antoni Jechalski, he shipped his sizeable *Geographical-Statistical Atlas of Poland* of 1916 not cold to Washington, but via neutral Sweden specially to the care of Bowman and the AGS in New York City (figure 2.1). Romer's goal was to gain support through back channels for Poland's cause and assert his position, ahead of Penck, as a leading expert (plate 1). He and Bowman mapped their trans-Atlantic partnership not on closed-door *Realpolitik*, or the politics of Vienna or Berlin, or of the kaiser's Mitteleuropa, but in the newfangled Wilsonianism of democracy and friendship. In remarkable letters between Bowman and Romer well into the late 1940s, they grafted "friendly" yet ambiguous twentieth-century norms onto a mythic plane of Europe's cooperative colonial traditions.

Woodrow Wilson's election in the U.S. in 1916 was very fortuitous for Bowman and his budding relationships. Wilson, a Virginian and the Democratic Party candidate, an academic (the only U.S. president with an earned Ph.D.) and former president of Princeton University from 1902 to 1910, defeated the Republican Charles Evans Hughes. He first gained support by trumpeting U.S. neutrality; ironically, his cartographic wedge in the American electorate was epitomized by the slogan, "He Kept Us Out of the War!" Once in office, Wilson reversed course. When he asked Congress to declare war in April 1917 (back when presidents did such things), it proved a major boon for Bowman and his career. Bowman, director of the AGS, was selected by Wilson's head politico Edward M. House (1858–1938) to become part of the U.S. Inquiry, which became the impressively named American Committee to Negotiate Peace. Touting himself as an expert on border disputes in North and South America, Bowman devoted all his energies to prepare for the postwar settlements. Trusted and selected as the U.S. chief territorial specialist, Bowman by April 1918 was charged with gathering up personnel to collect census records, gazetteers, and encyclopedias—and most of all, maps—for the application of national self-determination and border construction in 1918–19. Wilson's American Committee to Negotiate Peace would comprise men of character like Bowman, who in the former Princeton academic's words would be the "only disinterested people" when the Paris Conference began. Few realized how green the U.S. chief territorial specialist was, having only traveled in Latin America and never to any place in Europe.

OUT OF EURASIA

Between Budapest and the family's Pribékfalva estate, World War I also nationalized our Transylvanian count. It presented the expert with similar opportunities. Initially, Teleki served the Habsburg emperor Francis Joseph I in Austria-Hungary. On 18 July, just three weeks after the assassination of Archduke Franz Ferdinand, Teleki was drafted into the Austro-Hungarian army. On 31 July, he took leave of Budapest for Sarajevo and onto the Serbian and Romanian fronts. In December 1914, Teleki received a promotion to first lieutenant in the

Irregular Volunteer Service. Not surprisingly, life was miserable there. When Italy entered the war in May 1915, he was sent to the southeastern front to serve as a command officer. As the war dragged on, the Hungarian's enthusiasm for a German-led effort waned. The smallish Teleki was really in too frail condition to continue on the front. It was stalemated anyway. By fall 1915, the count had had enough of World War I. He hoped it would end soon. Sickly, bored, and ill-equipped for combat in the east, he was thrilled for the Hungarian cause when the military's general staff ordered him away from the battlefields to prepare military-topographical maps. This was boyish fantasy and a glorious opportunity. Teleki dedicated himself to producing cartographic knowledge, right up to his selection in 1918 into the Hungarian delegation. Count Albert Apponyi (1846–1933), the pro-Magyarization minister of education and a respected figure in Hungary (he was nominated multiple times for the Nobel Peace Prize), chose Teleki as his leading geographic expert in preparation for the Paris settlements.[27]

During World War I, Count Teleki of Austria-Hungary developed his outlook in political geography from earlier readings of Ratzel, adding Europe's biological prejudices about racial eugenics, assimilation, and urban development.[28] Teleki in the war's early years had an abiding interest in theories of the Turanian Society in Budapest (it lasted from 1910 to 1943), for which he served as president until 1916. He wrote regularly in its journal, *Turán*.[29] He was also the vice president of the Hungarian Eastern Cultural Center, where the explorer Count Béla Széchenyi (1837–1918) served as president. Among the Turanians were men preoccupied with Finno-Ugric languages and Magyar decline and dispersal. Studies varied from the population's settlement in the Danubian Basin to popular dreams of a Greater Hungary in the Balkans, in faraway "Eurasian" spaces from Turkey to Russia and Japan. The movement was protean and Teleki probably shared such latent fears. Yet he considered himself a scientist and therefore moderate, or at least this was how he reasoned. When he prepared "Landscape and Race" (*Táj és faj*) for the general meeting on 2 May 1916 of the Turanian Society, by then he was using geography as a tool of Greater Hungary research. He operated in the default mode of *Ostforschung* (German area-specific research on Europe's east) for mapping and studying populations by nationality in Transylvania and Romania and along Hungary's Balkan frontiers.[30] He sought a syncretism of *Landschaft* (landscape) and *Rasse* (race), cherry-picking ideas from familiar Germans as during his studies under Lóczy. He corresponded with foreign geographers, but he preferred not to get involved in military affairs. Instead, he served the Hungarian state in a civilian's capacity in its push for postwar independence. Teleki wrote on pedagogy in geography and continued as secretary general of the Hungarian Geographical Society (Magyar Földrajzi Társaság, or MFT), which remained a site of contact and intrigue for dilettantes to gain credence while securing their privilege.[31]

Offering his rejoinder to Herder's dire prediction about the disappearance of the Magyars in East Central Europe, Teleki in World War I appealed to Turanian exceptionalism across the Central Asian steppes.[32] From the prominent work of Paul Vidal de la Blache (1845–1918) on France's regions, he developed the notion of a Magyar-based "historic" space that was not purely linguistic or ethnographical, outside the framework of *Ostforschung*. In this way, Teleki slyly evaded the Anglophile empiricism of a Germanocentric generation in the 1880s and 1890s of Lóczy and Penck, turning to human geography that was not captive to the nationality principle. Magyars were the Danubian Basin's civilizers, having come from Eurasian climes into Europe, the Christian spaces of the St. Stephen crownlands. (Recall Teleki's 1908 marriage to Johanna, in St. Stephen Basilica in Budapest.) In "Turanianism as a Geographic Concept" (*A Turán földrajzi fogálom*), the expert appealed to authority and argued that the Turanian region from which the Magyar race emerged was a frontier.[33] Teleki found a useful spatial grammar, that Magyars were less a group than an entire European civilization, against outlying "primitives." Hungarians, like Europe's settlers, had a manifest destiny. They moved into the lands of St. Stephen, expanded out from Budapest, the Danubian Basin, and Transylvania well toward Romania and across the plains. Cognitive maps are muddled— where the U.S. West of the AGS 1912 excursion merged with the novelistic fantasies of Karl May, then was brought to modern Hungary via Africa, Japan, and Eurasia in 1916 by a Grand Canyon–loving Transylvanian count.

Borne of frustrations in 1915–16, Teleki's Turanian (or Eurasian) turn in geography was clearly infused by nationalism, as with so many fantasies of territorial gain in World War I. Nationalism does not explain everything, however. When the war turned worse, the count held to his illiberalism, which was more expansive. In August 1916, Romanian troops crossed onto the grounds of Teleki's Transylvania, prompting more fears of the loss of status, not to mention landholdings. There, in place, legacies of family lore kicked in, of the presence of Romanians since 1848–49, which pressed the clan further into compromises of office-holding politics, shaping their phobias and demographic obsessions. During the stalemates of 1916, Teleki took more interest in the *Geographical Bulletin* (*Földrajzi Közlemények*), the main publishing organ of the MFT. He wrote numerous articles, speeches, and reviews (from 1899 through 1929), principally of French and German books.[34] Pál favored the colonial MFT as a more respectable venue than the Turanian Society. He turned "scientist" by reviewing in 1916 a key German article in the history of the early modern cartography of northern Albania, published by Baron Franz Nopcsa von Felső-Szilvás (1877–1933), a professor of geology and paleontology in Vienna, for the journal of the Vienna Geological Society.[35] Concerning frontiers, in service to the state he looked as an academic for practical spatial models.[36]

Power grabs certainly motivated this Eurocentric geography. Pál in 1917 plunged himself into anticommunist politics, working with the Constitutional Party in Hungary. The aims of the aristocrat were not altogether coherent, but his passions were clear. He self-published *The History of Geographical Thought*, outlining theoretical aspirations for the discipline in Hungary's economic, educational, geopolitical, and social spheres.[37] In 1917, he delivered the first of many speeches before the parliament, in effect lobbying for government financing and the professionalization of Hungarian geography. Teleki even personalized his love of geography in a speech to the MFT main assembly in Budapest in 1918, which, he once admitted, was owed to a boyish obsession with Karl May. He noted, "Youth do not have literature on geography . . . [our] work is considered inferior . . . I can tell you, ladies and gentlemen, that I myself read all the geography books, from which I studied sketches of the earth. It is something I still recall most vividly from the books of Karl May. It is true that they are filled with fantasies (*fantaziaval*) and much that is not geography . . . [but] they persist in the child's memory."[38] Nevertheless, leaving emotions and youthful fictions behind, it was geography now that inspired the lifelong traveler's mid-wartime blueprints for a place out of Eurasia, from his late father's estates in Szatmár (Satu Mare) County south of the Tisza River, onto the Great Plains (Alföld) of Hungarian-Romanian frontiers, into a scientifically based European civilization.

FANTASY EASTS

As outlined in chapter 1, Romer in Poland probably had no greater rival than Stepan Rudnyts'kyi in Ukraine, at least before 1914. Starting in 1914, Rudnyts'kyi published new works in support of his country and the Central Powers. He was objective but never neutral, colonial in his knowledge, postcolonial in his politics. He aimed for creation of an independent federal Ukrainian republic in post-Habsburg space, and hoped in vain for international support.[39] When he left Lviv for Vienna at the outbreak of the war, together with many influential Ukrainian scientists, writers and activists, Rudnyts'kyi took part in the Union for the Liberation of Ukraine. "Levenko" was his preferred nom de plume. In Vienna, he reworked his *Short Geography of Ukraine* from Ukrainian to German, for use in and beyond the Central Powers' territories. His 1914 *Ukraina und die Ukrainer* was translated into English as *The Ukraine and the Ukrainians*, published by the Ukrainian National Council in New Jersey in 1915 (plate 2).[40] Working for the government-in-exile of a West Ukrainian National Republic, the *Short Geography* came out as *Ukraina: Land und Volk* (*Ukraine: Land and People*) in Berlin in 1915 and 1916. The vast 416-page *Ukraina* work, which featured forty illustrated tables and six maps, secured his scientific reputation in Berlin, Vienna, and beyond. It elevated his status as an expert on the east, especially in the eyes of Penck. His training in the natural sciences held great sway. A very Germanocentric map, the *Ethnographic Survey Map of Eastern Europe* (*Ethnographische Übersichtskarte von Osteuropa*), was appended to the 1916 edition of his general geography when it was published in Vienna (plate 3).[41]

Insistent on the unity of Ukraine's regions and Ruthenians and Ukrainians, Rudnyts'kyi's various work in World War I can be summarized in brief. First, if political geography was ever honorable as a science, his German-Ukrainian writings surely had an agenda. His books prioritized Ukraine's cross-border unity of populations with historic and demographic rights, along the Zbruch River and eastward across the Eurasian steppes. Rudnyts'kyi's preference was for a Ukrainian republic in Europe, with an electoral body and its capital in Lviv. Yet he too "spoke map" in common colonial codes. He reasoned from wartime Vienna that Poland and Russia were more significant dangers to Ukraine's future than the Hohenzollerns or Habsburgs. He advocated for "small" Ukrainian statehood as a mediating, safeguarded polity in Europe, proposing a Baltic to Black Sea federation (or bulwark) of small states stretching from Finland to Ukraine. Like Teleki, he foresaw the dangers of selectively applying language as a proxy for identity in order to carve out new nation-states. The Habsburg subject warned against ethnic antagonisms after empires, something he knew from Galicia. He forecast—correctly, as it turned out—that without this kind of cooperation, Ukraine could be reannexed by Soviet Russia. Communists in Ukraine, he thought, had the intent not just of revolution in Europe, but also the long-term prevention of Ukraine's statehood. His post-Habsburg plans for a buffered federation of sovereign states did not happen, for this was grander twentieth-century geopolitics. Yet Rudnyts'kyi predicted the future promise and dangers of unequally granting legal and cultural rights of national self-determination. Without incorporation into European structures or juridical backing by philo-Polish great powers in 1919, the transnational rights of minorities would remain unrecognized by many states.[42]

At this crossroads for making Ruthenia into modern Ukraine, Rudnyts'kyi's short political geography was printed in English, French, German, Italian, Hungarian, Czech, and Russian. Notably, it was never translated into Polish. Relative to the power of Poland, Germany, or Russia, Rudnyts'kyi's position on Habsburg diversity was never quite in a one-to-one "postcolonial" dialectic with a group or nationality. Yet as a statist, he was hardly liberal on the issue of minority rights or nationalities. He adapted Europe's maps to follow a sedentary logic like Penck and Romer, privileging a future Ukrainian kin-state, showing the maximum number of Ukrainian speakers on "ethnographic" frontiers for settlement. His caveat was that ideas of conational unity, what to other powers seemed like complicity with foreign powers or a Ukrainian Empire, from his Ukrainocentric standpoint would actually *prevent* a fratricidal war between Ruthenians in Habsburg lands and Little Russians in imperial Russia, who moved through and dwelled on opposite sides of the Zbruch River. To place Ukraine on Europe's map, he thought of Ukrainians not as local dialect-speakers or separate victims but as a group, communal sharers of high-cultural history, language, and space. In his 1915 pamphlet, *Why Do We Want an Independent Ukraine?* printed in Vienna, Berlin, Lviv, and Stockholm, he mixed science and propaganda to argue on behalf of these mobile rights.[43] In the 1916 article, *The Ukrainian Question from the Standpoint of Political Geography* (*Ukrains'ka sprava zi stanovyshcha politychnoi heohrafii*), in the Ukrainian daily *Dilo* printed in Lviv, he advanced national-geographic unity on open frontiers. NTSh members evidently distributed his works to Ukrainian POWs held by the Russian army, imprisoned in Austria-Hungary. In early 1917, the education-minded Galician followed the model of Romer's atlas of 1916 by preparing separate maps for a historical atlas of Ukraine. Due to political obstacles, financial constraints, and the lack of a clear patron abroad, these were never published as a single folio.

While the collegial exchanges of Penck and Romer in 1916 turned acrimonious in the first six months of 1917, Rudnyts'kyi insisted on Ukraine's sovereign territory. Geographers' maps intersected, but all nationalisms were not identical. Between empires and nations, Galicia was a Habsburg-Russian borderland, a zone of combat in World War I, and a wilder, more violent west on a mass scale than America west of the Mississippi.[44] Rudnyts'kyi got thrown aside easily, when Romer went public with Penck's review of his atlas on 5 June, on 1 January 1917 in an article for the *Kurier Lwówski*, and on 4 January in the Kraków-based *Głos narodu* (*Voice of the People*). In Berlin, Penck wrote to Romer from the Berlin Oceanographic Institute on 10 January, acknowledging his receipt of Romer's work, but not his agreement with Romer's claims. Penck's letter seems to have been delayed. It arrived too late. He asked Romer to make a copy of his private letter first so that he could "take it to the appropriate position," likely von Beseler in Warsaw. Romer replied in German on 15 January "with the expressions of my deepest devotion." He followed with an apology for how the exchanges had been handled. Whether it was sincere is not clear, but we do know that behind the scenes, Romer tried to discredit both Penck and Rudnyts'kyi in tandem. After all by 1917, these men shared in the vision of reducing Poland to its smaller Vienna borders of 1815, in the tsarist Kingdom of Poland.[45]

The next phase grew heated. Tensions mounted in a specific timeframe between the Two Emperors' Manifesto of 5 November 1916, the first Russian Revolution of February 1917, and the formal U.S. entry into World War I in April 1917. As the February Revolution broke out in Petrograd, the Ukrainian factor in the German-Polish conflict could not be ignored. An anonymous author wrote in defense of Ukraine's independence on 16 February in the publishing organ of Ukrainian Central Rada, *Ukrainische Korrespondenz*. The author may just as well have been Rudnyts'kyi; the article plagiarized his geographies.[46] In an angry response in defense of his Polish atlas and his *Military-Political Map of Poland on the Basis of the Emperors' Manifesto of 5 November 1916*, Romer on 16 March played the hard-line Galician card. He dismissed "Ukraine-centric thinking" (*Ukrainismus*) because it had a malignant effect on Poland's world image. He rejected multiethnic tolerance. He stressed the objective authority, over anything made by a Ukrainian, of his published maps in French and German, treating Ukrainian claims as baseless in Poland's "sphere of interests." He stressed that "the main goal of Polish sciences . . . [was] to marshall all their forces to the defense of truth and knowledge of . . . Polish interests." Romer thought any reference to a "Polish-Lithuanian-Ukrainian state" (*polnisch-litauisch-ukrainischer Staat*) was absurd. Hence, he rejected his father's old Habsburg project of Austro-Hungarian-Polish triloyalism as well. He concluded with an appeal to the like-minded, "Risum teneatis amici!" ("Friends, would you not keep from laughing!"). He could find no good reason to justify an independent Ukraine. Romer's maps were of a Greater Poland, not from 1815 but before 1772, in a Polish sphere of influence from the Carpathian Basin to Upper Silesia and Pomerania, and along the Baltic Sea corridor.[47]

Pretenses of objectivity seemed to vanish. Romer in anti-Mitteleuropa was guilty of doing to Rudnyts'kyi precisely what Penck had exacted from their broken relationship. In essence, Penck dismissed Romer as provincial, questioned his impartiality, and accused him of willful propaganda. On 27 March in the newspaper *Posener Tagblatt*, Penck published his "Polnisches" article in which he reviewed in full Romer's 1916 atlas. Penck wielded Ratzel's *Lebensraum* right back against Romer, that Poland's borders were "not fixed," and that "even among the Poles there are far-reaching differences of opinion in regard to the concept of 'Polish.'" He objected to "Romer's Poland . . . [that was] not only the province of West Prussia and Poznan, which coincided with Prussia during the Polish divisions, but also to East Prussia and most of Silesia." Instead, he argued, "Prussian Poland . . . [was] part of the German Reich with its nearly ten million inhabitants . . . Poles, Mazurians, and Kashubians." He judged as inferior the process of Romer's mapmaking, from the stage of financing, to technical production, to the visual effect on readers of its contents. He accused Romer of being backed by "Polish savings companies," targeting Romer's patron in Vienna who bankrolled production of the Polish atlas.[48] He blamed Romer for having produced the text not only in Polish (in the first place) and German (thirdly) but also in French, "the language of our enemies." He noted low counts of Poles on the edge of its frontiers, contrasting it with settlement "on German soil." *Völkisch* "American" frontiers in Ukraine for Penck's twentieth century appeared as sites of adventure, settlement, and feats of national strength—the unfurling destiny of *Volk, Raum, Kultur*, and *Boden* as attestations to German civilization. The Berliner refused to accept from a mere "Galician professor" the idea of an independent Poland.[49]

Emotions rise, like lava to the surface, in times of war and revolution. Penck from his Berlin chair of geography argued ad hominem against Romer. He accused "the Pole," in his words, not simply of being injudicious or too selective, for such distortions could well occur in any map generalization. Rather, Penck claimed, Romer *intentionally* falsified Poland's geography for the whole world to see. Penck mentioned the "memorable proclamation" of the two emperors on 5 November 1916, in line with the Vienna Congress of 1815. He cited as authoritative the Prussian Census of 1910. On the authority of German statistics, Penck wrote, "This share is higher than the participation of Poles in the general population, because in Polish eastern provinces of Prussia the Germans are more numerous offspring. Prof. Romer knows this well." He accused Romer of ignoring majority-German demographic presence on frontiers and German dominance along the Baltic. He then criticized Romer's technique: "In this way [Romer] manages to [show] a natural Polish ethnographic area considerably larger and well-rounded down to the sea, but [this] does not correspond to reality."[50] Politeness turned into hypocrisy. On 5 April, Penck thanked Romer for the delivery of his atlas in complete form, and noted his public review.[51] This exchange was the last the two men would have for the next seventeen years.[52]

Cartography often works as a kind of fringe science. In map wars of words and images, aspirational jealousy is one subtext. The worst insult either person could hurl was a dismissal of his opponent's career and ethical norms, an accusation of being prejudiced—in short, a dishonest man, a bad scholar, or a pseudoscientist.[53] On 4 May in Vienna, Romer published a detailed article in *Polen* in which he interpreted Penck's detailed public review as a charge that he had deliberately misused geographic science.[54] Romer's work was reprinted in *Kurier Lwówski* for Polish audiences on 10 May. He called attention to the reception of his atlas in the German press, pointing out that Penck had been slow ("ten months had passed") to raise *public* objections to the Polish atlas. Speaking on behalf of Polish geographers, Romer averred that "we regard as a joke" the idea that Ukrainians had historic rights to land, and that Polish lands existed for German development and settlement. Romer accused Penck in Polish of prizing himself as the "great protector of the Ukrainians," putting "Ukrainians" in scare quotes on the fantasy frontiers of Poland's historical geography.[55]

Rudnyts'kyi, wisely, stayed aloof from this muddle. East of Berlin, the Ukrainian geographer was exiled from Austro-Hungary and "Polish" Galicia by the war. He spent most of the latter stages after the Bolshevik Revolution in Vienna, with the exception of a trip to Berlin in February 1918 for Penck's *Festschrift*. His objectivity was on behalf of a transnational Ukraine, directed toward an audience of Europeans and Americans. It entailed playing multiple great-power sides of a continuing conflict. He expanded his German edition of *Ukraina: Land und Volk* in 1917 and 1918, which was translated into English and published in New Jersey as *Ukraine: Land and People* by the Union for the Liberation of Ukraine, a replica in the United States of the Vienna organization. More editions came out in Italian, Hungarian, English, Czech, and Russian. It raised Rudnyts'kyi's prestige worldwide. He became the leading expert and authority on Ukraine's geography. The book also appeared in French in 1918 and 1919, in advance of the Paris negotiations. Print runs of Rudnyts'kyi's books were phenomenal (recall here that the obscure Lan publisher had lost his original manuscript prior to 1914), exceeding two million copies. Objectivity mattered less by the war's catastrophic end. By the middle of 1918, the entire run of his *Short Geography of Ukraine* was again sold out.

APOTHEOSIS

So former map boys now had map wars. While impartiality was called into question, the advance politicization of German research on Europe's East

(*Ostforschung*) was reflected by Penck's life, status, and imperial position.[56] The professor gathered in allies during the 1917–18 academic year when he served as the interim rector of the University of Berlin (now Humboldt University).[57] In February 1918, a total of 162 of Penck's former and current students were invited to Berlin for a Festschrift, in honor of the great man's sixtieth birthday (he was actually born in September 1858). Selected among the fraternity of Europe's traveling scholars was Rudnyts'kyi. Romer was not invited.[58] The Ukrainian geographer arrived in Berlin from exile in Vienna and contributed a piece on geomorphology from his doctoral study of the Podolian flatland in Galicia.[59] Absent former students also included, notably, Jovan Cvijić (1865–1927), the Balkan ethnographic mapper and founder of the Serbian Geographical Society (est. 1910), who lectured during the war in Paris.[60]

The elevated tribute was a giant Germanophile 438-page book, across various disciplines of geography, published as part of the Bibliothek Geographische Handbücher series founded by Ratzel in 1882. Johann Engelhorn's firm in Stuttgart published the tome; it printed plenty of *völkisch* literature in the war, including Penck's account of his arrest by Scotland Yard in 1914. Penck's disciples deemed him a "keen observer" of natural phenomena, seeing "unmatched clarity" in his empirical research. They celebrated his "groundbreaking" labors and the announced his entrance into "the circle of men, in which geography has validity as an independent science." Everyone lauded the Berliner's warmth and generosity of spirit, how he stimulated their joy of discovery. From mineralogy to Alpine research and human geography, they emphasized the master's work and "the power of his personality and teachings." Our map man of Ostmitteleuropa, he who fell into post-1945 obscurity, was thanked in outsized terms for helping the world's geographers find solutions to their problems. As if Humboldt and Ritter came alive from bronze (both died in Berlin in 1859), Penck was apotheosized as "the wise and successful inspirer and organizer of geographical science."[61] Geographers became like holy figures, geography like wartime therapy.

Meanwhile, in July 1918 in German-occupied Warsaw, Penck's friend and colleague Erich Wunderlich (1889–1945), in charge of the Landeskundliche Commission, completed the foreword for the bulky *Handbuch von Polen*, the volume Penck was busy editing in Berlin. The younger Wunderlich was among Penck's trusted assistants during the war, at the university's Geographical Institute. The *Handbuch* survey project was undertaken in 1916 on the initiative of Colonel-General von Beseler, head of the Generalgouvernement Warschau.[62] Penck in Berlin was counted among the leading experts, in the traditions of Richthofen and Ratzel. Writing from Warsaw, Wunderlich praised the Berlin professor: "High respect is accorded to Privy Councilor Dr. Albrecht Penck, the geographer of Berlin University. He has drawn attention to interrelationships. He has stressed repeatedly the great importance of immediately carrying out regional studies of Poland from the German side." In the *Handbuch*, Wunderlich referred to Romer's geographic work eleven times, obsessively to counter it. He offered a litany of the expertise and titles of Germans, what fine empirical research was being done in Breslau (Wrocław), Krakau (Kraków), Lemberg (Lwów) and Warschau (Warszawa). He thanked Polish libraries, archives, and organizations as suppliers of information, but never in Polish. For the patronizing Wunderlich and Penck, Poland was made real by *German* academic *Ostforschung*, in labels by nationality, only insofar as "ethnicized" knowledge pertaining to a country's objective study was gathered and organized by their own scientists.[63]

Penck advanced this cause. On 3 August 1918, he spoke in the auditorium of the University of Berlin and invoked the history of Friedrich Wilhelm III of Prussia (r. 1797–1840), the university's nominal founder. He quoted with urgency the king's statement that "the state must replace through spiritual forces what it has physically lost."[64] He alluded to the role of nineteenth-century German sciences, which acquired knowledge about the earth. He noted the Prussian state's establishment of Ritter's chair, pointing out that Ritter was a university professor and a member of the Military Academy. Along with Humboldt, Ritter was the "master of country studies" (*Meister der Länderkunde*), a champion of comparative geography and "one of the two stars, who at the beginning of the nineteenth century shone with light in the geographical heaven of Germany." Penck lauded other Germans including Heinrich Kiepert (1818–99), the cartographer of Asia Minor and the Balkans, and Richthofen and Drygalski. He placed himself in the line of Richthofen, who moved from Leipzig to Berlin to become chair in 1886. He extolled the museum of the Oceanographic Institute, founded in 1900 and christened by Kaiser Wilhelm II in 1906. Even Humboldt and Ritter were recast as national defenders: "The great earthmoving (*weltbewegende*) war has revealed enormous benefits, and at the same time the directions in which further expansion (*Ausbau*) is necessary. The war has convinced us of the justness of the action to establish large state institutions dedicated to the cultivation of the sciences. . . . In that way the right path will be drawn for the future."[65] By such slick realignment, Penck canonized loyal Germans as the true torchbearers of *Wissenschaft*, promulgators of unity through science, frontiersmen and experts. They, like he, were Germany's elders, fathers coming home to educate their sons with the knowledge they had gained.

The patriarch never tired of insisting on his authority on these terms. A fundamentally insecure man, he always spoke with something to prove. He allowed himself to be drawn increasingly into esoteric debates over map generalization. In 1918, Penck informed like-minded geographers, viz., fellow German scholars writing in German-language publications, that Romer's maps were unscientific. They rested on shoddy Polish research techniques, misleading graphics, and unverifiable data. Inter-Allied reconstructors of Poland as a federal republic in 1918–19 were special targets of his envy, for German geographers were treated as instigators and locked out of preparations for the Paris talks (and in Paris itself). Penck charged that Romer's maps were false because the data used to claim frontiers along the Baltic Sea corridor failed to rely on the Prussian census of 1910, to which Romer actually had limited access for his 1916 atlas since the entire census data was not available.

Penck knew no Polish, however. In his basic xenophobia, Germany was always the center of learning. He remained suspicious of data reported by Poles or bilingual Polish-German schoolteachers. Upbringing of children also factored in here, for he accused Polish teachers of falsifying numbers of Polonophone pupils for gain. He denigrated the value of Romer's 1916 atlas of Poland by invalidating the 1911 *Schulstatistik* as a source, arguing that the German government never recognized Polish as an official language. He claimed the survey incorrectly included teachers from the lowest grades of schooling, where Germans were least represented. His charges again betrayed anxiety about migrants in the east, implying that because Poles had more children than the Germans, this did not entitle those peoples to rights beyond the 1815 borders. Penck prioritized the virtues of *Heimat*, *Volk*, and *Raum* in "German" global spaces for development and an expanding civilization.

By November 1918, Penck's cartographic anxieties were made manifest after the loss of the Second Reich and decolonization. Anger does not explain everything. It is true, however, that in the Weimar Republic, the conservative turned to the political right. Now lacking a militarized state in the Prussian tradition to be the patron of his academic work in geography, he took part in Berlin educational circles and made civic efforts to form *Volkshochschule*. Penck discredited Polish geographers by focusing on the 1910 Prussian state census against the 1911 *Schulstatistik* and denouncing pro-Polish maps such as Romer's and the one by the Vienna-based engineer Jakob Spett (see below) in 1918, *Nationalities Map of the Eastern Provinces of the German Empire*, published by the Justus Perthes commercial firm in Gotha. Penck accused both of violating the basics of German science (plate 4).[66] He picked apart the minutiae of many purported anti-German maps, noting that the 1910 census listed the Kashubian population as a separate nationality, in addition to Germans and Poles. Sometimes, he had a logical point. If Wilsonian self-determination were strictly applied, then the Kashubians, a minority among minorities, had the very same rights as Poles to the Baltic corridor. Penck moved toward interwar right-wing *Revisionspolitik* in this manner, away not only from Wilson's postwar League of Nations, but also from the spirit of objectivity that permeated the code of cooperative transnational geographers.[67]

How much of the shift was owed to Penck's ego is hard to ascertain. On 9 February 1919, Penck wrote an article in *Deutsche Allgemeine Zeitung* in which he announced, Romer-style, the start of a major cartographic project for the Weimar state.[68] Penck's work of February and March 1919 in their coloration, symbols, and content were designed as scientific texts. To raise the German public's literacy, he included a small-scale map to challenge Polish claims, but the map retained problems with the delimitation of frontiers. As Guntram Henrik Herb has shown in his superb study, Penck's map was also inconsistent in its graphic categorization of German and Polish majorities: the racial/ethnic/linguistic category of "pure German" (*rein deutsch*) in the legend was coded in such a way to appear Polish, exaggerating a Polish, and Slavic, demographic threat. The *völkisch* map created an impression that all the areas identified by cross-hatching were Polish. Penck's designs were finally reproduced in the weekly report of 14 March, and republished as a separate pamphlet. His maps expanded from earlier black-and-white work in the illustrated Leipzig periodical *Illustrierte Zeitung*. He used black dots to designate Germans, white circles with a black rim for Poles, and black-rimmed dots with a black center for the Kashubians. Since the symbols for the Kashubians and Germans seemed to merge, a color version was needed. In the one published in Berlin by the Geographical Society, blue dots designated the Germans and red dots the Poles, while Kashubians were represented by blue-gray dots, which unfortunately could not be easily distinguished from Germans.

This cartographic caviling showed a bigger trend. Obsessions with technique pose all sorts of dilemmas regarding modernity writ large, if modernity means rationality and usefully explains Europe's twentieth-century horrors.[69] Penck learned slowly what Romer grasped intuitively in a longer history of cartography, the subliminal codes of maps. The sixty-year-old Berlin professor in early 1919 could not find any rational, internal consistency in German maps. Maybe there was none. Nor could he give up his long-held belief in a Eurocentric progress of geography. Other limitations were less palpable. Penck's reputation as a renowned geomorphologist hurt him to the extent that he grew too comfortable, failing to gain technical training before 1914 in

statistical demography and cartography. Subfields of the new geography expanded faster than he could keep up. Penck was sensitive, even old-fashioned,
about charges of dilettantism and unfairness. After being one-upped technically by Romer's 1916 atlas of Poland, he now grew obsessed with how German
ethnographic maps had to be prepared by local and loyal experts in Berlin,
although this caused huge production delays.[70] Not until a full two months into
the Paris talks, on 19 March 1919, did Spett's 1918 map gain the attention of German geographers once a section of it appeared in *Le Temps*, the French daily.
Spett, a skilled Polish-Jewish engineer living in Vienna, was denounced for
his deceptions. National presses exaggerated sensitivities concerning the threat
of red revolution from the east in 1917–19 and Franco-Polish encirclement.
Mundane bourgeois comforts mattered more. For geographers in positions of
privilege, gendered norms about migrating Slavs or Jewish peoples "penetrating" borders factored in, out of fears that they would threaten a "German"
imagined civilization.[71]

Suspicions in Berlin further abounded as international proceedings of the
Polish Boundary Commission were carried out in March 1919, with Isaiah
Bowman in charge (see chapter 3). Here is where Penck's previous writings on
East European Jewry and on American-Jewish economic power, from his two
prewar trips to the United States in 1897 and 1908–9 and in his travel account
published in 1917, also mattered. There is little doubt that Penck's rational/emo-
tional maps of 1918–9 were informed by "judeobolshevik" equations. After the
Spett episode in March 1919, he spent the next two months pursuing a project
so gargantuan that it should be compared to Romer's grand 1916 atlas of Poland. Motives of men derived from fears of partition or dismemberment, that
a punitive Anglo-Franco-Polish alliance was being conjured against German
civilization. With Wunderlich and his own assistant Herbert Heyde in Berlin,
Penck assembled a team of experts to compile statistics, showing dispersals of
the Volk.[72] Frontiers encompassed some 11,000 communities of Polish, German, and bilingual speakers, based on the Prussian census of 1910. Penck and
Heyde represented a German majority with colored dots and squares, on an even-
tual 116 sheets for the *Map of the German Empire* (*Karte des deutsches Reiches*). It
was a reference to Penck and Hettner's *Das Deutsche Reich* of 1887, but by then
Penck's map was of an empire that no longer existed. The project could not be
done in time before the treaty settlements.

On 7 May 1919, when Weimar was first presented with the terms of the *Diktat*, Penck's team was still toiling away. Penck and Heyde managed to publish
just a part of the study in Berlin, the *Map of the Distribution of Germans and Poles
along the Warthe-line Networks and the Lower Weichsel, and on the Western Frontier in*

Posen.[73] The nineteen printed sheets covered only the area of the lower Vistula and the Warthe-Netze region. They handed over their maps to the state on 10 May. Penck promised more sheets to show "the predominantly German character of Western Posen," but Weimar was reluctant to turn to *Revisionspolitik* in Pomerania and refused financial backing. This lesson was not lost on *völkisch* men. In German debates in Berlin and Warsaw about postwar claims, two other friendly experts, Hans von Präsent and Dietrich Häberle, continued to revisit pre-1914 Prussian and Russian census data, Spett's map, Romer's work, and Penck's "objective" claims, content, and methods.[74] Despite these setbacks and his support of the kaiser in World War I, the Berlin chair of geography never lost his position during or after the war. Conventions of German academe persisted, as did regional academic respect for geography as an enterprise. When the so-called victors of World War I dispatched their plebiscite takers and sought a trusteeship guided by the League of Nations, Penck's German compatriots from Ostmitteleuropa were marginalized or forgotten.[75]

As if by accident, Penck thus became a post–World War I map man, if stubbornly via the modern geography of Germany's impossible borders, in a still unmappable East.[76] Without knowledge of Slavic languages, lands, peoples, or history (which he made no time to learn), the devotée of *Heimat* dreaded the loss of cherished illusions. German centers of knowledge and maps were universal, he thought, and could never lose their luster. Conversely, Polish designs and claims couldn't be justified even by science. The other side of Penck's settled Saxon identity of the 1870s–80s was his Prussian statism of the 1900s–10s, in which European geographers could simultaneously be inquiring scientists, civil servants, and handmaidens of empire. With the end of World War I and the German Revolution of 1918–19, however, the state had changed. The professor assisted in the drafting of an official petition of protest to Weimar, signed by more than 1,000 participants in the society.[77] By June 1919, Penck in Berlin became (ironically) what his rival Romer was in Galicia, a fretful defender of nineteenth-century Western civilization, now in the triply marginalized role of a decolonial European man, a prowar academic, and a revisionist.[78] When the freeze in German-Polish relations ensued, Germany's geographers were banished. They called for a boycott of the postwar International Geographers' Union (IGU). Colonial men of an expansionist order lost patience with the new zeal for democracy. Suspicious of Poland, Penck emphasized expertise for the purpose of defending Germans abroad, lobbying for territorial revision. Developing new modes and techniques for maps was a way to map the closed world and thereby get ahead. In this manner, Penck's isolation reflected the modes of *Ostforschung* practiced by men of the interwar period. For the rest of his life, the famous geomorphologist barely disguised his contempt for Romer and his map men, or for the Paris talks.

PAPRIKA GEOGRAPHY

In Teleki's turn from Turanianism toward Euro-nationalist anticommunism in Hungary's context of war and revolution, politics could be confusing. What worked was habit, in several ways. First, the count collected Europe's colonial maps, as he had done for Japan in the late 1900s, with "objective" future peace settlements in mind. This required a hunt for statistical data and graphics for soil fertility, population density, vegetation, livestock, orography (areal differentiation), and hydrography. When he reviewed an atlas with thirty maps published by the firm of Dietrich Reimer, he warned of the rival productive capacity of Russia across the Eurasian plains. Teleki elevated the colonial enterprise of geographical knowledge of the East, for the 1918 atlas extensively used German research on the Russian Empire as well as India, "The Crops of India in Their Geographical Distribution," published at the kaiser's Kolonialinstitut in Hamburg in 1914.[79] He held up as examples the institutes established in Budapest and the German ones in Istanbul, precisely where Penck's son Walther worked as a geologist. Second, Teleki admired Ostmitteleuropa geography but had to adapt it to national purposes. In his April 1918 MFT address, he bemoaned Hungarians' lack of self-confidence, scarce resources, and sparse support for research. Germany's Second Reich as it "organized now a network [of scientists] all over the empire" set the standard due to its "influence and willpower . . . [and] intellectual cross-fertilization," and developed broad programs for research in statistics, geology, cartography, and geomorphology.

Third, the aristocrat retained his passion for travel in the cause of modern science. He mused about trips similar to Penck's from the 1890s onward, to New Guinea and the Balkans, arguing that Hungary would reap results from more funding. He alluded to the prewar budget, since back in 1913 then the MFT could grow its library collections and acquire foreign journals, as all major societies in Europe had done. The bibliophile praised a favorite book by the adventurer Baron Nopcsa on the early modern mapping of Northern Albania (which he reviewed), also pointing out the Austro-Hungarian state's prominent work in mapping out Serbia and Ukraine (he excluded Romania). Fourth, Teleki handled his moodiness (he was prone to depression) and shielded feelings through science. He resolved his ambiguity toward the Central Powers' war aims by an academic's zeal for science and patriotic, duty-bound expertise.

He thought soldiers "would be involved only partly" while ethnographers and geologists conducted research in Europe and Asia Minor in the tradition of Lóczy and Cholnoky. Geography became Teleki's fused life-and-work quest, a job forged by war and taken with utter seriousness, for "national and scientific prestige . . . our appreciation and self-empowerment . . . vital and appropriate economic interests."[80]

These four subliminal contexts help place the "Carte Rouge" of Count Teleki and Baron Nopcsa in perspective. Teleki saw Hungary's need to retain the values and practices of scientific progress in geography. The map's full title was *The Ethnographical Map of Hungary According to Population Density* (*Magyaroszág néprajzi térképe a népsürüség alapján*; *Carte ethnographique de la Hongrie construite en accordance avec la densitè* [*sic*] *de la population*), published in 1918 and 1919 (plate 5).[81] It depicted the Magyar population in a bright red (*vörös*) (whence its informal name, "Carte Rouge"), moving into Europe and settling into a central area in Budapest and the Danubian Basin. Teleki's map betrayed nation-state fears of multiethnic society and diversity, often seen post-1918 and misleadingly by territorial architects as a previous Habsburg weakness. By September 1918, in preparation for Paris, he concluded similarly to Penck that "Slavic" and/or "democratic" influence over the peace talks was too great. Given Hungary's contested borders with Czechoslovakia, Romania, and Yugoslavia, he had scarce hope that postwar settlements would be arranged in its favor. In early October 1918, Teleki worked to form a league for Hungary's territorial defense, a carbon copy of the Germans' revisionist Deutsche Schutzbund (see chapter 3). As Austria-Hungary dissolved, dozens of experts, geographers, and statisticians rushed to collect census data, often according to the categorical focus after 1848–49 on mutually exclusive nationality and confession. Teleki took part in this as well, gathering up more maps of Hungary in November, illustrating ethnographic minorities and districts with Magyar speakers, distribution by confession, and levels of literacy.[82]

Fretting that they lacked time to make an entire atlas for the delegation, Teleki and Nopcsa made the "Carte Rouge" of late 1918 and early 1919 their showpiece while a postwar Hungarian state was cobbled together. Just as the armistice was signed on 11 November, Teleki fell ill with a serious case of influenza. On 13 November, the Romanian, Czechoslovak, and Serbian forces invaded Hungary with France's support. On 16 November, the government under Count Mihály Károlyi (1875–1955) was hastily established and Hungary's postwar republic was declared. On the defensive, mulling over how to improve the map at his Pribékfalva estate in December 1918, Teleki asserted population density even ahead of nationality. He followed the linear logic of modernization, that the density of assimilated Magyars increased as peasants moved to the city, became literate, and settled naturally into St. Stephen enclaves.

Magyars were the privileged European ethnocentric norm. They appeared in bright paprika red from core to periphery, across frontiers under threat of "dismemberment" by hostile small-power neighbors and unhelpful great-power victors. Entire districts, based on 1910 imperial census data, were shown bereft of populations that were actually present.[83] One square millimeter represented 100 inhabitants.[84] Teleki and Nopcsa thereby proved that nationalities could all be measured by density, all in the same way. As if one were mapping the Balkans and Asia Minor in the 1870s and 1880s, as German experts had done, or the U.S. West, they left deserted the uplands, marshes, plains, mountainous areas, lowlands, and frontiers with few inhabitants. Persons were of course present, including entire Native American nations, living and claiming lands there; in U.S. frontier history, one might recall President Andrew Jackson's signing of the Indian Removal Act of 1830.[85]

The "Carte Rouge" was printed in December 1918 in English, French, and Hungarian prepared on a scale of 1:200,000, adjusted to 1:300:000 for the February 1919 version. Teleki solicited the assistance of Nopcsa, whom he needed along with men from the MFT, the MTA, the Bureau of Statistics, and the Ministry of Commerce. Sources included a 1903 Hungarian ethnographic map and censuses of Austria-Hungary, up to and including the 1910 survey. It rebuked Emmanuel de Martonne (1873–1955), the leading expert of France's delegation in 1919, who skillfully used a similar technique for showing population density in competing maps of Transylvania. Martonne favored Romanian statehood across "eastern Hungary" (what Teleki called it), while insisting on French great-power sponsorship of the Little Entente. Martonne essentially concurred against Hungarian designs with Tomáš Masaryk (1850–1937) and Robert Seton-Watson (1879–1951). Probably, Teleki realized the collusion earlier when Masaryk turned to Seton-Watson, his ally in London in 1915, to publish maps in the *New Europe* periodical against the Central Powers and Hungary in particular.[86] Masaryk and the Scotsman Seton-Watson, supporters of federal solutions for Czechoslovakia and Yugoslavia, a kind of Middle Europe as Middle America, hardly considered all nationalities in East Central Europe to be on a level playing field.[87] Racially for them, Magyars were quite low on the totem, judged somewhere along with peoples of Africa, viewed as incapable of their own governance. In *Racial Problems in Hungary* (1908), Seton-Watson penned a study of Hungary's oppressive Magyarization, objectively as he saw it, of its minorities including the Czechs, Slovaks, and Romanians.

Men of the great powers were immersed in these studies. In fact, England's David Lloyd George (1863–1945) grew so irritated by the floods of materials from his experts on East Central Europe, which he could hardly decipher, that he asked his delegation in frustration, "Who are the Slovaks?"[88] (He honestly did not know.) When Seton-Watson arrived as a private citizen in Paris in 1919, the prime minister refused to appoint him to the British Empire's delegation, or allow him to take part in an official capacity. Seton-Watson's linguistic expertise made him England's most knowledgeable expert on nationalities. A kind of academic-as-lobbyist, he was influential in the development of the School of Slavonic and East European Studies in London, founded by Masaryk in 1915. Yet Seton-Watson was mostly ignored by his own ranks. Lloyd George was more concerned with Great Britain's commercial spheres than with Wilson's holy rhetoric of national self-determination, or a vaguely defined struggle for twentieth-century democratic rights against communism. Mapping East Central Europe thus fell to other neocolonial men.[89]

On 24 December 1918, Romanian troops occupied the city of Koloszvár/Cluj, including the university where Teleki's compatriot Cholnoky taught. On 1 January 1919, the same day the Red Army took over Minsk in Belarus, Czechoslovak legions occupied Pozsony/Bratislava. On 11 January, one week before Paris, Károlyi became president of Hungary's republic. Oszkár Jászi (1875–1957), the historian and minister of nationalities, tried in vain to persuade the Romanians to remain in Romania, under a revised election law.[90] In Budapest on 21 March, Béla Kun's communists seized power and for 133 days in the capital, a Hungarian SSR was set up. The anticommunist Teleki appealed to the science he learned under Lóczy, as his "Carte Rouge" reflected and aggravated border conflicts. Placed on the defensive after the Central Powers lost the war, he ran specially afoul of the Czechoslovaks, namely Edvard Beneš (1884–1948) and Masaryk who opposed Greater Hungary and went abroad as "democrats" to the United States to campaign for self-determination. Masaryk, guided by the Czechoslovak National Council, traveled all the way from Moscow via Vladivostok to Vancouver and Chicago. He had pro-Czech and pro-American leanings (he was married to an American, Charlotte Garrigue) and grass-roots support in "democratic" hyphenated diaspora spaces of an American electorate, but his anti-German and anti-Magyar prejudices were grounded in colonial Europe.[91] To make "historic" claims, Teleki faced competing graphic arguments by Czechs, Romanians, Poles, Ukrainians, South Slavs, and others, added to the dismissive silence of the Great Powers. All competed in an agonistic space for maps.

Observers of Ostmitteleuropa ignored maps at their peril. Yet this did not deter smart insiders, pressure groups, or lobbyists like the count and the baron from smartly carrying them around in their briefcases. When Count Apponyi, head of the Hungarian delegation, tried to give the "Carte Rouge" to Lloyd George, the British prime minister roundly rejected it. The French prime minister, Georges Clemenceau (1841–1929), in support of the Little Entente, likewise shut down the Council of Four's session before discussions on Hungary's borders could begin.[92] Teleki appealed, haplessly, to the legacy of objectivity in nineteenth-century science, as evidenced by his review of a compilation, *Die Geographie als Wissenschaft und Lehrfach*, in the *Geographical Bulletin* in 1919. He re-examined geomorphology schools since the 1890s, comparing Penck and Davis. He idealized geography as *Wissenschaft* and promoted its uses in Hungarian universities and society. He imagined a line of progress from ancients Ptolemy and Strabo to moderns such as Humboldt, Richthofen, and Vidal de la Blache. He stressed the "valuable new ideas . . . [for] a young generation of Hungarian geographers . . . [with] their methodological and conceptual questions." Geography was a unity (*Einheit*) of interrelated knowledge. Cartography was humanity's collective achievement. Positioning himself as Hungary's premier expert, the aristocrat used geography and cartography to reinvent his objectivity and Europeanness.[93]

––––––––

At the start of World War I, our map men were not just national heroes or villains. They were agents of a transnational, moving history. Once Penck faced the threat of defeat and loss of status in Germany during World War I, he sought to exact revenge not on England or modernity, but on Romer personally in 1916–17. Penck's alerts about his former Polish student, just short of arrest, went fatefully unheeded. After Colonel-General von Beseler intervened, Dr. Józef Reinländer, the chief of police in Lwów who despite his German name identified as Polish, refused to detain the Polish geographer. General Alfred Hübl, head of the Military Geographical Institute in Vienna, also guaranteed Romer's safety. Hübl approved of the atlas's publication by insisting that the work was "pure science" and that its uses as propaganda were incidental.[94] Romer had beaten the Germans at their own task—he learned to draw maps that are more persuasive. He went on to lay Polish frontier claims to Galicia, Upper Silesia, Pomerania, and the Danzig corridor. Romer's Greater Poland was then advertised to the Western victors as civilized, peaceable, and democratic. Although Romer rejected Mitteleuropa as an economic and political idea, he also generated Polish maps in the "East Central" tradition before 1914. Penck and Romer were men who carried on Europe's dynastic empires, far beyond World War I. In their symbolic outlooks and "scientific" *Ostforschung* (or *Westforschung*), they marked the visceral nature of border violence during the two world wars.[95]

Yet the lines after 1919 are equally blurry. After having a full century to think about it, to merely remark that maps during the 1914 to 1919 period were nationally constructed is not a great insight, but a truism that elites, state leaders, diplomats, and geographers of that milieu likely already knew. Indeed, some grasped it intuitively. Long celebrated by patriots and associated by Teleki's acolytes with his suicide/martyrdom on the eve of the Nazi invasion of Yugoslavia through Hungary in April 1941, the "Carte Rouge" must be resituated within the (in)glorious norms of Europe's pre-1914 men—those who loved geography and maps a bit too much.[96] The "Carte Rouge" was a tool of the *nineteenth* century. In the map's surreal and subliminal codes, its message was the defeat of the Little Entente, marginalization of Romanians and Jews, and omission of rural lands and indigenous peoples, whisked out of history. Non-Magyars were left out, a blindness in white toward denizens of mountainous areas and neighboring valleys. The true-false aspect of Teleki's map, long a preoccupation of the fiction of nation-states, is less important than its design as a fantasy text, or its aura as a visual artifact. If such maps were arguments in graphic form, they could be poor and terrifying ones. Pál's paprika geography is a perfect capstone to a long story of transformation, minus nationality or modernity, of Europe's "civilizing" tribes and clans. In the garb of objectivity, the "Carte Rouge," like the Penck-Romer conflict, was *affective* as the transnational geography of textured lives.

CHAPTER THREE

COURTIERS

With the Treaty of Versailles in June 1919, the Weimar Republic relinquished Germany's colonies and signed Article 231, the war guilt clause. Saddled in heavy debt, Germany paid reparations to the victors. Penck too paid a personal price. He was left out of the postwar world of geographic science, to which he owed his reputation and emergence out of provincial Saxony in the 1870s and 1880s. He railed against decolonization not on liberal or universal grounds, but by damning the treaty and defending German munificence ahead of England. In his presidential address of 1919 to the Berlin Geographical Society, he welcomed home returning German troops as *Kulturträger*, the West's great explorers. He praised struggles of such men "able to return home in an honorable way, with weapons in hand, as German victors." Colonial East Africa was a European mission, he declared, for "each one of them has become an Africa-traveler (*Afrikareisender*) and can offer recollections . . . The geographer can learn from them individually . . . In full suspense one awaits first reports on the grandest and most original African expedition that has been ever made. The Berlin Geographical Society feels proud and . . . satisfied."[1] Penck, however, had lost his court to serve. The war's victors prohibited the Germans from participation in international congresses, and they in turn boycotted the new Versailles order in the IGU. Penck's life into the early 1920s therefore became a decolonial one, in which he and other German geographers were left isolated, making finely persuasive maps and holding geopolitical conversations mostly among themselves.

IN SEARCH OF PATRONS

In the Paris of 1919, Bowman and Romer grew ever closer. Each man dreamt of large-scale cross-border coordination of geographic data for a global world. They reprised Penck's original role in Austria, in Germany, and abroad. They esteemed professionalization but in different ways, Romer out of a defeated aristocracy, Bowman from rural poverty into a highly status-conscious U.S. East Coast bourgeoisie. They produced knowledge in quests for security and social advancement. They encoded a mission for Western civilization as objectivity and professional, civic-minded struggle. Bowman, the Wilsonian proponent of friendship and Anglo-American free markets, was seen by his Galician friend as a kindred spirit; in turn, Romer sought his favor to defend Poland's territorial integrity. Romer sent an endless stream of maps, atlases, and letters across the Atlantic to the AGS building in New York City, hoping to influence U.S. experts. For Bowman and Romer, their German-inspired relationship (about which both were silent) was never one of equals, and certainly not of two democracies. Catalyzed by the war, it became a tense courtship in which socially performed selves broke down.[2] The "friendship" of these two globally minded scientists out of old Ostmitteleuropa comes out in two still critically unexamined sets of sources: Bowman's Paris diary of 1919 and the Bowman-Romer letters over the course of four decades, from the 1900s to the end of the 1940s.[3]

Isaiah Bowman was gifted at finding patrons. He impressed the right men at the right time, academics like Davis, Penck, Jefferson, and Wilson himself. Yet the American lacked experience—his first trip abroad to Europe was for the Paris talks. We pick up the story on 27 January 1919, when he and the Harvard Polonist Robert Howard Lord (1885–1954) had lunch at 2 Quai de Tokio. That evening, he met with Lord and "a dozen of our men" in the Hôtel de Crillon office, where they discussed "ethnography, combat situation, [and] boundary geography . . . [with] good results." Lord was among those who "stayed until midnight discussing their special problems."[4] Such gatherings were business meetings, in which our men stood and pored over maps. The geographers had many intense late-night study sessions at the hotel where the U.S. delegation stayed, at tables and on their hands and knees, gazing at maps as sources.[5] These Ivy League explorers shared a zeal for the research seminar, to critically examine documents as Leopold von Ranke (1795–1886) had done with his Berlin students in the 1820s. Insofar as geography or history were sciences, many practitioners were text-centric religious men, although not always in a typical way. Lord was an Episcopalian who converted, after his Paris and Polish experiences, to Catholicism in 1919; he resigned later from his professorship of history at Harvard, in 1924, to become a priest in Boston.

When Isaiah Bowman came to the French capital, he was not on a free-thinker's quest. He was not looking to seek out urban space or the phantasmagorias of a modern *flâneur*. Instead, he arrived with a mission and found respite at a famous site. On 16 February, he took a walking tour of Paris and strolled around the Notre Dame Cathedral. It was a kind of psychogeography at work, a space for the streetwalker's wonderment.[6] The precocious Harvard student who once devoured William Wordsworth and Ralph Waldo Emerson figured the church as a spiritual voyage. He wrote, "Half way to the cathedral the clouds broke and a full moon came out, setting the cathedral into relief all the more plainly because of the low clouds which hung about. We [with Frank L. Warrin, Jr., of the State Department] walked around it and finally passed into the archbishop's garden at the rear. From the terrace at the edge of the garden we looked down upon the town, which was half obscured by mist, with an occasional light peeping through." The mere site of the cathedral stuck a chord. Bowman lost himself in reverie,

> We heard its history from the guide, and examined quite minutely the various parts which had been built at different periods. It was a wonderful day, for the sun had come out and the sunlight lay on the landscape for miles around, in fact clear to the horizon except toward the southwest where a rainstorm was gathering. We talked of the cathedral's great age and of the great stretch of French history which it had witnessed, until even the present war seems a very brief event. It was begun several hundred years before Columbus set sail for America, and saw the Hundred Years' War, the Thirty Years' War, and all the intervening history of France. We come back to Paris with our minds filled with the beauty and the majesty of it.

The myth of the explorer's discovery—here Bowman indulged in a favorite chronotope of colonial science. He explored frontiers in reverse from the New World. At Notre Dame, he found Tintern Abbey in 1919, in an era of Bolshevism and postwar disorientation. Bowman prepared himself for Europe with this intimation of immortality. He discovered his appointed role as a providential American, and a true Wilsonian map man.

As chief territorial specialist in the American Committee to Negotiate Peace since April 1918, Bowman served the state through "Colonel" Edward M. House (he was never an actual colonel). Meanwhile Romer felt left out in the new Poland, frustrated from early on with the divisiveness of its two leading wartime politicians, Józef Piłsudski and Roman Dmowski (1864–1939). He preferred the high road, appealing to science and the Americans. Romer became valuable as a conduit to Bowman, through the sympathetic Lord and his bilingual Polish-American assistants Henryk Arctowski (1871–1958) and S. J. Zowski (1880–1940) on the U.S. Inquiry. Bowman was an expert who could gain the ear of Wilson, who like Lloyd George knew embarrassingly little about East Central Europe and had to rely on a "secret team of experts" to study maps sent to them by the Americans.[7] In 1917, the Allied Supreme Council was set up by Lloyd George to coordinate strategy. It came up with the Curzon Line (after Lord Curzon, 1859–1925), an absolutely arbitrary line, east of Białystok and Lwów and west of Brześć and Stanisławów in its first 1919 version, as the Polish Republic's future border with the USSR.

In the later stages of World War I, the council proposed a special International Boundary Commission (IBC) on Polish affairs, to prepare for Paris and work on Poland's eastern frontiers (figure 3.1). Bowman was appointed its head. Romer, together with Lord, became Bowman's European wild card, a go-to expert. In fact, Bowman saved all the Polish maps, atlases, pamphlets, and cartograms he received "with some corrections" from Romer personally via the Polish National Committee. On 19 February, Romer wrote, "My dear Bowman, I have the honour to send You my book on the Polish NW-borderlands in Prussia with some corrections in the tables . . . in order to make similar corrections in the copy given yesterday to the mission's library. At the same time I ask You, Dear Bowman, to set me by the soldier which comes with the book to You the two volumes of Prussian Gemeinde/Lexikon, it is Posen and Schlesien. I will return this book at the same day in the afternoon-hours. Cordial handshakes from Yours, E. Romer."[8]

Courting Bowman, Romer smartly anticipated the American's powerful role on the IBC, on which Bowman served with the diplomats Jules-Martin Cambon (1845–1935) of France, Baron William Tyrell (1866–1947) of Britain, and Marquis della Torretta (1873–1962) of Italy. The next day, 20 February, its members met for the first time. Bowman noted that they "discussed Polish correspondence and agreed to recommend to Supreme War Council the occupation of Dantzig [sic] as a base for shipment to Poland of arms and

FIGURE 3.1. The International Boundary Committee (IBC) on Polish Affairs in 1919. Bowman is in the front row, second from the left. Romer, whose maps were most used, was not on the committee and is therefore not pictured. Courtesy of Special Collections, The Milton S. Eisenhower Library, The Johns Hopkins University, Baltimore, Maryland. Isaiah Bowman Papers (IB-P Ms. 58), Series XIII (Paris Peace Conference).

ammunition. . . . Also to occupy later the strategic points between Danzig and Warsaw because line established by [Marshal Ferdinand] Foch in North Western Poland, in armistice of last Sunday, does not run to sea."[9] Bowman never mentioned that it was the disingenuously cordial Romer who furnished the cartographic data in the first place.

By late February 1919, Polish and Ukrainian forces remained at war in defunct Habsburg imperial Galicia, on the outskirts of contested Lemberg/ Lwów/Lviv. Polish delegates in Paris pressured Bowman to take on their cause, but Piłsudski and Dmowski lacked any consensus on Poland's future borders.[10] Since the Allied victors did not allow the Poles to attend closed-door meetings, suspicions intensified. Dmowski called Bowman in order to set up a lavish dinner for 27 February.[11] It was conveniently timed. The day before, Cora Bowman arrived in Paris on 26 February, leaving their two sons with her aunt and uncle in the U.S.[12] Isaiah and Cora, who had her own patrician style, were absolutely in awe of the Polish banquet. The German-turned-Anglophile Bowman reported that he met the "descendant of our Revolutionary hero [Pulaski]" and "other Poles including the geographer Römer [he spelled the surname this way] from Lwow." Bowman was impressed by the fact that twenty different Poles in the "delegation of experts" gave well-rehearsed scientific reports. He did not think it was objectionable that the Poles spoke for other nationalities, or that no Germans were present. Bowman inquisitively "asked particularly about the desires of the Protestant Poles of Mazuria in s. East Prussia." He found Dmowski appealing and was unperturbed by his anti-Semitism, of which the British and French delegations were at least aware. "In evening [I] dined with M. Dmowski the Polish plenipotentiary," he noted, "and we talked intimately of all Polish problems. He is quite philosophical and broad and strikes me as a very able man." In his handwritten account of the dinner, Bowman Germanized his subordinate European friend's surname as "Römer." An innocent abroad was clearly charmed.[13]

Still, the dutiful pan-American careerist knew he had to put on a transcendent veneer of Wilsonian impartiality, just as the British and French delegations extended their agendas for a settled long-term peace.[14] On 28 February, Bowman "scientifically" focused on Europe's East. He met with two British colonels to review demarcations for Poland's western frontier with Germany.[15] The next day, 1 March, he and British experts went over the Teschen/Cieszyn issue. Bowman boasted that they came to his office and "agreed on everything easily." The IBC by then had added a fifth expert, Ambassador Kentaro Otchiai (1870–1926) of Japan, whom Bowman entirely ignored. At the third meeting, as if already at the League of Nations, he listened to "the French, British and American Commissioners regarding the German frontier of Poland."[16] When Bowman met that afternoon at the Quai d'Orsay "to settle on Western frontiers of Poland," the British started the discussion and Bowman represented the U.S. position, emphasizing the priority of Czechoslovak and Polish democracy.[17] Bowman claimed that they "agreed in principle all around but referred it and the Teschen problems to the subcommittee," comprising General Henri Le Rond, Colonel Frederick Kisch, and himself.[18] On 3 March, Bowman met at 10 am with Le Rond and Kisch again, to settle Teschen/Cieszyn and western Polish frontiers.[19] On 4 March, Colonels Robert McCormack and Edmond Taylor came in to get instructions for their assignments in Poland. They were assigned to go with French general Fernand de Langle de Cary and join the Polish inter-Allied mission in Warsaw. Bowman noted with added intensity that they "talked over the whole Polish problem."[20]

Here the crucial point from Bowman's diary is less his day-to-day chronicle than the fact that these March 1919 debates occurred with emotional undercurrents in Wilson's new world of "open diplomacy." Europe's military great powers were a closed membership community. This and the continuing Polish-Ukrainian War of 1918–19 over Galicia caused Polish delegates in Paris terrible dread. Of course, they were not the only ones locked out. In the longue durée, the aristocratic Romer's diverse background, the civil service tradition of his forefathers in East Central Europe, and his positivist understanding of geography/cartography as useful to governments all shaped his next move. Convinced that his expertise was not being properly used, he judged Dmowski no better than an ideologue, a kind of Bolshevik on the right, inflexible and dismissive of intelligence that did not match his outlook. So on 8 March, Romer bolted straight to the American geographer and his team of map men, circumventing Polish politics again. He wrote, "My dear Bowman, Because of a terrible occupation by our own Commandor [*sic*] Dmowski . . . was absolutely impossible to prepare all reports which we have promised sent to You. Now I have a great part of them and will be to You very obliged, if I could bring them to You and speak with You by this occasion."[21] By then, things got desperate. With the Polish-Ukrainian War escalating, the Romers' home in Lwów was under threat. Romer's own brother Jan, a general in Piłsudski's legions, was in active combat on the city's outskirts.

Meeting regularly, Bowman and Lord traveled regularly to Warsaw to find solutions on the premise of the Curzon Line. Postdynastic solutions according to language dated back to sources gathered by the U.S. Inquiry in 1917–18. These maps were embarrassingly shoddy and stored as classified for a long time (figures 3.2a and 3.2b). On 15 March, Bowman received Lord's telegram and jotted in his diary, "If Ukr. capture city will massacre whole population."[22] Bowman really needed Romer, whose tactics worked in a clever and overly personal way. In his family, his own sister-in-law Stefania, married to Jan the general, was a Polonized Ukrainian. By insisting on Polish-European assimilation and pointing to kinship and families within nations, as Penck had done in 1914, he managed to secure his friend and patron's sympathy. Bowman's defense of civilization concurred with Romer that all Ukrainians were undeserving of self-determination. When the American met with Ukrainian delegates on 8 April, it had no effect whatsoever on his long-held prejudices.[23] They dismissed Ukrainian "ethnographic" claims in East Galicia in favor of Poland's "historical" claims to statehood. The primitive trope was so perilous because it was highly effective. It could be invoked by anyone, anywhere. Starting from within the U.S. Inquiry and its maps, many of which were drawn hastily, Bowman, Lord, House, and Wilson saw in Europe's already unobjective cartography whatever they too wanted to see.

Put in the starkest of nineteenth-century terms, behind all talk of self-determination were Europe's colonial empires and a world of Realpolitik. Maps were not democratic but drawn by colonial experts, to make sub-Saharan Africa and the Balkans seem legible. Map men were like Columbuses or Napoleons,

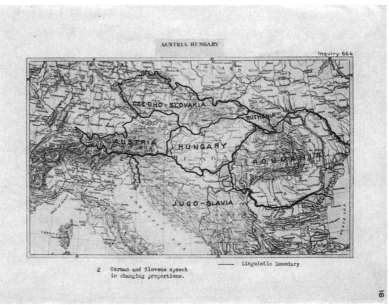

FIGURE 3.2A AND 3.2B. These maps were classified and never before published: "Poland, Lithuania and Western Ukraine" (U.S. Inquiry, no. 657), and "Austria-Hungary" (U.S. Inquiry, no. 664). Attempts to define linguistic boundaries led to unmappable areas which were designated as "mixed speech." The area including Lemberg/Lwów/Lviv is marked after Romer's work as a "possible addition to Poland." ("East Prussia" is misspelled.) No single author, but probably the "Polish" team of Bowman, Lord, Arctowski, and Zowski compiled these for the U.S. Inquiry (American Committee to Negotiate Peace) and East Central Europe in 1917–18. Courtesy of Special Collections, The Milton S. Eisenhower Library, The Johns Hopkins University, Baltimore, Maryland. Isaiah Bowman Papers (IB-P Ms. 58), Series XIII (Paris Peace Conference), U.S. Inquiry File.

or Kipling's men of character. On 5 April, Bowman noted the presence of Polish and Czecholovak commissions in the first meeting at the Quai d'Orsay, to discuss the Teschen/Cieszyn issue. Bowman proposed that Czechoslovakia relinquish the Teschen and Karwin areas to Poland. The British and French objected. The American ignored the democratic Masaryk and explained, with Lord, how the Czech leader had "compromised by proposing the administrative frontiers of Teschen and Karwin rather than new line to be made by a bndry com'n with all the attendant delays and possible disorders." Bowman in defense of Poland noted that the commission at his urging "however did stipulate a very large measure of autonomy for the Czecho-Slovak position and this was agreed to." At dinner he declared victory, "at Polish delegation headquarters . . . with the scientific staff including Gen'l Romer who commanded the Polish army against the Bolsheviks about 60 miles north of Lemberg." Bowman recorded a photo of "our experts with Col. House in his office."[24] The American was pretty good at posing as the centerpiece of his own homespun tales.

Although tensions of empire persisted in the transnational AGS and pre-1914 academic tradition, the "disinterested" Bowman and Lord saw no contradiction in befriending Romer, supporting Polish interests, and fraternizing in private while mediating Wilson's so-called scientific peace for East Central Europe. On 15 April, he and Lord met again for lunch with "General Romer and Dr. E. Romer [for] a general discussion of [the] Polish problem." Bowman performed impartiality by *appearing* to hear out ethnonational grievances. He met on the same day with a Ukrainian delegation to discuss the Polish-Ukrainian armistice, insisting that he "would be glad to be a medium of com'n between them [the Ukrainians] and the Conference on national questions." After hearing complaints, he shared Romer's rejection of their political goals. The Poles had grievances as well. But a different lesson was learned. Even President Wilson himself grew weary of Polish violence when anti-Jewish pogroms in Lwów were carried out by Poles, and the Polish-Ukrainian War in Galicia escalated.[25] Bowman quickly echoed the sentiment of his patron, losing patience with Romer. In his entry of 27 April, he wrote that Romer once more "came to talk Polish conditions 12:00 to 1:00. I brought him back to questions of an armistice. A flood of irrelevant words. Wants to tie up territorial questions with armistice."[26] Pathologies of romance, domination, and alienation were the stuff of politics. As the Polish poet Czesław Miłosz astutely observed in his commentary *The Captive Mind* (1953) nearly half a century later, a psychology of love and hate was at work. Tumultuous courtships from a mapped-out East could not be explained rationally by nationalities issues, schemes of economic development, or geopolitics.[27] Each Wilsonian friendship had limits, and the Bowman-Romer tryst was no different.

AMONG THE DEFEATED

In Berlin in 1919, Penck was a disappointed lover who fell back into the company of the like-minded. Major articles during and after the war appeared in the journal of the Berlin Geographical Society. His German writings came at an anxious time, when there was a new flurry of German maps and cartography-related literature.[28] Against Romer's Poland, Penck insisted on several points: (1) the data of Polish maps showing access to the corridor was faulty and not scientific; (2) Romer had been incorrect in claiming borderland Kashubians, Mazurians, and bilingual groups as Polish (or all-Polish); (3) landholdings remained in the hands of German property owners; (4) German statistics, diagrams, and tables were superior and correct; and (5) there was no proof of "pure Polish" or Polish majority groups by language in contested provinces of Poznań and West Prussia. He viewed Poland as a "landlocked country" (*Binnenstaat*) in its 1815 borders. In defense of German *Boden* (soil), he claimed it was possible "to go from Berlin to Königsberg and not cross or step into a village of mostly Poles."[29] He invoked frontier diversity to diffuse Polish influence, always as he saw fit. This was a *divide et impera* tactic once again, from at least Rome to colonial Europe. Penck learned from the Galician Romer—the "Polish language area . . . the larger space . . . has a strong admixture everywhere: Germans to the West, Ukrainians in the East, and Jews in the center and East."[30] The time was ripe again for depicting Poland as an existential threat to German civilization, as Bismarck had done in a failed *Kulturkampf*, but this time in open lands across the Oder (Odra), a kind of Mississippi River for German resource management and economic development.

Penck's initial obsession over borders after 1919 was to "prove" that German experts' maps were technically advanced and had better data. Therefore, they were true. Objecting to the league's treatment of the corridor issue, his works supported German revision on grounds that Germans from time immemorial were present in Upper Silesia. They alone were torchbearers of Europe's civilization.[31] Penck reworked the Kashubian question more broadly by arguing for the reversal, on scientific grounds, of claims of Romer and Spett, who in their maps counted the Kashubians and Mazurians as Poles. He criticized Romer for using the 1911 *Schulstatistik* and accused Spett of deliberate falsification. Penck appointed his own assistant, Herbert Heyde, for fact-checking empirical purposes. He averred that the Kashubians were not related to "Slavic" Poles and stated, "A Polish corridor to the [Baltic] Sea in the sense of a language area has never existed. It is a slogan for political aspirations, and ignores the difference between Poles and Kaschubes."[32] Albrecht thought it was inaccurate for Poles to claim bilinguals as ethnic Poles, since such speakers were located on German soil and had been forced to admit bilingualism. Penck's logic resorted to essentialist discourse, attribution of malicious intent, and cartographies of fictive German unity, all for the cause of its settled lands and peoples across the globe.

In Weimar's Foreign Office from March to May 1919 (as mentioned), Penck's grand project on the scale of Romer's atlas of 1916 was ignored. *Völkisch* pressure tactics intensified over the issue of unity in Upper Silesia in 1919–21. Penck's colleague Walter Stahlberg, who had worked at the Oceanographic Institute's museum, stressed to compliant social democrats like Max Eckert the need to use Penck's maps for countering Romer's harmful effect on inter-Allied negotiations. Arguments soon grew technical. In Upper Silesia, they claimed, it was more proper to consider population density of language groups by the choropleth method, to measure the vitality of German culture on its settled, advanced Eastern frontiers. Choropleth maps, Penck claimed, worked better than Romer's isopleth maps because Romer misrepresented regional distribution of language speakers by districts. Penck continued to belittle Romer and the "forged" Spett map of 1918 by appealing to accuracy and *Wissenschaft*. He always defended objectivity as something cross-border Germans possessed, on the bases of the "correct" plebiscite results he and his associates discovered. In the March 1921 plebiscite in Upper Silesia, Romer "proved" that 300,000 Polish-speaking persons voted to remain in the Weimar state, while only 5,000 German speakers wished to stay in Poland's Second Republic. In this fact-based universe but now without a set of pre-1914 true/false standards for verification, the Berlin geomorphologist had to wait for his revenge.

Wissenschaft was the preoccupation of insecure map men in this way. Old colonial practices remained. Prejudices were reworked into positivist defenses of German authority and the superiority of German geographic science. Five years too late, Penck zeroed in on Romer's ethnographic maps from the 1916 atlas of Poland. He objected again on technical grounds: first, that Romer's style of representation was not suited to the map scale he used; second, that Romer represented mixed areas by shading in colors of different languages, basing claims on faulty 1911 data; and third, that Romer used language to draw irregular minority "islands" inside a German-majority area, thus giving the erroneous impression that Poland's frontiers were Polish. He dismissed Romer's isarithmic method, proposing instead a method through the graduated variation of a dot map, drawing from his 1919 work with Heyde. Penck showed that the absolute number of individuals for each language group could be represented correctly, color-coded by language and the fundament of place. The man from Saxony favored German regionalism to restrict Poles to a landlocked ethnic "kingdom," in effect a defense of *Volksgruppe* in an imagined German-unified East.[33] Penck's prepossessions of domination from a nineteenth-century past were made evident in this way. Mapping Germany's East, the *völkisch* Berlin scientist and writer turned with hope and despair toward a world of post-Versailles revisionism in the 1920s.

RUMP STATE

As for Teleki, in the Paris negotiations of 1919 the victorious Entente regarded Hungary as a rump successor state to the Austro-Hungarian Empire, a punishable belligerent power in alliance with Germany's aims of 1914. This impelled Teleki to lobby and utilize all his contacts from Europe's nineteenth-century world. Starting in fall 1918, he worked to create the League for the Protection of Hungary's Territorial Integrity, or TEVÉL.[34] The irredentist league, made up of elites and experts like Teleki's mentor Lajos Lóczy, hosted influential foreign visitors including Major Lawrence Martin from the U.S., a friend and ally of Bowman. The moderate left already in December 1918 pilloried TEVÉL in Hungary. For instance, the socialist daily *Népszava* (*People's Voice*) accused it of damaging the new country's image in the world.[35] Teleki himself was regarded with suspicion. On 14 February 1919, while the closed Paris talks went on, the Council of Ministers prevented his participation. Other threats, real and imagined, came from Kun's class-belligerent Leninist revolution in Budapest, which fractured Teleki's Anglo-American dreams of bourgeois Magyar modernization.

The count certainly feared when the Third International (Comintern), a successor to the First (1864–76) and Second (1889–1914) International, was set up in Moscow in early March 1919. Teleki was certainly no Marxist. Nor was he a bourgeois or "white" internationalist, since that might mean the war's victors and collaborating forces could enter Hungarian lands and overthrow Kun's communists by force. He was supportive of counterrevolution, however, when it suited his purposes. Holding more faith in European diplomacy, he insisted that foreign troops and especially Serbs, Czechs, and Romanians should not be involved on Hungarian soil. He hoped that Horthy's resistance would defeat the communists, just as he hoped the Bolsheviks would somehow lose. But Teleki was an idiosyncratic conservative in East Central Europe, anticommunist and transnational, not quite a Wilsonian either. He rejected self-determination, since that meant reduction of Hungary's "historic" St. Stephen lands. Yet he reasoned that an Entente-led mandate system, guided by multilingual experts from pre-1914 diplomacy to redress grievances through the League of Nations and a Eurocentric neocolonial trusteeship, was a generally *good* idea.[36] Both leagues—one Wilsonian, one unapologetically revisionist and protofascist—were useful for Teleki's ingenious Transylvanian balancing act, for they could ultimately salvage what remained of Ostmitteleuropa geographic science before 1914. Anticommunist Hungary looked like a protectorate and a bastion of Christendom and Western or European civilization, in general conformity with his noble clan's Transylvanianism and early modern history.[37]

Yet square pegs like Teleki fit badly in round twentieth-century holes. Really, none of the delegations or politicians of Washington, Paris, or London in 1919 favored any nuanced treatment for Hungary or dispersed Magyars across borders after World War I. Lacking objective science or a useful court of appeal, the Anglo- and Francophile Teleki called for justice in the nationalist language of a pro-Western, but not quite Wilsonian or Leninist, self-determination. The last choice was revisionism as an ideology.[38] Hardly prolific in cartography, Teleki barely produced three world maps before 1914, but beginning with the "Carte Rouge" of 1918–19, he learned modern mapping on the fly, how to make and use maps in choice polite appeals to the civilizing men in power. He extended this strategy in his writings, for instance in the *Short Notes on the Economic and Political Geography of Hungary* in 1919, which served as a first draft for lectures on American economic geography in 1921.[39]

The count gave it personally to Bowman as a gift for the AGS library and collections. Since we lack many surviving letters from Teleki's life, this fact coupled with his mental maps in *Short Notes* offers us three essential clues. First, he emphasized a visual distillation of the textual, for the Hungarian language was daunting for outsiders to access. Seven maps and diagrams appeared at the end of the article, including three prepared by Teleki himself. These were a "Physical Map of Hungary: The Natural Regions," "The Density of Population: The Same Natural Regions Marked," and "Origin of the Towns of Hungary according to the Classification of Prof. E. Cholnoky." Second, Teleki wrote the article in English. Appealing to authority, he peppered it with references to a Eurocentric canon. On the very first page, he cited another hero, Vidal de la Blache, before 1914, "the world-renown master of modern French geography . . . [for whom] Hungary represents one of the most striking morphological unities on the physical maps of Europe." Teleki drew authority from "Les Divisions Regionales de la France," stressing Hungary's ecological integrity and "morphological unity . . . situated in the three great parts of Europe." Third, the count drew parallels of Magyars who migrated from Eurasian climes to become settlers, as in the U.S. West, on the Great Plains (*Alföld*) as a kind of imagined heartland space. He wrote that Hungary "has been in human history a meeting-place and war ground of the Northwest, South and East. The people, which took possession of the great plain, the dominating region of the country . . . had to defend Hungary in the course of history against Eastern, Southern and Western invasions: against Tartars, Turks and Germans."[40] Hungarians had abandoned their old Eurasian climes, just as Teleki seems to have done in the Budapest of 1916, in favor of "Western instead of Byzantine civilization and friendship."

Teleki's serious ideas were half-baked all the same. He redrew Hungary as integral to Western civilization against "primitives." Hungarians spoke a Finno-Ugric language, but together with Finns and Estonians, he mapped the Magyars into Europe. He also looked south to Italy for his spiritual geography, as Penck looked north to Sweden. Teleki valorized Hungary's Renaissance period above all. He stressed the high culture of the Hungarian nation, those who now distanced the country in 1919 from past "German" (and Austrian) influence. The playboy count, not too religious in his early life, now found cause to recast counter-Reformation Transylvania as a bulwark of Europe. He wrote, "The temperament of the Hungarians resembles that of the Southern people, his comprehension is quick, his speech laconic. . . . Hungarian culture has had its best individual and national development when it met with the culture of people, having the same temperament."[41] Teleki added that in the Danubian Basin, Budapest became the locus of Euro-Magyar migration and settlement. When non-Magyars such as Romanians, Jews, Slavs, and other minorities came into Hungary's historic space and assimilated to high Magyar culture and language, they too would reach the plateau of Europeanness. He wrote, "Budapest is situated on a spot of exceptional geographical weight. . . . Amongst all capitals of Europe, Budapest stands almost unrivalled as to her strength of centralization. This is a law of nature and by any means not the consequence of Magyar chauvinistic politics, which can be easily proved by the map."[42] In defense of unity, he appealed to "natural" ecology: the Danube and Tisza rivers were interlocked, as were trade networks of interconnected roads and railways. "The division of Hungary," he wrote, "would separate geographical regions which are dependent on each other. It would create boundaries where nature laid down the foundations of commercial intercourse and mutual economical life."[43] Following this logic, Trianon Hungary was therefore unscientific, retrograde, and immoral.

Teleki's preferred arguments for Hungary were also arguments for Europe's totality and unity—of a genuinely concerned, if pathetic and wounded sort. They were made in the post–World War I context of the Paris Peace Talks on the one hand, beginning 18 January 1919, and the communist revolution of a Hungarian SSR on the other, from 21 March to 1 August. He combined a nineteenth-century intelligentsia's high culture with positivist work for nation and state building, to educate a map-literate public into a new character. He played multiple sides by politely sending maps and statistical pamphlets to America, which the Czechs and Poles had been doing for years. His charisma and contacts made him influential in his own insistent, peculiar way. He rejected Eurasia for the West. Notably in the Hungarian-American diaspora, Eugene Piványi, the author of *Hungarians in the American Civil War*, prepared a pamphlet inspired by Teleki called *Some Facts about the Proposed Dismemberment of Hungary, with a Map, Statistical Table and Two Appendices*. The work included a "Map of Hungary Showing Proposed Dismemberment," with lands "claimed by Czechs . . . claimed by Rumania . . . claimed by Servia in dispute between Servia and Rumania." Statistics for this trans-Atlantic lobbying effort were supplied by the Hungarian Statistical Office in Budapest and published for electoral purposes by the Hungarian American Federation.[44]

Teleki's conservative ambitions for the role of maps in politics were not all sweetness and light. In his postwar public speeches in 1919, the count referred to Hungary's Jewish population in loaded political language, as the "Jewish question," or *zsidokerdes*.[45] In a thinly veiled anti-Semitic speech in Szeged (it was published by TEVÉL) on 14 December, Teleki appealed to the new Christian Unity party in which he was involved, announcing that "we will fight for the hegemony of the Hungarian nation."[46] Maps preceded territory (the French philosopher Jean Baudrillard rehashed this East European cliché as a grand postmodern insight in 1981) toward the century's end.[47] Though he had thought to show himself as European, therefore civilized and not chauvinist, the Europe he imagined was surely an intolerant place. It cast out "others," including non-Christian migrants. Teleki's Jewish question was framed in Horthy's Hungary through an ethnocentric lens in 1919. He fixed on the idea that the Little Entente of Romanians, Yugoslavs (for him principally Serbs), and Czechoslovaks presented grave threats.

Not wanting Hungary reduced to a rump nation-state, his fantasy of the St. Stephen concept of historical unity was coupled with "scientific" proof of Hungary's economic geography. The disillusioned Transylvanian Francophile thus saw Galician Jewish immigrants as Bowman the U.S. Midwesterner viewed Native Americans from his childhood days. If one can follow the mood, or logic: he dismissed Galician migrants' rights and visibility; he drew lines around "good" assimilated Jews by language who supported the Magyar nation; he saw those "bad" Jews of Habsburg and Russian imperial lands, which included parts of Poland, Ukraine, and Romania, as tainted by Bolshevism. He used this "Ostjuden" stereotype wherever it suited him, instrumentally as a political wedge to mobilize a modern, Christian, anti-Semitic, and privileged titular Magyar bourgeoisie. By advancing Christian unity as fact, Teleki's tool of anti-Semitism from the heartland owed as much to Enlightenment science and anthropology as to nineteenth-century norms of economic, social, religious, and biological racism.[48] Whether the count believed what he wrote and meant what he said about Jews is surely relevant. The point here is that his prejudices solidly advanced his career, and not just in the making of interwar Hungary's academic maps.

MELOTRAUMA

The Hungarian delegation had no role in peace negotiations after World War I until 1 December 1919. By that time, the rump borders of Hungary were decided on generally pro-French terms, in favor of the Little Entente. Only after the German question was settled and Versailles was formalized were Hungarians allowed to take part in the negotiations in January–March 1920, which led to the Treaty of Trianon. When the treaty finally was signed on 4 June 1920, Hungarians commonly saw it as a *Diktat*. In Teleki's politically victim-centered narrative of trauma, Hungary was encircled, occupied, and "dismembered" after World War I.[49]

Terms for the defeated are visible today. The country lost roughly two-thirds of its population, one-third of its territory, half of its most populous cities, and its access to the sea. Reduced from 282,000 square kilometers to 93,000 square kilometers, its population decreased from 18.2 million to 7.9 million, making it central Europe's smallest country. Its army was limited to 35,000 men. The country paid reparations. Czechoslovakia, Yugoslavia, and Romania, which fought against a Greater Hungary and Habsburg legitimists, became home to 3.3 million Hungarians, more than half of which were in solid blocs directly adjacent to Trianon Hungary's borders. Containment logic followed from indelicately applied self-determination: that Magyars were the country's titular nationality, and they, like colonists, would repopulate into cities and the Budapest-centric space of the Danubian Basin. The "Carte Rouge" was made into a tool of the victors' containment, of the fear of Magyarization and Hungarians beyond nation-state borders, instead of flexible frontier space. Few groups or parties considered the treaty peaceful or acceptable. Calls for justice and revenge had lasting effects on all the minority peoples of Hungary and former Habsburg lands. Reactions ranged from nonviolent diplomatic revision and reform of nationality policy to the paramilitary reconquest of Hungary's lost territories.[50] Geographers failed spectacularly in 1919–20 when the country lost all of its "historic" and "ethnic" claims.[51]

Physically drained, Count Teleki took the Trianon injustice personally. In June 1920, it was for him the catastrophic result of a national search for wholeness, as if he had lost an internal organ. Yet if we look closer in the months preceding, the "trauma" posed many opportunities for him and others' careers, at last to nationalize and professionalize a corps of Ostmitteleuropa map men and science. He assembled maps and statistics for the Paris talks and preliminary Trianon negotiations from January to March 1920, resulting in the Treaty of Neuilly.[52] In the first months of 1920, Teleki was made a full professor of geography at Budapest University's new school of economics and public administration, appointed by the university's regent.[53] The Ministry of Commerce published *The Economies of Hungary in Maps* early that year, a vast oeuvre, together with the MFT and MTK. There would be many editions revised and enlarged with more sophisticated maps by Francis de Heinrich, the minister of commerce, and compiled by Aladár Edvi Illés, the ministerial councilor and chief of the department, along with Albert Halász, a chemical engineer, trade inspector, and friend of Teleki. The count penned the edition's preface in English:

> This atlas was drawn during the work of preparation for the peace conference. From the moment we saw the way in which peace was settled with Germany, we had not the least hope of changing the minds and decisions of the conference taken and fixed without asking much about the conditions or the will of populations. Still we had to put our argument before them, even when our memoranda and maps remained closed and folded. It was our duty towards the nation, towards future generations and—towards our foes and judges. . . . We have not worked for the moment of the peace conference and we have not used exaggerations which at those times would have been of use. We worked for the tranquilised minds of the future. There is no exaggeration, to improve their knowledge. We wanted always to contribute to the development of that Europe which we defended during centuries. This

atlas shows not only what Hungary lost, what she retained, but shows quite abstractly to the foreigner, to the neutral how an economic unity was cut to pieces.[54]

Such "surgical" efforts had dubious effect. When the count returned to cartography, the *Ethnographical Map of Hungary Based on the Density of Population* in 1920 was based primarily on the "Carte Rouge." It was published at the Hungarian Geographical Institute in Budapest, in Hungarian and English, and reprinted in The Hague.[55] He issued in Paris and Budapest a series of maps and pamphlets on Hungary's frontiers, including *La Hongrie du Sud: Questions de l'Europe Orientale* and *La Hongrie Occidental*.[56] With Jenő Cholnoky and Ferenc Fodor (1887–1962), he put together the *Economic-Geographical Map of Hungary* in Hungarian, English, and French. Along with the "Carte Rouge," it is one of the two of Teleki's maps displayed prominently today in the permanent gallery of the Hungarian National Museum.[57]

On 10 May 1920, Teleki gave one of his many national speeches against the tragedy of Trianon, but he was not a garden-variety nativist. He appealed to "historic" Hungary's St. Stephen crownlands from the ninth century onward, on through the Battle of Mohács in 1526, to Hungary's defense of Europe. It prompted raucous applause in the National Assembly. Calling attention to the folly, he lauded John Maynard Keynes and the economic critics of 1919. He invoked U.S. anticommunist politicians and experts like Bowman, a supporter of the League of Nations, and William C. Bullitt, the diplomat who testified in the U.S. Senate against Versailles. Teleki praised such Americans, whom he saw as men of honor, because they "today unreservedly admit . . . the flaws that were the product of the peace conference." He appealed to Hungarian "moral duty" to support revision. Teleki looked for international allies to end "war psychosis," to assist Hungary against communism, and "free . . . the nation from bondage."[58] In the meantime, elites' frustrations over Trianon boiled over in the Hungarian parliament from June to November 1920.

From 19 July 1920 to 14 April 1921, Teleki was tapped by Admiral Horthy to serve his first stint as prime minister of Hungary, thus exceeding the wildest dreams of his father Géza, who had died in 1913. Teleki kept one foot in academe and the other in politics. On 5 October, immediately after the department of economic sciences at Budapest University was founded, he became centrally engaged in it. Notoriously as prime minister, he introduced the *numerus clausus* in Hungary in 1920, the first such discriminatory act in Europe. This restricted Jewish admission to university based on race and religion, and limited Jewish numbers in public service and the echelons of Hungary's emerging bourgeoisie. Teleki essentially forced Hungary's wildly diverse Jewish classes, including secular academics, to "choose" loyalties by modern categories of belonging to the body politic. Laws enabled Teleki to deploy the judeobolshevik stereotype. He invited pathologies of anti-Semitism in 1919 into the retributive production of more maps.[59] Hungary's fearful interwar right now could prompt panic in an electorate against alleged Jewish cultural, political, and economic preponderance in the "Judapest" capital. By prohibiting the promotion of East European Jews into urban professions and positions of state, Horthy and Teleki reserved powerful seats for privileged men like themselves. Officials thereby remade themselves from victims into patrons, bourgeois experts and travelers in service to states, courtiers of regimes with favored kin, allies, and underlings. Even the Trianoners—especially the Trianoners—were Europeans.

VICTORS IN ARMS

Once Versailles was signed, it had an impact on Bowman and Romer immediately and for the rest of their lives. By the end of July 1919, the Poles seized full control of the badly damaged city of Lwów in the Polish-Ukrainian War, along with Ukrainophone areas of former East Galicia. Bowman, done as the U.S. chief territorial specialist, sailed back safely across the Atlantic to resume his AGS directorship in New York City. Romer stayed on in Paris to develop contacts and complete reports for the Polish Sejm (parliament) on East Galicia, Lower Silesia, and future Polish-Czech and Polish-Lithuanian borders. Romer's own diary characteristically noted his heavy work discipline. He depicted his wife Jadwiga as always nurturing and supportive, a calming influence in stressful situations. She came to Paris on 27 July and stayed on for at least two more months.[60] Meanwhile in Warsaw, the Sejm debated the finer points of Versailles and finally ratified it on 31 July.

Not until October 1919 did Romer leave Paris by train. He stopped to help set up the Polish Military Geographical Institute (Wojskowy Instytut Geograficzny or WIG) in Warsaw, which began without essentials of equipment or personnel. On 17 November, he brought his belongings from Vienna to resettle permanently with his family in Lwów, now as a citizen of Poland's republic. In Lwów, Romer became chair of geography at the rebuilt, renamed Jan Kazimierz University. He was adamant that Polish mapmaking, drawing from Vienna's commercial rather than St. Petersburg's military traditions (at least after Peter the Great), must remain in civilian hands. The WIG was modeled on Vienna's wartime Military-Geographical Institute, where Romer gathered resources for the 1916 atlas he sent to the United States. He brought other data and maps from the Central Statistical Committee, the House of Trade and Industry, the libraries of Vienna University and Jagiellonian University, and the statistical office in Kraków. Romer received contributions also from Austrian ministers' collections as well as individual Polish professors, priests, and antiquarians.

Versailles may have quieted the Western front after the war, but it is oddly overlooked that Poland's political borders by the end of 1919 were as yet unresolved. Romer was cued to Polish-Ukrainian and Polish-Soviet tensions. On 17 December, he pleaded with Bowman, in a handwritten letter, on the theme of un-self-governable Ukrainians and the League of Nations' decision on East Galicia:

> My hearty wish was to return completely into the science and the school . . . the "entente" dont [sic] allowed! In some weeks came the news about the Eastern Galicia, the arrangement of which by the S. Court was an attempt on the solidity of the Polish state as hard as the solution of Danzig or Upper Silesia. I ask for the reason of such a solution. Going from the exact principles of nationalities was the return of Galicia to Poland possible under the supposition only, that the Rythenians [sic] are not able to govern hisself [sic]. The idea of administration under the authorities of the L. of Nations was without ground, because this principle is to apply for the natives peoples [sic] only, and the territory of eastern Galicia was since centuries by European societies administrated. The fact is, that eastern G. was given to Poland in administration for 25 years, then in the supposition, that after this period the Ruthenian people will be able to govern himself [sic]. This hypothesis is in my opinion false.[61]

Romer appealed to "exact principles" in the colonial Penckian way. In the letter he sent, he penciled in "extra-European" in place of "natives peoples," in effect casting "Rythenians" out of Polish claims. He located "science and the school" exclusively among Poles. The logic was Eurocentric, that since the Ukrainians lacked high culture they were undeserving of statehood. Romer's prejudices must have resonated with Bowman's Anglophile frontier ideology, and if not with his Wilsonian racism, then with the Yale geographer's previous empirical work on indigenous Andean peoples in Peru and Ecuador.[62]

Romer continued to ship photos across the Atlantic, along with newly produced Polish books, maps, and atlases, to Bowman at the AGS. With Poland's eastern flank exposed, he kept up the pressure on Bowman.[63] In a postcard of 2 March 1920, he requested the addresses of Lord, the geologist Reginald Daly (1871–1957), and Major (soon to be Colonel) Martin, to send them more materials on Poland. In the same breath, he passed along "kindest regards to Madame Bowman and yourself very sincerely."[64] Bowman, again happy to show off his connections, reciprocated. He wrote on 20 April that Daly and Lord were now working at Harvard (the Canadian-born Daly was head of the geology department from 1912 to 1942), and Martin was in the military intelligence division of the U.S. Army in D.C. The scientist Arctowski, who worked under Lord in the U.S. Inquiry, had just returned to Lwów; Bowman praised him as "a strong man who has many fruitful years ahead of him" and hoped he would return for "an extended stay" in the United States.[65]

The geographers' language became more coded and intimate as they faltered and fretted in 1920, as they did in Paris in 1919, that their expertise was being ignored by politicians. Bowman confessed to Romer, "You must be sure to send me word whenever any geographical matters come up in your country of which we ought to know here at the Society, and then in addition to that you must write frequently about your personal matters, for I shall always be deeply interested in them after our pleasant days together at Paris. So long as I live I shall remember the great drama of the peace conference and our frequent discussions on Polish problems."[66] On 30 April, Bowman wrote, "Remember that we shall always be glad to get any material that you know of in Poland, whether in Polish, Russian, German, or French, . . . we are able to pay for such materials . . . the white eagle is soaring." Bowman urged his friend to finesse a kind of PR account of his 1916 atlas of Poland, wanting him to narrate its path to the U.S. and how the atlas was utilized by the Inquiry. "By all means print the letter which I wrote to you . . . regarding the way in which the Atlas of Poland came to America and the use that was made of it by the American Peace Delegation and the 'Inquiry,'" he wrote.[67] Years after its publication, Bowman charged himself with finding someone to write a strong review of it for *Geographical Review*, the AGS journal. Seeing no conflict of interest, he handpicked Lord, his trusted ally in the U.S. Inquiry and the IBC on Polish affairs. Lord, Romer's friend, was even invited by the diplomat-pianist Ignacy Paderewski (1860–1941) to come for a visit to Lwów, just after the Versailles treaty was signed.

Of all our men, Romer was a relentless charmer. Indeed, one of the least appreciated aspects of his cartography was his timing and back-channeling skills. Romer sent his newly revised atlas of Poland with text in English, Polish, and French for Bowman's own research on his forthcoming book, *The*

New World. From Lwów he wrote on 30 April, "Today we are 80 kilom. distant from Kijów-Kiev, and we work with the highest seal on the 'Ukraine' and perhaps You will be interested that I am not against it, whatever I not conceal the dangers wich [*sic*] are connected with this question for Poland. The transitory character of the country of lower Dniepr/Ucraina is lasting on the psychology of the people which are the product of this physiographical position!" Romer mapped Ukraine in 1920 as a transitional zone, the very notion he *rejected* for Poland in 1912 when its integrity was denied by Nałkowski. Romer in 1919–20 denied historical agency to Ukrainians and consigned to the dustbin the political maps Rudnyts'kyi drew in wartime Vienna. He hoped "to be free from all foreign and intern[ational] public illnesses and go back to the scientific occupation, [and] to look for some rest."[68]

To moderate his overbearing streak, Romer had a distinct way of making Bowman nervous while appealing to his vanity and mundane frustrations. In a letter of 19 June, Bowman mentioned two packages the AGS had sent to Lemberg (as he referred to the city), but were returned to the office in New York. In an exchange about the incompetence of their postal services, Bowman confessed his post-league irritations, "I beg of you, my dear Romer, to interest yourself in government matters and not to allow your new government and your new officials to become so stupid and inefficient as those who run this country at the present time."[69] They rehashed the old dream of objectivity, *plus* anticommunism: "I hope that you will make writing to me a regular task, for I enjoy your letters exceedingly and only wish, in spite of your protests as to English literary style, that I could write Polish one-tenth as well, for your sentences are always perfectly clear. . . . Mrs. Bowman joins me in best wishes and regards to you and yours. Long live Poland! (but look out that you do not get too deep into Russian territory!)" In a letter sent to Lwów on 23 July, Bowman noted that Lord was writing the review of the 1916 atlas. Seeing what he wanted to see, he offered sympathy for the Wilsonian struggle: "You must be deeply engaged at this time on account of the struggle on your eastern front. It would be ever so much better if there could be agreed upon some line on which you could stand with Allied support. Doubtless in the near future something of this sort will be arranged, in order than you shall not be left alone to fight this battle with the Reds."[70]

Once Piłsudski's forces were victorious in Warsaw over the Bolsheviks in August 1920, Romer sent two parcels to Bowman as a gift. One had copies of a new wall map defining the borders of the Second Republic; the other had the "territorial problems of Poland" with a focus on Polish rights, and a copy of thirty-one maps for the second edition of Romer's "congress atlas" in preparation for Riga in 1921. This was important, yet still on the surface of what the map men shared. Romer knew how to sweeten things. Bowman privately thanked Romer in December for sending "a panorama of the Tatra," which in Poland evoked national unity and iconic landscape beauty. He stated, "I am going to put it on the wall of my room at the house as a reminder of yourself and of this very interesting new part of the Polish realm."[71] Bowman could not "impartially" hang a Polish political map in his AGS office in New York City, but he could place a landscape portrait in his home. Romer's Tatras had meaningful appeal, a symbolic "Polish" site for rejuvenation, the retreat where Eugeniusz and Jadwiga had a prized summer home and transmitted aristocratic and family values to their children.[72] Our "friendly" victors were starting to intuit a great deal more about each other's interior lives.

NEW WORLDS, NEW MEN

Back in imperial America, another friendship was falling apart. In 1899, Davis in America postulated—correctly, Bowman and Penck once thought—a "universal" deductive cycle of erosion. It represented a breakthrough in geomorphology before World War I, and provided fruitful grounds for German-American exchange. Called Davis's erosion cycle, it was "an ideal cycle of geological uplift, erosion, and deposition, which wore down uplifted landscapes into . . . peneplains and produced a sequence of young, mature, and old landscape forms."[73] Prewar exchanges between Davis and Penck in Berlin in the 1908–9 academic year resulted profitably in a translation of Davis's lectures, and the printing in Berlin in 1912 of his work, as *Die Erklärende Beschreibung der Landformen.*[74] Prior to 1914, Davis's theory was central, though not essential, to Penck's own scientific work. Now after Versailles, it suddenly became an affront to the esteemed contributions of Germans to Europe's geography. Penck appears to have had no objection until *after* World War I, when Siegfried Passarge (1866–1958) and Alfred Hettner (1859–1941), two prominent geographers who inspired his own son Walther's research in geology, seem to have convinced him that the American's model was of limited use, just one of multiple explanations. Penck took it upon himself to argue against Davis for his German audience, tearing apart his general theory cycle as far too dogmatic. The chair of geography in the Berlin of 1919 took the high road, adhering again to the virtues of objectivity and field-based empirical methods.

The war changed this, too. When science and political geography intersected in the early 1920s in the new geopolitical way, it was not just for Penck who refused either to disavow or to assume responsibility for Germany's foreign policy. Who "arrogantly" started the Davis-Penck quarrel is hard to determine, perhaps not even relevant. Having made their fame in the 1890s and 1900s, these men now charged each other with letting prejudice interfere with free inquiry.[75] Davis in the *American Geographical Review* responded with a snide nationalist review of Penck's *Festschrift* of 1918 and chastised the Berlin chair for not believing in his "universal" cycle.[76] Instead of courteous dialogue, these elders indulged in the pettiness of ad hominem ivory-tower quarrels. Davis soon solicited the help of his protégé Bowman at the AGS; the two map men exchanged at least eighty-six letters between January 1917 and December 1932.[77] In a letter of 2 February 1920, the Harvard professor complained about Penck's militarism. "Several papers from Penck came lately," he wrote. "In his article on Die erdk. Wissenschaften and der Unv Berlin, 1918, he opens by recalling that four years before, Germany was driven to war by Russia's breaking her word, on the east, and that a few days later England attacked Germany on the west. . . . Not a word does he say about German's breaking her treaty with Belgium! To what extremity is a partisan driven!"[78] With high reputations and egos that required stroking, Bowman found himself in the middle of a postwar of words between his "fathers."[79] As U.S.-German geographers' relations deteriorated, reflecting wider Franco-German and German-Polish patterns in the 1920s, such outbursts became commonplace.

Across an ocean in Poland's Second Republic, Romer was the driving force behind the country's maps. In preparation for Polish-Soviet negotiations at Riga in March 1921, Romer was in charge of mapping Poland's "east" and the Soviet "west," arranging geopolitically the settlement (from the disputed Curzon Line of 1919) of borders in Ukraine, Belarus, and the Baltic states in the aftermath of the Polish-Soviet War. He published his *Congress Atlas of Poland* in 1921, in effect the second edition of his 1916 atlas, with new and more persuasive thematic maps. He literally supplanted Penck in England from summer 1914, becoming an honorary member of the Royal Geographical Society of London. Once a frenetically busy Romer returned from Riga, he set about restoring geography in Lwów and at the newly Polonized Jan Kazimierz University.

Romer established the Książnica-Atlas Cartographic Institute for Poland, a firm modeled on Vienna's commercial traditions and Justus Perthes in Gotha. Książnica-Atlas opened its first branch in Lwów and an affiliate in Warsaw. It became Poland's most technically advanced organ for map production in the interwar period. Romer recruited junior map men just as the great powers and nations of East Central Europe had gathered their experts for Paris. He worked closely with Stanisław Pawłowski (1882–1940), who became professor of geography at University of Poznań, and Teofil Szumański (1875–1944), a talented cartographer adept at drawing maps for the Polish National Committee.[80] Romer organized all the lithography, graphic design, and print shops. He arranged excursions for students to Poland's new seacoasts in the north and south. He promoted geomorphology and glaciology. He showed off Poland's reborn landscapes, from the Eastern Carpathians to the Tatras and the Holy Cross (Świętokrzyskie) Mountains to Podolia in the Subcarpathians. Romer achieved something few others could manage in strained postwar economies, the "German" mobilization in the 1920s of a "Polish" national-professional corps of geographers and cartographers.[81] He even trained his own Penckian sons Witold and Edmund to intern with talented Polish associates at the institute.

Since maps could be expensive to produce, none of this was easy. Even Bowman faced similar challenges with *The New World: Problems in Political Geography*, the book for which he became best known in spring 1921.[82] A brief summary is useful here. Significantly, Bowman's book was one of the first American policy texts in the genre of U.S. exceptionalism and the Anglo-American "heartland" logic of hegemony vis-à-vis Russia/the USSR. As Sir Halford J. Mackinder (1861–1947) famously put it, to control East Central Europe and Eurasia was to rule the world.[83] It was illustrated with an astounding number of maps—to be exact, 215 (figure 3.3). Like his Berlin mentor, Bowman envisaged the *whole* world from a single Archimedean point. He drew explicitly from the U.S. project for the 1:1,000,000 map of Hispanic America, which commenced in 1920 (after Penck's International World Map of 1913, proposed in 1891), and of course from Romer.

On the most basic level, Bowman asserted the U.S. right, based on its revolutionary principles, to extend its hegemonic global interests in the name of liberal democracy.[84] He combined frontier ideology with a dose of frustrated Wilsonianism, his assessment of an adversarial world left by war and revolution and overseen by the League of Nations. Bowman applied many of Romer's Greater Poland views to the Polish-German, Polish-Czechoslovak, and Polish-Ukrainian frontiers he once adjudicated. He depicted Poland as a bastion of Western civilization, the Midwesterner's own lost kind of mini-America in Europe. *The New World* was tribute to Wilsonian democracy, though as we shall see, Bowman was by no means democratic in the dealings of his family or interpersonal life. Yet it came at a key juncture in his friendship with Romer. Best sellers are not necessarily the best books. In its time, nevertheless, the 1921 treatise was widely read and translated into multiple languages, including German, Polish, Russian, and Chinese.

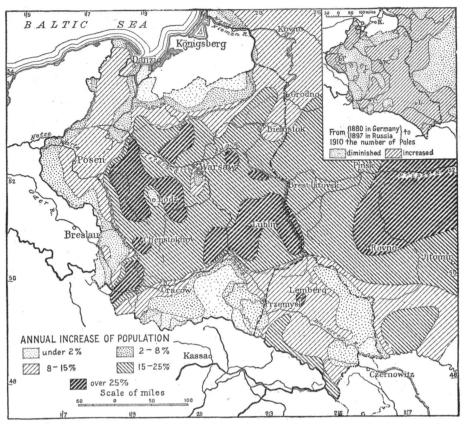

FIG. 182. Note that Russian Poland had the greatest increase of population. It is significant in relation to Polish claims and ambitions on the east that Poles have increased in Eastern Galicia and in general in the belt of country just east of the line recommended by the Peace Conference at Paris (heavy broken line). Such a tendency might ultimately change the ethnic situation. Based on Romer, *Atlas géographique et statistique de la Pologne*, 1916, Pls. 7 and 11.

FIGURE 3.3. Map (fig. 182) from the first edition of Bowman's *The New World: Problems in Political Geography* (Yonkers-on-Hudson, N.Y.: World Book Company, 1921), 353. Based on Romer's *Geographical-Statistical Atlas of Poland* (1916), the map relies on Romer to note "the ethnic situation" and Polish population growth in "Eastern Galicia." It was one of 215 explanatory maps in total, used by Bowman as illustrations for his book.

Bowman liked to think that his career took off because of intelligence and merit. In Washington, D.C., he was elected in 1921 to the board of directors for the newly founded U.S. Council on Foreign Relations. He became a member of the editorial advisory board of *Foreign Affairs*, the council's policymaking journal of record. He remained on as director of the AGS in New York City until 1935, when he was chosen as president of Johns Hopkins University. Actual reception of his treatise was mixed. German and Hungarian geographers were among the least pleased. Teleki reviewed it, as we will see, quietly and much later, in Hungarian in 1929. Germans regarded him as anti-German, due to his support for Polish (Romer) and French (Martonne and Gallois) geographers. Berlin-based geopolitical thinkers led by Karl Haushofer (1869–1946) in the *Zeitschrift für Geopolitik* gave pointed rebukes. *Völkisch* geographers of an older sort read it as a stalwart pro-Polish book penned by the Allied victors, evidence of a huge betrayal.

Even French geographers, holding his Anglo-Saxon name in suspicion, often suspected Bowman of being pro-German or British. The scientist clung to impartiality and found these dismissals amusing.[85] He insisted, breaking from a German past, that the twentieth-century world he explored was a new one, the continuation of Wilson's "scientific peace" that everyone in fact sought mutually in the Paris of 1919. When Bowman set about revising it for a new edition, Romer offered to send European materials to him from Warsaw.[86] Theirs was less a lament for Ostmitteleuropa than an imagined rupture, led by map men of stature in the 1920s. Continuities were very significant. Bowman preserved an illusion of political geography as international *Wissenschaft*, beyond geopolitics and revisionist German, Austrian, Hungarian, and Bulgarian geographers in their boycott of (or banning from) the IGU, the organization of which he became vice president in 1928 and then president in 1931, with Romer as his vice president.

STRINGS TO PULL

Having begun to doubt Penck's geography thanks to Davis, the new-worlder Bowman next jumped into a kind of map-loving bromance (forgive the buzzword) with his Polish ally Romer. In a letter from Lwów on 1 October 1921, Romer curiously defined their friendship as "credit." "My dear Friend," he wrote, "It is very right, if You use such a capital, calling me, but it is very much proof for Your friendship . . . If You use this call to me, You give me a credit, for which I am proud, but never know, if I came in the position to prove, that I merited this credit. I assure You that I am very much obliged for Your credit, and that in my sentiments is a large place for reminiscences, which are tied with You. . . . You force me to think over Your attention and kindness for me."[87] The new-worlders coinvested in Romer's maps and plans, working in European tandem to rebuild the fraternity of global experts on geographic matters. That was exhausting. In Romer's words to his "dear friend" in this important document:

I am morally completely tired, I could no longer work in conditions, which are for men thinking real and honest, no more tolerable. Much of the blame falls on our government's backs, on our incompetent managers, but not all blames it remains enough blames for the world, for the big countries, which came in the straightest controversy with the principles confessed. I came to this kind of work never more, with the one exception, I will in the next summer write my memoirs not for today, but for the future, they are too sad. Since six years I have had my first holidays this summer . . . this recovered me and my family well. The wife of me and the children which have seen and suffered some during the last three years, but which lived also something the ucrainian and bolsheviks war needed also a rest, which was well given. Our industrial establishment of the cartographic inst. goes more slowly because of financial difficulties, but never is stopped and we work and produce also in worst conditions, which will in any case be better in the next six weeks. As the thing will be mature, I will You report. I am, as always, Sincerely Yours, Eug. Romer. My brother which You and Madame B remember has also a very hard life . . . His life is much harder than mine . . . he is completely in the public work, and can find no consolation in the private work.[88]

The letter outlined the two geographers' public/private sensibilities and shared male bonds. In the stateside role of experts, they worried about being ignored, Bowman by the failure of the U.S. to join the League of Nations, Romer by the sidelining of his expertise by Dmowski and Piłsudski. Themes of health, work discipline, and convalescence shaped the manly selves they dramatically performed—and how their vitality was sapped (Romer alluded to his brother, Jan the general) by public affairs and labors. Meaningful reflections such as these were written to each other while on vacation, Bowman from the family's cottage in New Hampshire, Romer from the Tatras in postcards from Zakopane with its spectacular sanatoria and mountain views, spaces for the (Penckian) study of glaciology.[89] In several exchanges from 1921 to 1925, the Tatras (but not the Alps) represented Romer's quasi-mystical place for outdoor work and regeneration, away from Polish politics.[90]

The profession these men loved was more than a mere résumé or *Lebenslauf*; it was a gendered construction throughout the 1920s. Romer guided a new generation of service-oriented men, and some women, into a didactic European project, as Penck had partly accomplished when his students gathered for his *Festschrift* in 1918.[91] The study of colonial geography, they thought, bred men's character as in a world of past empires. Romer wrote to Bowman, "I assure I have never had such a great number of moral excellent men in the laboratory, but all these men, exactly all men and girl[s] are more or less ruined by the war. Among 8 males is 6 over 30 year old and 3 of them are married and have to nourish their family. With only one exception all students was exactly 7 years on the front and 2 of them made acquaintances with the Bolshevik-regime. About a similar prehistory [I] have my girl-students . . . going through the Bolsheviks' hell."[92]

Romer hoped a corps of *Kulturträger* would emerge. Pawłowski and Szumański followed their mentor's goal of extending Polish geography further into education and research training for new cadres of national geographers and cartographers. Romer reassembled a canon of knowledge, from before Humboldt and Ritter to the Polish Enlightenment in the writings of Hugo Kołłątaj (1750–1812), Stanisław Staszic (1755–1826), and Jan Śniadecki (1756–1830).[93] He sought to introduce Polish geography in the very first class of gymnasium. He returned to geomorphological research. He produced maps and atlases for schools. Romer led students on expeditions in order to transform studies from the rote-based, schoolmasterly style of imperial Vienna into research at a higher level. He introduced students to geology, cartometrics, and learning methods. He assigned weekly repertories for all years. He gave students literature in Polish, German, French, English, and other major European languages. Romer's training introduced novel pedagogy to maps.

Although Romer became a twentieth-century map man by the early 1920s and his pre-1914 and Parisian networks ensured a wider reputation, it was hardly easy to get his vaster projects off the ground in Poland or turn a profit. In his Książnica-Polska cartographic firm, as it was originally called, overhead was a perennial problem. To stay solvent, Romer sought the help of Jan Treter (1889–1966), the Lwów-based economist at the Polish Academy of Commerce, where he had taught before World War I. Because the Second Republic lacked funds for expensive atlases and textbooks, he relied less on the state than on private patrons and donors and on the Stefczyk bank in Poland for loans, as he had done during the 1910s. Romer found help, through Treter, from the Polish banking sector. Aware of the poor condition of cooperative firms, to which banks could refuse loans based on an assessment of low or unreliable credit, Romer sought to finance map production through a joint-stock company guided by Treter.

Treter's savvy string-pulling and problem-solving ability, much underappreciated, proved a giant asset. He was Romer's ally in business and a reliable mainstay. Treter was House and Bowman rolled into one, a Polish recruiter of talented men, a resource manager. Recognizing that postwar popular demand for maps was growing, he shuttled back and forth to Vienna from the old Military-Geographical Institute. He bought cannibalized lithography machines and convinced specialists from the former Habsburg Empire to come to Lwów and work alongside Romer. Romer relied on Treter's know-how to raise funds from the Polish Industrial Bank and the Bank Krajowy. Romer was elected chairman of the supervisory board of the Spółka Akcyjna Atlas in 1923; Spółka merged in 1924 with Książnica-Polska to form the Książnica-Atlas firm. Książnica-Atlas quickly became a success. It turned out to be the largest commercial firm for maps in interwar Poland, operating from within the Cartographic Institute on the premises of Jan Kazimierz University in Lwów, inside the larger Geographical Institute. It printed maps for schoolbooks and academic books, lesson plans, scientific publications, and popular science texts. All maps produced by Książnica-Atlas acquired a logo (the "map-as-logo," literally in Benedict Anderson's term), associating the "Atlas" trademark with the institute and Romer. The title "im. Romera" was added; the Romer Cartographic Institute was named after its founder. The firm was Romer's brainchild, a League of Nations for the latest map designs. It lasted in Lwów until the invasion of Poland in September 1939.[94]

To further reconnect Poland to world geography, in 1923 Romer set up the journal *Polish Cartographical Review* (*Polski Przegląd Kartograficzny*) to emphasize geographic knowledge and rigorous study *in Polish* for the Second Republic's citizens. Romer, Pawłowski, and Szumański reproached the shortcomings of German maps and the tendentiousness of Prussian censuses and compilations. (Poland's state developed its first census in 1921.) Contributors reviewed maps, articles, and books of Germans in depth. Romer became involved in the Association of Polish Geography Teachers and its new quarterly, *Czasopismo Geograficzne*. He disapproved of abrogations of parliamentary democracy: when the Military-Geographical Institute (WIG) completed its topographical survey of Poland on the scale of 1:840,000 in 1923, its officials demanded to inspect maps that were produced by the Romer Institute. Seeing this as interference, Romer would not sacrifice civilian authority to military censorship. After prickly reviews in the *Polish Cartographical Review* in 1924, he and the WIG kept at an uneasy working distance.[95] From his institute, the Lwów geographer sent thematic maps around the world and received numerous accolades.[96] He was made an honorary member of new geographical societies, and awarded the prestigious Prix Gallois by the Paris Geographical Society. Appreciation even extended to the USSR, for in the relative thaw in Polish-Soviet relations that followed the Treaty of Riga in 1921, Romer earned the title of corresponding member of the (former imperial) Geographical Society in Leningrad and honorary member of the Soviet Geographical Society in Moscow. A cordial correspondence ensued with Veniamin Semyonov (1870–1942), the accomplished urban geographer and son of Pyotr Semyonov (1827–1914), imperial Russia's most famous geographer. Between 1920 and 1937, Romer received honors from scientific societies and associations in Warsaw, Prague, Poznań, Toruń, Belgrade, Bratislava, Chicago, Leningrad, London, Moscow, Neuchâtel, Paris, Rome, Sofia, Stockholm, and Toronto. None were in Germany or Austria.[97]

In this way, Romer at last achieved his Penckian stardom. He was invited for the 1922–23 academic year to come to America on exchange, to teach at Clark University in Worcester, Massachusetts, but he regretted that he was not able to leave Lwów because of his many duties.[98] Bowman wished all success to Romer and the former U.S. Inquiry member Arctowski for building up his institute. He politely added that "if there is anything that we can do [in America] at any time we shall be happy to help you."[99] Romer surely grasped the importance of studying transnationally to acquire expertise, resources, and technology for cartography, as he had done in Germany and Austria prior to World War I. In his letter to Bowman of 14 April 1923, the Pole voiced concern: "Now I have some troubles with my older son [Witold], who received from our Polytechnical School the diploma of engineer chemist, and desires to receive a practice in one of the greater factory of wood-pulp/cellulose in America, for learning the kind of work in this country and for learning an industry, which is very important for Poland." An outsider to the American ruling class, Romer was clearly asking for a favor. He told Bowman that he had written to Henry Noble MacCracken, the president of Vassar College (from 1915 to 1946), who had traveled to Lwów and "was so kind to show interest in this question and promised to help my son in realizing his desires." Romer wrote also to Stanisław Łubieński, secretary of the Polish-American Chamber of Commerce in New York City, and Colonel A. B. Barber, "who was a long time in Poland." Romer shrewdly pursued study abroad hopes for his two sons Witold and Edmund: "I have too much to do in my home, I can no more dream to leave my country, but I will be very happy to see my children learn in the wide world. . . . I have sons, of whom work and character give me much hope, they will not waste the time in Poland. But peace we need much and for a longer time, because our government is young and must grow to be ripe."[100] Glad to show off, Bowman replied in a letter of 7 May, "I am writing to Dr. MacCracken today . . . to arrange for common action in regard to your son. As a college president he undoubtedly knows the way in which to obtain the best information regarding possibilities for your son. . . . Certainly I shall do everything I can."[101]

This intriguing episode of 1923 showed the power of an old boys' network of privilege and colonial patronage, the psychology of space behind surfaces of maps and according to personalized norms of *Wissenschaft* in Europe and America. The class- and region-traveling Bowman had similar ambitions for his sons Walter and Robert. He immediately dialed up prominent friends from New York to assist Romer. (Bowman did not regard his daughter Olive as worthy of the same professional training.) In May 1923, he wrote to MacCracken at Vassar and to Ferris J. Meigs in New York, a paper industry magnate and chairman of Vassar's board of trustees.[102] MacCracken replied in the affirmative: "I am very much interested in what you write about the son of Dr. Romer, and am now taking up with some of my friends who are in the pulp and paper industry the question of trying to locate some practical position for him. I am not now connected with any such industry. . . . However, I am delighted to do anything I can for you in this matter."[103] Bowman read this as good news and forwarded it to Romer, appending the letters from Meigs and MacCracken.[104] Tying geography and cartography as a joint enterprise to professional lives and grander projects, our patriarchs shared in the determination to gain access, by any means necessary, to scientific institutions for the advancement of biological and intellectual "sons" into a technical intelligentsia.

SCENES FROM A BREAKUP

Walther Penck, son of Albrecht, was a geographer of precocious talent. He was his father's pride and joy. Educated to become a geologist in Germany and Switzerland, he studied abroad at Yale in 1908–9 at the time that Bowman taught there, also while his father was invited from Columbia, on exchange with Davis from Harvard, to deliver the Silliman Lectures. Walther did extensive research on landforms in South America and had enviable knowledge of the Andes Mountains. He defended his doctorate in geology in 1914, at the University of Heidelberg. In Germany's war on the Western front, he served in the army and saw combat in Alsace. In fall 1915, he was promoted to professor of geology and mineralogy at the University of Constantinople, provided with this opportunity by the Second Empire and its academic establishment. He completed research in 1916–18 in Anatolia and along the Dardanelles. He then taught at the Halkaly Agricultural College, where he trained many Turkish geographers at the Dârülfünûn, the only research institute in the Ottoman Empire in which geology was taught as an independent discipline.[105] Back in Germany in 1918–19, Walther Penck collaborated with his father's colleagues

Hettner and Passarge. Yet he came to regard Davis's "universal" cycle of erosion as an American prejudice and egregious scientific error.[106]

William Morris Davis in America refused to believe that new German methods rendered his general theory invalid. Nor, following the "empirical" proofs of Passarge and Hettner, could he accept that his theory was merely one possible explanation for erosion of the world's landforms. He complained privately in letters to Bowman that the Pencks were surely blinded by World War I. Now they would never acknowledge their bias, nor accept the preeminence of American theoretical work. The Harvard man put Bowman smack in the middle of a personal moment of crisis and professional quarrel. In July 1923, Davis's second wife died, sending the seventy-three-year-old geographer into a depression. He grieved in an unusual manner, by writing to Bowman and taking open aim at Berlin. In his letter, he mentioned his wife's death but only after six vituperative paragraphs against so-called scientific geomorphologists.[107] Bowman, retreating to a Wilsonian pose of disinterestedness, had of course known and corresponded on cordial terms with both Davis and Penck separately for nearly two decades.[108]

Nevertheless, in the old polite form, Bowman took patient heed of his father-mentor. He reviewed Walther's seminal 1922 book, *The Puna of Atacama*, about Andean landforms and border disputes between Chile and Bolivia in the War of the Pacific (1879–83). The review was safe and formulaic, appearing in *Geographical Review*. Bowman praised Davis's stellar career and the harmony of his American and worldwide reputation.[109] Ever the diplomat, the Wilsonian wrote privately to Penck the elder on 26 July: "I have just gone through your son's book on the Puna of Atacama and find it a most admirable piece of work. It is one of the best regional studies that I know of anywhere. I only regret that he should have misinterpreted (as it seems to me) the idea of the erosion cycle. . . . I hope that in our criticism of this matter we shall not fall into a narrowly nationalistic attitude. It would be a great misfortune to civilization if we should do so."[110]

Against "narrowly nationalistic" charges on the ground and on both sides, a lesson learned from Paris in 1919, Bowman chose Western civilization. Regardless, by summer 1923 at the AGS, he gave a full impression that he no longer aligned with the Pencks. He wrote behind Penck's back to Davis on that same day, "Let me say at once that the book by young Penck is most admirable, taken as a whole. . . . [But] when it comes to the erosion cycle, he appears to me to be very 'young and foolish,' if I may so put it. . . . It seems to me that he is going out of his way to take a crack at the idea of the cycle, while yet being compelled to employ it in explanation of the landscapes he found."[111] In advancing U.S.

global power, the author of *The New World* of 1921 always thought to align what was best for geographic science with his own career aspirations. On 27 July, a perceptibly angry Davis in a next-day rant damned Walther's "very limited, rigid, special scheme" and took umbrage with "the father [who] is as exasperating as the son." Davis reinterpreted the elder Penck's body of work since the 1890s, seeing their history of exchange as an intra-Western standoff. Davis attested to Penck's willingness to concede, on the symbolic ground of frontiers they once explored, for when "[he] was over here in 1897, and we had a trip across Canada westward across the U.S. eastward together, . . . he became much more imbued with the cycle idea."[112] From New York, Bowman agreed with Davis: "I thought every line of it deserved to be said. I could scarcely have understood the attitude of a group of German scientists in constantly stalking or waylaying the scheme of the cycle were I not going through a similar experience in regard to our Hispanic America Program."[113]

What the Americans did not know in this showdown on the West's (academic) global frontiers of the early 1920s was that Walther Penck, just thirty-five in 1923, was terminally ill with cancer. Economically and personally, Albrecht's life was unraveling. On 22 August in a letter to Bowman, the professor referred gently to his son's suffering in Stuttgart. He proceeded to defend geographic science on Walther's behalf and took up the mantle of his son's findings, offering what sounded like a modest empirical refutation of Davis's cycle:

> I am very pleased at your opinion of Walther's Puna book. He wrote it with devotion after having gathered the material with difficulty, and he was grateful that he was able to procure funds for its publication from the United States . . . I am so familiar with his [Walther's] views that I know exactly what his attitude is toward the geographical cycle. In principle it coincides with mine . . . my son and I do not think of injecting any nationalistic "moment" into this question. To me in scientific matters Davis is a scientist and not an American. I regret that several of my compatriots [Hettner and Passarge] have felt differently. My son would write the same thing if he were not severely ill in Stuttgart. Since Christmas he has not written any scientific paper and since Easter no letter.[114]

On 25 September, four days before Walther died, Bowman from New York responded in detail to Penck in Berlin. "I am very sorry to hear of your son's illness," he wrote, "and I can only express the hope that he will have improved by this time and that my work of appreciation regarding his book will have reached him." Bowman cited his Harvard education as proof that he, but not

Walther at Yale, was personally far more familiar with Davis's theory, recalling his time with Davis as his teacher; that Davis's theory was not some "law of inevitable sequence" that eliminated complexity, but nuanced and flexible, like models in evolutionary biology. Davis was a critical scientist and a problem solver, not a dogmatist, the "master . . . [with] penetrating questions and comments [that] so far outran my own abilities [was] not to be criticized for the deficiencies of his students." Not willing to jettison the theory, Bowman stressed the nineteenth-century myth of progress passed down from Davis and the elder Penck to himself and Walther in the twentieth century. He wrote, "The proponent of any scientific theory necessarily expressed the theory in simple terms in the first instance. . . . [He] has written a textbook in which not only the theory but its bearings and applications are set forth. As time goes on he enlarges and amplifies, and his students do likewise." Bowman wrote, "Because I know you so well I cannot impute insincerity to you. I have great respect for your work and I feel that you have great respect for Professor Davis' work."[115] Though Penck had broken with Romer in 1917 and Davis with Penck by 1923, Bowman now finally sided with Davis, while appealing to polite norms in the same breath.

On 29 September 1923, Walther Penck, the geologist, only son and heir to Albrecht Penck's scientific legacy, died of throat cancer. Upon hearing of Walther's death, Bowman wrote on 18 October, "I want to tell you at once how very deeply grieved we are here at the Geographical Society to learn of the sad death of your son, Dr. Walther Penck." Bowman informed the German master that Yale's Department of Geology and Geography and its Sheffield Scientific School remembered him fondly. Speaking institutionally, the AGS director expressed a "feeling of deep regret" and offered "our deepest sympathy." He asked the elder Penck to send his son's posthumous works to be published in the *Geographical Review*.[116] Yet when the mourning father told Bowman of Walther's death, the two men read between the lines and tried not to dwell on it. They continued to resolve Penck's falling out with Davis and American geography. Penck began his reply of 30 October "in a friendly spirit" and "to welcome our exchange of views about Davis' cycle theory." By then, life had really intervened. He told Bowman of the harsh economic conditions in Germany, how he was "anxious as to what my daughter-in-law and her children can live on. Times are very hard for us here. An American dollar is today worth 60 billion marks."[117] The social status of the Pencks had plummeted. The patron Bowman offered condolences and added that the AGS was willing to purchase any of Walther's manuscripts, maps, or other papers of scholarly or archival value, including his writings on South America.[118] At an emotional impasse, they retreated to norms each of them deeply respected. Bowman waited "until your son's book has been printed so that I may study it very closely and carefully and thus reach a more mature conclusion regarding his conception of the cycle . . . I shall reserve my personal judgment on the matter until I can read the argument in extenso."[119]

———

After 1919, conflicts involving our map men, political and psychological, burrowed into the grounds they made. Groundlessness persisted, however. Two years before he died in 1924, Franz Kafka wrote *The Castle* (*Das Schloss*) as a work of dystopian fiction. It featured a topographer trapped in a kind of peculiar matrix, in a dark bureaucracy. Yet the place for the fantasy, though pessimistic and rational in the sense of an iron cage, was not exactly a gridded or modern planned city. The Prague writer set the story elsewhere, on frontiers, in the borderland village of Spindlermühle (Špindlerův Mlýn), today a ski resort town in the Czech Republic. The flimsy unreliable narrator's all-seeing (but not all-knowing) voice emerged at the tale's outset, "It was late evening when K. arrived. The village lay under deep snow. There was no sign of the Castle hill, fog and darkness surrounded it, not even the faintest gleam of light suggested the large Castle. K. stood a long time on the wooden bridge that leads from the main road to the village, gazing upward into the seeming emptiness."[120] Kafka's deterritorialized "K." character stood on interwar East Central Europe's shifting plates. In the sleepy Sudeten Mountains where Germany, Poland, and Czechoslovakia met, the locale had an absence of overseeing gods.[121] The lodestar in the heavens was an empty overlook. The place was an atlas obscura, foggy, off the grid, absent of clear sight or definition.

Our men lived similarly by map fantasies. Identities were elusive, performed in the transnational places where they slipped through and lobbied for power. They lived not in matrices but among cartography's paradoxes—emotion as reason, science as fiction, truth as propaganda. As some grew alienated and apart, others reattached themselves. After World War I, they too were Kafkas who had a great deal at stake—the loss of a country, the loss of status, the loss of family, the loss of a successful career. Out of frustration, they hoped for power, for a just order, to be regarded as experts, for history to have meaning and geography to move humanity forward again. Human character had changed, not in December 1910 but a little later, in or about June 1919. Simmering up to the surface, their maps emotionally depicted the world's most potent and surreal anxieties. Objectivity became subjectivity. Modernity was not yet visible.

CHAPTER FOUR

BERUF

In Max Weber's *Politics as a Vocation* (*Politik als Beruf*), written in 1919, the sociologist offered his definition of the state as a human community that claims a monopoly on the legitimate use of physical force in a given territory. Weber delivered it during a German revolution in Munich when Bavaria briefly became a socialist republic, and it was printed in January just as the Paris talks got under way. The essay enters our spotlight more for its *Politik* than for its *Beruf*, a harder-to-translate term meaning "profession," yet implying a higher civic sense of service and moral duty. Our map men were men of *Beruf*, builders of institutions, but they were also traveling men of science. They went abroad in search of friends in wider communities, to lobby for and defend territory. Their ethnocentric maps like the "Carte Rouge," in the garb of reason, expressed personal pain or injury—partition, mutilation, dismemberment, emasculation, lost honor and status, unfulfilled hopes, co-opted *Wissenschaft*, defeated defenses of civilization.[1] After the war, Penck found himself among revisionists. He tried to regain his spatial bearings, but neither he, a geomorphologist, nor his son Walther, a geologist who died in 1923, was ever a geopolitician.[2] Penck's scientific pupils used nebulous signifiers of *Volk, Boden, Raum, Kultur,* and *Grenzen*, borders that could be open or closed. They envisioned a world of science that could be studied in its frontiers, mapped by hard-working family-loving men.[3] Lives of such professional experts were transnational. Lives involved more than nations, or bold lines by a state around a territory.

Although Albrecht Penck in his *Beruf* dabbled in the classical geopolitics of a first generation from the 1890s to the 1920s, he still adhered to the norms of a growing nineteenth-century academic guild.[4] He deeply supported empirical work.[5] The elder Penck in Berlin and Pál Teleki in Budapest were tempted by Karl Haushofer, a transitional figure who borrowed loosely from Mackinder, Mahan, and Kjellén alongside Humboldt, Ritter, and Ratzel in his writings. The founder of the Institut für Geopolitik in Munich in 1922 and the *Zeitschrift für Geopolitik* in Berlin in 1923, he eventually became a tutor to Rudolf Hess and was long thought, incorrectly, to have had a strong influence on Hitler. The Pencks, however, were not human geographers or Haushoferians.[6] In fact, Penck wrote little that was original after his son's death in 1923, and almost nothing in 1924 when the three-volume compilation *Macht and Erde* (*Power and Earth*), spearheaded by Haushofer, was printed in Berlin. Haushofer's geopolitical circles dubbed Versailles a *Diktat* in the revanchist 1924 volume, a brazen German rebuttal to Bowman's *New World* for fear of increased French or Polish influence. They criticized the victors and the U.S. Dismissive of non-Germans' rights and rights in general, appeals to objectivity seemed old-fashioned. While Romer gained traction with his maps, Penck in Weimar Germany found himself marginalized. Teleki was somewhere in between. When pretenses of *Wissenschaft* quickly faded in interwar scenarios, there remained one last thing these Western map men could do that Rudnyts'kyi in Ukraine could not—they could return home. It is to our Ukrainian's anxious sense of *Beruf* in postdynastic Ostmitteleuropa that we shall now turn.

VIENNA-PRAGUE-KHARKOV

For Stepan Rudnyts'kyi, the period between Europe's two world wars began in a tragic and horribly inauspicious way. In summer 1918, his wife Sybilla died of heart illness and complications from pneumonia. She left behind three children. A geographer in exile, he could not care for Emilia, Levko, and Orysia on his own. Sofia Dnistrians'ka, Stepan's sister in Vienna, assumed all responsibilities of childcare. Through the cataclysms that followed—Germany's surrender, multiethnic Habsburg devolution, and the Polish-Ukrainian War from November 1918 to July 1919—Rudnyts'kyi lived and worked in Vienna.

While the Paris talks went on in the first months of 1919, Polish military forces under Piłsudski ignored inter-Allied calls for a cease-fire and stretched Poland's pre-1772 limits in the *kresy* of Poland-Lithuania and Ukraine. Ukrainians made diplomatic overtures to halt a Polish empire, but having been de facto on the side of the defeated Central Powers of Germany and Austria-Hungary, they had to contend with lingering French anti-German sentiment and German and Anglo-American suspicions of Slavs, Jews, and communists. The U.S., Britain, and France roundly ignored Rudnyts'kyi's expertise, prized by German scientists and Penck even more than Romer's during wartime. The former subject of Austria-Hungary worked with Hrushevs'kyi and consulted with the government-in-exile of the Ukrainian National Republic, seeking its independence as a small federal nation-state in East Central Europe. Rudnyts'kyi's maps claimed lands on the conational basis of shared history, culture, and language on both sides of the Zbruch River, the now-defunct imperial Habsburg-Russian border.

In the postrevolutionary USSR and Soviet Europe, the Bolsheviks badly needed colonial geographers and map men like Rudnyts'kyi in the West to disseminate knowledge and serve the tasks of modernization. As Francine Hirsch has pointed out, Lenin considered scientists in service to planners to function as bourgeois experts, though they had been part of capitalist powers and pre-1914 Europe's exploitative order.[7] The knowledge such non-Marxist academics possessed was useful, even though Moscow held their political loyalties in question.[8] True believers in socialism existed. In general, however, not all were fully on board with revolutionary violence, a one-party state's claims to democratic socialism, or Moscow-centric communism in the Comintern (started in March 1919) of Lenin and Trotsky.

By the time the Polish-Ukrainian War ended in July 1919, three weeks after Versailles was signed, Rudnyts'kyi and other Ukrainian scientists could not retreat to their former lives. Marshal Piłsudski's so-called pacification of East Galicia removed the possibility of forming a Ukrainian university there. This was a major goal since the 1890s of the polymath activist Ivan Franko (1856–1916) and the Ukrainian Radical Party, and of the Shevchenko Scientific Society (est. 1873), the Ukrainian association that worked as an unofficial academy of sciences. On 22 November, the Polish government claimed the university in Lemberg/Lwów/Lviv and renamed it Jan Kazimierz University (today's Ivan Franko National University). In "Polish" Galicia, Rudnyts'kyi earned his doctorate in geography. In "German" exile, he spoke, wrote, and did research in multiple languages, including Czech. Ostmitteleuropa endured after 1918–19 through his transnational life and expertise. Rudnyts'kyi was employed as a professor of geography at the Ukrainian Free University in Vienna in the early 1920s. He taught in Vienna until he moved to Prague in 1923, invited to Charles University to lecture there and at the Ukrainian Higher Pedagogical Institute.[9]

Having lost his professional status in Polish Lwów, the Galician took on organic work among the Ukrainian intelligentsia, in exile from 1919 to 1923. He continued Habsburg Galician positivist traditions since the 1860s (recall that Galicia became autonomous in the Austria-Hungary of 1867) of working within and sometimes for the imperial partitioning powers on behalf of a national cause. In 1923, Rudnyts'kyi wrote his seminal work on Ukraine's political geography, the *Survey of the National Territory of Ukraine* (*Ohliad natsional'noi terytorii Ukrainy*).[10] He drew from his German training in physical and human geography to argue for the unity and integrity of Ukraine's lands. It too was a kind of new-worldism beyond the nation, for Rudnyts'kyi was critical of the diplomatic handling of the Ukrainian question by the League of Nations. He worried that Ukraine would be stuck in permanent limbo, between the nationalizing policies of Poland's state and Soviet nationality policies in the years of its New Economic Policy (NEP) from 1921 to 1928. Hrushevs'kyi, who hoped to make Ukrainian history a scientific and autonomous discipline, was in full support of Rudnyts'kyi and other Galician repatriates who elected to relocate from away from Poland and Central Europe "back" to Soviet Ukraine. Hrushevs'kyi did it himself in 1924. It made perfect sense for Ukrainian educated professionals who considered themselves by moral duty (*Beruf*) and outlook to be developers of Europe's most vital technologies across borders. Moreover, communist Moscow and Kiev appeared more welcoming than Piłsudski's Warsaw after Poland's violent annexations of Lwów in 1919 and Wilno in 1920. Soviet federalism and affirmative action prior to the start of Stalin's passport system seemed a better compromise, at least initially, in support of Ukrainian institutions and rights against the dangers of chauvinism. Rudnyts'kyi was hardly a Leninist, but he was not wrong to surmise that his European work in geography, coupled with decades of research and teaching, could turn out to be useful.[11]

Power scenarios changed fast, however. Lenin suffered the first of three strokes in 1922 and died in 1924. When the Ukrainian SSR in 1925 extended an invitation to Rudnyts'kyi, they extended it not just to him but to his family, to live, work, and settle "colonially" in a privileged Soviet Ukrainian ethnoterritorial space. For the time being, Rudnyts'kyi concluded that he could not return to Lviv or live unperturbed as an academic in Poland's Second Republic. He chose to accept the invitation, together with many other Galician friends and peers. Expat colleagues who returned included fellow Ukrainian geographers from the Galician technical intelligentsia, such as Hrihorii Velychko (1863–1933) and Volodymyr Herynovych (1883–1949). They had similar backgrounds, knew German, and held high hopes for the development of Ukrainian geography.[12]

Rudnyts'kyi was very enticed by the prospect of organizing research in the Ukrainian SSR. On his family's prospects starting in 1926, Rudnyts'kyi corresponded with his sister Sofia and brother-in-law Stanislav (Stash) Dnistrians'kyi, probably his most trusted friend. Stash even wrote a book in 1919, reprinted in English and French, on Ukraine's future and the Paris Peace Conference.[13] When Rudnyts'kyi arrived, the communist authorities appointed him to work as professor of geography at the Kharkov Institute of People's Education. He also became director of the Ukrainian Scientific Research Institute of Geography and Cartography, which he helped administer. Turning down competitive offers from Charles University in Prague and from Vienna and Berlin, Rudnyts'kyi elected by 1925–26 to settle and work as a scholar in Soviet Ukraine. Political support was fragile, however, more than he realized.

When the Ukrainian Scientific Research Institute was launched, Rudnyts'kyi had to rely on Mykola Skrypnyk (1872–1933), the leading Ukrainian Bolshevik and national communist, who walked a delicate line with Moscow in the 1920s. Skrypnyk turned out to be unreliable as a patron. Rudnyts'kyi intended to recruit other experts from abroad, a longstanding practice in European empires, including Russia under Peter the Great, but it did not happen. In part, the Soviet world, including the world of revolutionary émigrés, had already turned "geopolitical." Eurasianist geopolitical discourses gained more credence in the 1920s, though this movement of spatial thinkers was more protean than has been thought.[14] Rudnyts'kyi's aims for an independent pre-1914 school of Ukrainian geography and scientific cartography fell short. He may have been blind politically, but as a nationalizer in Europe, he was not misguided. Between 1926 and 1929, Ukrainian projects, already held in suspicion, were underfunded and then defunded. Rudnyts'kyi was surprised that authorities in Moscow did not allow him to form a geographical society, or even have a geographical journal. They allowed him to serve as the geography subject editor of the Ukrainian Soviet Encyclopedia, but that work never appeared in print.

Rudnyts'kyi's expressly transnational patriotic work and naïve Beruf raises troublesome issues. How does one judge a scientific geographer? His "geopolitics," if it can be called that, was from maps not made, from sources that were destroyed or never reached the light of day. Did his "nationalism," if it can be called that, make him culpable in the 1930s and 1940s, along with the fascism of the Organization of Ukrainian Nationalists (OUN), or ethnonationalist movements that succeeded him? These are different questions: we must be careful about engaging in Soviet Russian and Ukrainian history only in retrospect from World War II.[15] Rather like the Galician Romer in communist Poland after 1944–45, as we'll see in chapter 7, Rudnyts'kyi's positivism and conservatism were nineteenth-century holdovers. They did not make him an anti-Semite, but rather open to federalism. By the mid to late 1920s, the *homo geographicus* was geopoliticized into *homo sovieticus*. The scientist could not conduct research as he pleased, since Soviet bans extended to the printing of political Ukrainian maps in the entire country. In wall maps of Ukraine, for instance, he could only show ethnographic settlement of a united Ukrainophone population (in search of a high culture, he like other Europeans ignored dialectal variants), but the fixation on unity was typical of European nation building. He colonially depicted Ukrainians eastward, dispersed across frontiers. He substituted Ukrainians for Germans, a privileged titular nationality, just as Teleki had done in his famous "Carte Rouge" of 1918–19.

In essence, many maps in Rudnyts'kyi's mind could not be prepared or disseminated. Soviet Marxist agendas in Ukraine met limits in political geography when NEP came to an abrupt end with Stalin's First Five-Year Plan in 1928. At that time, Soviet Ukraine shifted out toward a primordial and Russocentric idea of nationality.[16] In Soviet Kharkov the "German" professor Rudnyts'kyi became vulnerable even earlier, when he became director in 1926 of the new Ukrainian Scientific Research Institute, and in 1929 the first-ever chair of the geography department of the All-Ukrainian SSR Academy of Sciences (VUAN).[17] With the party-state's collectivization and industrialization through the First Five-Year Plan of 1928–32 came so-called socialism in one country and Stalin's use of force, intimidation, arrest, and deportation directed against a mobile, highly educated technical intelligentsia. The fate that awaited the Habsburg subject as a geographer was common to fellow Galicians. Border-crossing elites who once belonged to empires were given passports, sent to live territorial lives according to "ethnic" nationality in Stalin's 1930s. The geographer's fate would be left to a suspicious and murderous NKVD.[18]

BODILY WORK

In the first few months of 1921, the world-traveling prime minister of Hungary could not have been pleased with Bowman's fame, but Count Teleki always hoped to keep up personal diplomatic relations in the ways of the AGS 1912 excursion. In April 1921, Charles IV von Habsburg made the first of two failed attempts to regain the throne of Hungary. In Austria, the king had been refused and delegitimized by the Habsburg Law, passed by the new republic's parliament in 1919. Teleki's government did not survive the first restoration coup, though Admiral Miklós Horthy defeated both attempts at restoration in April and November 1921, when he refused to back the legitimists, or royalists, with the armed forces. On 14 April, Teleki resigned. He immediately returned to working as a geographer in the department of economic sciences at Budapest University. Teleki was elated to return to his driving passion. He sought to continue lobbying for Hungary's struggling pre-1914 institutions in Budapest, such as the MFT and MTA. He also served as the head of irredentist organizations, which disseminated maps for education at home and abroad, to promote "justice for Hungary."

The count kept up a whirlwind itinerary. During 1–29 August 1921, he was in the United States, invited by Colonel Lawrence Martin in Washington, D.C., to give eight lectures at Williams College in Massachusetts.[19] He addressed an audience of eager geographers, including his friend Bowman. The new Republican president, Warren G. Harding (in office 1921–23), invited him to the White House. He met with U.S. Secretary of State Charles Evans Hughes (1862–1948), the candidate who opposed Wilson in the 1916 election. This second trip in the U.S. turned out to be another formative event of Teleki's life. When he got back to Budapest, Teleki published "The Nationalities Question from a Geographer's Point of View," which was important in three respects.

First, he highlighted population density as a marker of civilization, since Magyars in "various natural landscapes" had "adapted to economic realities," a mentality that was ideal for frontier exploration and colonial settlement. Teleki presupposed a kind of French assimilatory practice for Magyars as Europe's civilizers, outside of "Herderian" mutually exclusive nationalities, following Vidal de la Blache's prewar work on French regionalism. Second, Teleki defined the fantasy of Hungarian unity by everything imaginable: race, region, language, culture, and religion—above all, religion. Hungarians could be Roman Catholics, Calvinists, Unitarians, and Lutherans "without exception." He prioritized his family's baptismal heritage but marginalized Jews, Muslims, and Romanian Orthodox. Third, Teleki appealed to the canonical authority of Europe's geographers. Since the St. Stephen borders were "historical," they rested beyond ordinary politics. They permitted the multilingual count to blame anyone and anything but himself for Hungary's woes—victors of World War I, self-determination, the Little Entente, Jews from the East, non-Magyar nationalities—all the while updating the Transylvania of his noble family's heritage into revisionist dreams of a future unity.[20]

Teleki's geospatial logic may have been confused, but he clearly looked across the Atlantic to America for a Bowman-like success story. In lectures on trans-Atlantic geography, he employed a favorite quote from Bernhardus Varenius's 1650 *Geographia Generalis*: "When part of the ocean is moved, the whole is moved." (He quoted the Latin incorrectly, as "cum pars Oceanus movetur, totus movetur," but it should be "cum pars Oceani movetur, totus Oceanus movetur.") The count added another motto to this, one of his rallying cries after Trianon, "Always study with a map by your side! Learn from the map! Practice map reading not just with one map but with several types of maps!"[21] In 1922, he published his American- and German-inspired "Statistics and the Map in Economic Geography" and wrote didactically about map generalization using statistics. An ideal economic geographer, he claimed, would "monitor and research materials." Geographers had to know boundaries as well as the physical size, names, and form of settlements, routes, rivers, lakes, swamps, mountains, and all landscapes for economic development. Unity came by way of hydrography, geomorphology, and climatology. Knowledge of landscapes was essential, as in nineteenth-century etching of cartograms in the U.S. for the government's exploitation of land in the South and West.[22]

For agriculture, industry, and commerce, one had to research how wheat, cotton, corn, and other major staples were all keys to expansion, in the United States as in Hungary. In addition, his dreams for geography were on Prussian-German terms.[23] Geographers offered expertise to the state. A geographer's "human mind" by way of "synthesis" was responsible for integrating the physical, chemical, physiological, and social aspects of the earth. The earth's entire surface was "culture and civilization," and the "geography of life sciences . . . and geographer's thinking should be genetic."[24] Teleki's moving fundament was a spiritual and material *föld*, meaning earth or ground, in the gender neutrality of the Magyar language.

When Teleki published these thoughts together in *Economic Geography of America, with Special Reference to the United States*, it allowed him to transpose his boyhood Karl May obsessions into science. It also continued the observational tradition of Europe's aristocratic foray travel writing on America, men such as Hector St. Jean de Crèvecœur (1735–1813), Alexis de Tocqueville (1805–59), and Béla Széchenyi. The 1922 work was Teleki's first major work in the genre, before his American lectures were published in 1923.[25] His staunchly anticommunist book, *The Evolution of Hungary and Its Place in European History*, was published in New York in 1923. Lawrence Martin endorsed it in an introduction.[26] There,

Teleki associated anarchy with Ukraine and East European Jewry with communism, and identified Galicians and Jewish migrants to Hungary as groups not yet assimilated to Magyar Christendom. This played well on Hungarians' popular European anti-Semitic fears, the bodily loss of territory, health, and wholeness, modern Jewish and/or communist hidden hands in the supposedly conspiratorial power expressed by maps.[27] In Budapest as in the United States during its red scares, some Jews were communists. Of course, many were not. Teleki's interior geography of Magyar settlement and a "healthy" civilization in 1921–23 connected the dots for political gain and yoked all of these emotional subtexts to the project of post-Trianon revision.

In Hungarian society, post-Trianon revisionist maps in the early 1920s were never just an academic matter (figure 4.1).[28] Cartography involved bodies and minds. Teleki's diplomatic travels took a drastic toll on his health. In June 1922, after being made dean of the school of economics at the new Corvinus University of Budapest (est. 1920), he fell quite seriously ill. He complained to his friend Cholnoky about an infection and high fever. When the outdoorsman finally relented and went to a doctor, the diagnosis was not good. Teleki was afflicted with renal tuberculosis. That November, he had an operation to have one of his kidneys removed. His recovery at the turn of 1923 was painful and took several months. Countess Teleki blamed the Habsburgs, apparently believing he had contracted the disease during World War I, but this was beside the point. The indefatigable Teleki, prone throughout his life to manic overwork and brooding depressions, knew he was in trouble. He thought that he might retire from his Budapest professorship. At one point, he even recommended his protégé Ferenc Fodor (1887–1962) as his successor.[29] However,

FIGURE 4.1. Graphic design by Rezsö Balázsfi or Balázsfy (1885–1973) in Budapest, likely 1921 or 1922. A good example of aesthetic styles of irredentist politics in Horthy's Hungary, in common circulation after the Treaty of Trianon (1920). The artist was also known for his pastels and erotic prints in the interwar period.

the administration of the Ministry of Commerce and school of economics refused to consider it. Though Teleki received a leave of absence for the 1923–24 academic year, he never took time off for recuperation.[30] For his last nineteen years, the count had to use a catheter three times a day just to avert total kidney failure. He was in constant pain and had chronic bowel troubles. No cure yet existed for his condition.[31]

OF GLACIERS AND MEN

On the losing side of World War I, Teleki abroad and Rudnyts'kyi in Ukraine tried vigorously to keep up *Beruf*, but little by little, they saw actual sponsorship for *Politik* evaporate. Romer's fame grew. Penck became more isolated. It did not ever hurt to have a trusty patron. In 1924, the U.S. State Department named the "Romer Glacier" in Glacier Bay, Alaska, after the Polish geographer's contributions to glaciology and his travels there. In the Last Frontier, Romer's reputation soared. He was rewarded at home for his exploits abroad. For example, he was elected, all at once in 1924, a full member of Polish Geographical Society (est. 1918), the head of the editorial board of Książnica-Atlas, the firm he founded for producing most of Poland's maps, and president of the National Committee of Geographers. He retained the last of these until 1938. Penck meanwhile burrowed into his private life in 1924 and 1925. Nearing retirement, he wrote just one piece in 1924, on Mittenwald as a border marker in the German Alps, and another in 1925 in Haushofer's *Zeitschrift für Geopolitik*, not on geopolitics but on physical anthropology and the subsciences of geography.[32]

Bowman in post-Wilsonian America kept to the geographer's task of diplomacy. He worked to fulfill his promise to the elder Penck after Walther's death in 1923, to publish an English translation of *Die morphologische Analyse*. It had been printed posthumously in Stuttgart in 1924, edited and with a preface by Albrecht.[33] Davis strongly disapproved. He even needled Bowman to write a hostile review of it. When Bowman and Penck renewed their epistles in February 1925, Penck opened on a glad note, informing Bowman that a copy of Walther's book was being sent to New York. Bowman replied on 25 February 1925, "I shall give personal attention to the review of Walther's book. It is going to be of great interest to me. It may not be possible to finish it before the summer gives me a little breathing spell." Thus, their affection for *Wissenschaft* endured despite Davis's burning animosity. In a reminiscence of their first encounter at Harvard in 1904, Bowman wrote, "Naturally, it goes a little slower in the German and there is also the fact that it is so condensed and *scientific* [emphasis added] in treatment that it has to be read slowly and thoughtfully."[34]

Poland was still a ground for friendship, but it was also where our map men learned about the limits of European history. Statistical demography posed difficulty in Poland's almost unmappable *kresy* (borderlands) and Germany's now-closed frontiers.[35] Ever aware of the fine line between science and propaganda, Romer shipped abroad the newest maps his workshop produced. He informed Bowman that the *Polish Cartographical Review* was available to anyone in America who wanted to read it. Romer received "through [Henryk] Arctowski some diagram-maps of the war-territories in Europe" to send to Bowman. These went straightaway to the AGS, and a few weeks later, Bowman sent Romer additional maps on Chile, Africa, Hispanic America, and Alaska.[36] Romer was thrilled to receive these, without asking. He returned new Polish works from his *General Geographical Atlas*, which he began in 1925 and reissued throughout the 1930s, to illustrate Poland's borders and regional unity. He let his enthusiasm for maps often get the best of him. For one thing, he seemed not to grasp the boundaries of personal space when he wrote a P.S. in a letter of 30 November, "Please, Bowman, to give the order, that the new publications of the A.G.S. arrive to me without special demand. I desire them all!"[37] Romer may have intended this as a figure of speech. Bowman read it literally and replied coolly, "Unfortunately, the regulations of the Society do not permit us to send gratis all of our publications. The cost is quite prohibitive and we have to treat people more or less the same." He followed with a reprimand, "A word about the geographical publications now appearing in Polish. Of course you must know that publication in Polish means that no professional geographer in the United States reads them, not one. Is it not imperative that you should always publish abstracts, either in English or in French, and preferably in both?"[38]

It took time for any veteran of Paris to be more careful, but the American knew how to be circumspect. In fall 1925, when Bowman finished his review article of the posthumously published work of Walther Penck, Davis applied back channel pressure again. In several letters of November–December 1925, Davis reviewed his "intimate association" with Penck and their interwoven career trajectories.[39] Once Bowman completed his review under Davis's furious instructions and requests, he sent a copy across the Atlantic, soliciting the elder Penck's reply for print in a future issue of *Geographical Review*.[40] Bowman published a generally constructive review, *before* Penck's was in print, of "The Analysis of Land Forms: Walther Penck on the Topographic Cycle" in the January 1926 issue of *Geographical Review*. Only afterward was Penck's private correspondence to Bowman from 30 December, with the author's permission, published in the April 1926 issue of *Geographical Review*.[41] Davis had many clever stratagems. In fact, the translated book of Walther Penck, a seminal critique of Davis's "universal" cycle of erosion, was blocked out of the American geographical establishment effectively until 1953.[42]

AN AMERICAN IN MOSUL

The worldly Count Teleki espoused an unusual sort of internationalism, a kind of American frontier selfhood steeped in his clan's history of Transylvania, European colonialism, and the German ideology of revision. After another diplomatic trip to Finland in July 1924, he placed himself in the vanguard of research in Hungary. He produced his syntheses of economic, political, social, and regional geography in the 1920s and 1930s. Like Romer, Teleki had young, talented colleagues at his beckon, such as his protégé Fodor as well as Franz Koch (1901–74), who each wrote geography textbooks for academic use. In post-Trianon Hungary, the revisionist illiberalism of European geography was streamlined into the school system. Teleki approved the creation of the new Sociographical Institute (Szociografiai Intézet) under the auspices of the Hungarian Academy of Sciences (MTA). Most curious for our history of transnational lives, the globetrotting count accepted a mission to northern Iraq, which was to take place from 22 December 1924 to 1 July 1925.

Given his reputation beyond Hungary as a moderate, Teleki was selected by the League of Nations (which Hungary joined in September 1922) to adjudicate a boundary dispute in the Mosul district, between Turkey and Iraq. The problem seemed intractable. Following Kemal Ataturk's creation of a secular republic in 1923, Turkey's political elites were making demands on the oil-rich city of Mosul, part of the former Ottoman Empire. The league formed a three-person international commission and charged it with resolving the Kurdish boundary settlement under Lieutenant-Colonel Wallace A. Lyon (1892–1977), the head provincial administrator in Northern Iraq and Kurdistan.[43] Selected with Teleki were the Swedish attaché Einar af Wirsén (1875–1946), the chair, and the Belgian colonel Albert Paulis (1875–1933), a veteran of World War I. Wirsén was stationed in the Ottoman Empire in World War I and had witnessed the Armenian genocide, while Paulis had deep sympathies with the plight of Assyrian Christians in Northern Iraq.[44]

The Transylvanian was hardly an objective choice, however. He, like the others, was supportive of a mandate system led by Europe's powers in the Middle East, a glorified imperial trusteeship for the twentieth century. Moreover, he was swayed by the *Beruf* of a twentieth-century geographic profession, for neocolonial development and privilege. In preparations for the Mosul trip, Teleki followed the habits and procedures of his activity for making the "Carte Rouge," gathering available men, maps, and information. Since the U.S. was not in the league, he felt free to ask for its "independent" information. He and his MFT compatriots insisted on receiving American literature and maps principally from the AGS in New York City and the U.S. Geological Survey (USGS) in Washington, D.C. In a decolonial environment, by then the MFT was in dire financial shape, a sign of things to come for other geographical societies of its nineteenth-century kind, even Bowman's cherished AGS (following the 1929 economic crash). Relations ensued between the librarian of the MFT, István Dubovitz, and the U.S. Department of the Interior in 1924 to acquire information on geological resources.[45]

Teleki's and Dubovitz's efforts at securing a friendly scientific exchange of expertise between the U.S. and Hungary seem to have worked. Teleki corresponded with the experts of Paris and his friends Bowman and Martin in America.[46] Martin supplied Teleki with U.S. maps of Turkey, drawn secretly in the early stages of World War I. He corresponded personally with him in 1925.[47] Based on the repute of his lectures at Williamstown in August 1921 and the international standing of Martin, who introduced Teleki's book favorably in 1923, Teleki received from Carl O. Sauer (1889–1975) a letter concerning the new geography program at University of California, Berkeley. The German-trained Sauer asked him for help "to secure a young geographer who can speak English" and looked forward to "the opportunity of becoming acquainted increasingly with our European colleagues to come and teach there in 1925–26, since a vacancy was coming available."[48]

Bowman, too, was a veteran of boundary disputes. He enthused about "objective" Mosul work, especially because Teleki planned to use American maps. On 17 September 1925, Bowman wrote approvingly from New York,

> Some weeks ago your report on the frontier between Turkey and Iraq came to my office but I was away at the time and only today have I been able to set about a serious study of the report. I find both the text and the maps fascinating to the highest degree. As soon as I have finished with the reading of it I shall prepare a note for the *Geographical Review* to be published in January. It seems to be an admirable piece of work and I congratulate you on the part you played in it and thank you for your kindness in remembering me and the Society with a copy. How much simpler it would be to settle questions of this sort if a firm geographical basis were laid in the first instance, as you have

done here. With continued good wishes and regards, and with compliments to Countess Teleki.[49]

These explorer men of *Beruf* were in high demand throughout the 1920s. Plugged into an American establishment, it was the first time a geographer from the defeated Central Powers was asked to serve on an international commission. Nevertheless, German and Hungarian geographers were sidelined from the IGU after Versailles. Teleki's sojourn was an opportune moment to talk about boundary adjustments, to consider Allied territorial injustices, and to rework the prejudices and policies of the Little Entente toward Hungary's rump state.

In the sphere-of-influence system in the Middle East set by the Sykes-Picot Agreement of 1916 and overseen by the league, brute interests had to be disguised. Petroleum resources in the 1920s were of central concern to the victors, especially the U.K. and France. Wilsonianism had limits, but unlike Bowman, Teleki was a European and not a Wilsonian (or Leninist). Teleki and the commission were charged with making an impartial recommendation in 1925. To do this, they had to determine in the Middle East, by ethnic language and confession in the style of maps by nationality for Eastern Europe and Africa: first, whether the population of Mosul belonged to Turkey or to Iraq in 1924; second, whether the region should stay part of Iraq; and third, as a result, whether there should be a British mandate for twenty-five years to safeguard Kurdish rights. By the alchemy of Europe's maps, they cloaked disunity as unity, emotions as science, and interests as ideals. Teleki wrote on 21 December 1924 to the League of Nations office in Geneva, noting that they were having some trouble finding a translator.[50] The three experts concluded that what the people of Mosul essentially wanted (they were never asked) was to belong neither to Turkey nor to Iraq. From the point of view of the Geneva office, Iraq effectively under U.K. mandate (1920–32) was the lesser of two bad choices.

The League of Nations ultimately accepted the three-person Mosul Commission's report, against Turkey's objections. It decided in December 1925 to award the oil-rich city to Iraq under control of the United Kingdom. Teleki, his political clan being no friend of Muslims or of Ataturk's emergent secular republic, celebrated it as the achievement of a scientist and professional expert. Like Bowman and Penck, his *Beruf* and German Anglophilia endured. What these map men saw as objective, others came to regard as collusion, the defense of great-power privilege in the guise of Western civilization, or Europeans clinging to their extraterritorial resources—all things that the resolution of the Mosul question was.[51]

While Penck's life can be read in hindsight as a story of geopolitics or Germany's lost honor, or an antecedent of 1945, 1933, or 1914, his maps were similarly rooted in the colonial geographic traditions of East Central Europe before 1914.[52] He reasoned that Germans abroad in Bohemia, Moravia, and Silesia would reap the benefits of a quick neocolonial revision.[53] Penck's "Volks- und Kulturboden" map of 1925 was drawn by the graphic artist Arnold Hillen Ziegfeld (b. 1894), who also contributed to the volume in which Penck's conceptual text appeared (plate 6). It illustrated convergence in Weimar of three trends within German geographical discourse: (1) *Geopolitik* or *Weltpolitik* as a new language of power; (2) the *völkisch* tradition as a revived neoromantic project of unity and totality, for fantasies of a provincial, antiurban and "natural" premodern past; and (3) the modern rise of anti-Slavic *Ostforschung* as a twentieth-century discipline for expertise, against perceived threats of an educated non-German bourgeoisie and demographically significant Jewish and Slavic-speaking populations in East Central Europe.

Revenge politics are the gears that grind out maps. Penck asserted his expert authority against his prodigal son, the Galician Romer. It was significant that Penck's 1925 map appeared in Leipzig, in the Deutsche Schutzbund (DSB), or League for the Protection of Germans. The work was prepared for a compilation of essays, *Volk unter Völkern*, published in German in Breslau/Wrocław (figure 4.2).[54] Envious of Romer's gains in the West, Penck envisaged the fantasy of a German-dominant economic, political, and cultural earth. He saw Germans as unified, colonially settled in the East, pioneers in new world lands. The graphology of "Volks- und Kulturboden" had a subliminal logic to it, which the historian Ingo Haar has translated as "German ethnicity and land cultivation."[55] Penck's work was also emotional politics, replete with frustrations directed at Poland, the end of the status for which he, as a German expert, had traveled the world in search of data. There is no evidence he ever went to Poland, until the last trip of his life (in chapter 7).[56] Penck's *völkisch* map was a Rorschach blot of lost lands, lost fame, lost bourgeois-professional status—in short, a lost transnational life.

Rational/emotional, objective/subjective, real/surreal: reframed this way, the map laid out honor codes among men for scientific study of German populations and frontiers to the east. It was made possible by Versailles and the economic depression that followed. Penck's influence was reflected by the Leipzig organization founded on 19 February, the Deutsche Mittelstelle (German Central Agency) für Volks- und Kulturbodenforschung.[57] This

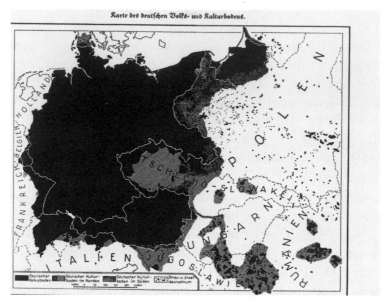

FIGURE 4.2. Original "Karte des deutschen Volks- und Kulturboden" by Albrecht Penck, included as a supplement to his essay "Deutscher Volks- und Kulturboden," in Dr. K[arl] C[hristian] von Loesch, *Volk unter Völkern: Bücher des Deutschtums*, vol. 1, für den Deutschen Schutzbund (Breslau: Ferdinand Hirt, 1925), 62–73. The designer Arnold Hillen Ziegfeld was not credited here. Penck's own copy is in the Albrecht Penck Papers (AP-P), Leibniz-Institut für Landeskunde, Archiv für Geographie, K. 877, S. 12.

state-supported organization fell under the jurisdiction of the Prussian Ministry of the Interior. The first issue of the map was printed in March 1925. The Leipziger Stiftung, as it was called, called for new boundaries determined by Penck and his associate Wilhelm Volz (1870–1958), based on the 1910 Prussian census. Volz received approval from the Ministry of the Interior to become the resource manager in Leipzig, and he was given funds for research and an office headquarters in the city. His task was to provide reports on all Germans beyond Germany's borders. Colonial science was repackaged by pressure groups for nationalist revision, another bitter lesson Penck had learned from Romer's wartime maneuvering in Vienna, and in the immediate years after World War I.

All of these map men invented cross-border forms of Polonophobia. Penck and Volz, arguing in the early 1920s against the liberal internationalism of Gustav Stresemann (1878–1929) and the League of Nations, asserted natural rights of Germans abroad, in place of juridical ones for minorities in the unacknowledged new nation-states. The argument continued an end run around Czechoslovak or Polish democracy, so-called. They claimed the Vistula had been German for two thousand years. Primitive Slavs, so went the logic, being incapable of surviving conflict with effectively organized, superior Germans, lived there only temporarily. Apologists for German colonial power, conservative *Ostforscher* supported authority and objected to modernity, i.e., any notion of rule of law or "Slavic" democracy, based on the notion that Slavic peoples could not run a democracy, or the pretext that Germans territorially were under the unlawful occupation of Czechoslovakia and Poland. They drew up *Volksgruppen* and so-called *Volkslisten* for their map not by sensible logic but by homology, with categories of Germanizable populations. They spoke as experts in prevalent *postcolonial* victim discourse, to bring back a national empire. The alchemy of borders (*Grenzen*) as boundaries on maps in the 1920s was that they could be open and closed simultaneously, sites of danger or bodily penetration, to forge ever more maps showing territorial injustice and claims to German superiority.[58]

For Penck's 1925 "Volks- und Kulturboden" map, any territorial division of the area that privileged non-Germans was unacceptable. He defined *Volksboden* in terms of "an ethnic group [that] establishes itself in a fixed zone of settlement whose peripheries dissolve into ethnically mixed zones." German *Volksboden*, the soil in which an ethnic group is anchored, was to be protected in areas bordering Poland and Czechoslovakia by "internal colonization" measures, i.e., erecting a "settlement wall" of German peasants there.[59] Based on population politics, experts like him arrived to reorder the lands. The *Volkskörper*, like the recuperating soldier after World War I, would be restored to health by German high culture and the state. *Kulturboden* referred to areas beyond the cities and inhabited not by German but by Slavic majorities, in which German explorers, settlers, developers, and conquerors played civilizing roles. In the text to the 1925 map, Penck wrote, "The German *Kulturboden* is the greatest achievement of the German Volk. . . . Wherever Germans live sociably and use the Earth's surface, it [the *Kulturboden*] appears."[60] Structured in this hazily gendered manner, the loosest definition was most advantageous, for one could suppose an open frontier to allow German scientists to "penetrate" into the Soviet Eurasian agricultural hinterland.[61] Any development since medieval times could be chalked up to German land cultivation and settler development, for frontier Germans had the sturdiest bonds of all the modern leveled nationalities. Their civilization, Penck supposed, was present in all aspects of European modernization, above all in agriculture, commerce, and transport networks.

Penck's map launched a petty PR movement of sorts, in liminal space between political geography and geopolitics. More than 120 German politicians and researchers were attracted to Penck's designs of 1925. Eager careerists such as Karl Christian von Loesch (1880–1951), Friedrich Metz (1890–1969), and

Max Hildebert Boehm (1891–1968) found rapt audiences to signal German achievements as a world-historical people.[62] Progress of civilization by romantic nationalism is a soothing tale for uncertain times. Penck noted, contrary to Romer, the low yields and record of subpar performance of the Polish economy (*polnische Wirtschaft*) in the agricultural sector.[63] He lauded culture as the greatest achievement (*Leistung*) for Germans and argued that patterns of settlement meant that the *Volk* were history's exceptional *Kulturträger* with frontiers extending into Europe's east. Penetration presupposed dominance. Unbounded frontiers went "deep into Hungary," beyond the "limits of the old German Reich" into the Beskids, Moravia, Slovak Považie, the Zips in Slovakia, to Transylvania in Romania, the Volga, the Caucasus, the Crimea, Bessarabia, and Volhynia. In the Ratzelian sense, *Raum* was planetary—and therefore limitless.

In his last philosophical work, Walter Benjamin before his suicide in 1940 famously criticized this historicism by writing about messianic time, rejecting the past as emblematic of progress, the past as retrievable in what it actually was, with all history rolled into the present. But Penck's frontier space of 1925 was messianic space—open in global frontiers, wherever Germans carried knowledge, spread community like seed, cultivated others' lives, grew things in the soil, and reproduced culture for all of humanity's benefit. Time was space. The timeless *Raum* of the nineteenth-century explorer was terrestrial, therefore *not* really mappable.[64] In a more mundane sense, Penck's paper map of 1925 acquired a psychic life of its own, picked up by many German associations to suit their agendas.[65] Penck, who had long lobbied for improvements in German civic education, witnessed his "Volks- und Kulturboden" earn the status of a buzzword, popularized by more maps and atlases, adapted into school curricula for geographic literacy to cultivate young minds (figure 4.3).

Naturally, this was very dangerous. The Germans' sense of anger and loss mirrored the tragic death of Albrecht and Ida's son Walther in 1923—the real and perceived threats of his depletion of knowledge and authority, manly vitality and virility.[66] At the end of the 1925–26 academic year, at the age of sixty-five, he left a mixed career behind. Penck's chair in geography at the University of Berlin was given to Norbert Krebs (1876–1947), his protégé and handpicked successor, ensuring continuity in the Richthofen school. Albrecht and Ida quietly moved back to Leipzig, the place of his birth in 1858. He continued to work outside academe, on behalf of the Verein für das Deutschtum im Ausland (Association for Germans Abroad) and the Foundation for Research on Ethnicity and Land Cultivation (Stiftung für deutsche Volks- und Kulturbodenforschung), founded in the Saxon capital in 1926. He sought affinity among

FIGURE 4.3. Map showing Penck's influence in Saxony, also a mark of complex continuity out of Ostmitteleuropa suggesting "natural" frontiers to the east. Prepared after Penck by H. Fischer, "Der Deutsche Volks- und Kulturboden in Mittel- und Osteuropa," on a scale of 1:16:000,000 (Leipzig: Georg. Anstalt von Wagner & Debes, around 1930). Sent to Bowman and the AGS in New York City. Printed on behalf of the Verein für das Deutschtums im Ausland or VDA, founded in Berlin in 1908 from German culture/education associations in Vienna and Berlin in the 1880s, renamed and repurposed as the Volksbund für das Deutschtum im Ausland by the Nazi Party in 1933. Courtesy of the American Geographical Society Library, University of Wisconsin-Milwaukee Libraries, 600-C Europe (Central) C-193.

young objectionists to the Locarno Pact. Funding for this Leipziger Stiftung was guaranteed by the Prussian Interior Ministry, which financed the purchase of a building and contributed costs for technical infrastructure.[67] Maps allowed them to layer a new Poland, by a kind of symbolic European transference, into a "lost" U.S. West or German (East) Africa. Maps gave Penck's corps of new map men a flattened sense of geography and geopolitics as mission (*Beruf*), all while the earth was still spinning.

A SORT OF HEIMAT-COMING

In relocating to Leipzig in 1926, Penck's transnational choice was a little different from Rudnyts'kyi's. At least he had somewhere familiar to go. As he assumed the need for living space in Germany's east, Leipzig institutions for revision opposed the German-Polish borders of 1919, Polish access to the corridor, the plebiscite of 1921, and the status quo supported by Stresemann and his 1926 Nobel Prize colaureate, the French prime minister Aristide Briand (1862–1932). Penck's anti-Slavic, neocolonial men were architects of an illiberal Europe, with him on the political right. Leading irredentists such as Loesch and Haushofer put no faith in Stresemann or the League of Nations, which demanded German acceptance of war guilt, territorial loss, and reparations.[68] They concurred with Penck that Germans had a natural right to settle on all the frontiers of the earth's surface.

Loesch boisterously demanded that Versailles be scrapped entirely, Austria be annexed by Germany, and even Czechoslovakia, Poland, and Yugoslavia be held accountable for overpopulation and the unsavory conditions for Germans in their borders. The new *völkisch* geographers prized authority over democracy, homogeneity over diversity. They sought out "heartland" space and politicized demography to a much greater and more modern extent than Penck and Romer had done in the 1916–21 period.[69] Yet there was continuity, for Penck's civilizing mission was a Prussocentric European model, German pseudohistory from the nineteenth century, that asserted that it was the job of the state to foster the nation and unite a central civilization. Scientists contributed to *Wissenschaft* while advancing the governmentality of a lost century. Practitioners of *Beruf* formed an administrative order for settlement of lands; to manage space down to the lowest level in villages, towns, and cities; and to delineate borders for the maximum rationalization and exploitation of all resources.[70]

On these Saxon regional grounds, Penck's frontier fantasy was not entirely fascist or, for that matter, even modern. There was no effective state, at least not yet, in the place where he found scientific purpose for training new "sons" out of Ostmitteleuropa. In the years of the Leipziger Stiftung from 1926 to 1931, Penck's most enthusiastic allies were young men like Friedrich Metz and Emil Meynen (1902–94). They aligned behind him to advance the "Volks- und Kulturboden" while developing research projects and extraterritorial defense of *Volksgruppe* stock abroad.[71] Old *völkisch* geographers by then had little more than a symbolic nexus of power, but they were distillers of vast compendia of knowledge. Experts were a step below the ministerial level in Leipzig. There was wide diversity among conservative German academics and expansionists, who obtained geographic knowledge in the old style. Schools of geopolitics cannot accurately reflect this eclectic milieu of men spawned by other map men, for revisionist reports on borders were penned by German academics such as Hans Rothfels (1891–1976), the sociologist Gunter Ipsen (1899–1984) from the University of Königsberg, and the anti-Semitic Theodor Schieder

(1908–84) and Werner Conze (1910–86). They apotheosized the *Volk* in Ratzel- or Haushofer-inspired discourses of race and space; they affiliated with right- wing parties against Versailles; and they claimed that German settlers lived on contested frontiers, were global victims, and had special cultural missions to uphold. Above all, they angrily resisted the loss of German civilization. Wei- mar, in their view, had compromised too far with "Slavic" successor democra- cies in Czechoslovakia and Poland, supported by the West's victors, in which Germans were de-privileged into a minority and had their rights dependent on alien state policies.

The Leipziger Stiftung from 1926 to 1931 was neocolonial, racist, anti- Semitic, anti-Western, even Eurasianist in this respect. That it laid some of the groundwork for the Third Reich and men such as Hess or plans like Himmler's Generalplan Ost is beyond a doubt. Not all historiographical roads, however, lead to a German catastrophe of 1933 or 1941. Turns in Penck's transnational life *preceded* Haushofer's agendas in the 1920s, or the Stiftung.[72] Take 1928, for instance, when Penck helped to organize the centenary celebrations of the Ber- lin Geographical Society. On the society's one-hundredth anniversary, he was feted for his seventieth birthday. By invitation, Rudnyts'kyi took the train to Berlin from Kharkov, and just as in 1918, Romer was not invited. Penck, once famous and now retired, hardly changed his geographer's life or his political mind. He recycled a lot of academic boilerplate: in a 1928 article penned from Leipzig, he declared that the unity of geography should be based on the spatial arrangement of phenomena on the earth's surface, with respect to both physi- cal and human processes, insofar as such processes helped to characterize the unique areas of the earth's surface.[73] Reiterating his work in geomorphology in the 1890s and 1900s and writings from wartime Berlin in the 1910s, Penck repeated the powerful myth of the authority of nineteenth-century German geographic science, again the straight line (with Davis left out) from Humboldt and Ritter to Richthofen. The greatest German geographers, he surmised, earned their scientific acclaim not by geopolitics or studying the diversity of urban life, but by getting outdoors, finding patterns of sameness, and observ- ing the earth's terrain and phenomena empirically.

REVISION INSTITUTIONALIZED

At the height of his diplomatic fame, Teleki resisted getting involved in geo- politics, at least at first. Seeing himself more as a political scientist and an eco- nomic geographer, he turned toward Germany on terms similar to Penck. He was elected to the Hungarian Academy of Sciences (MTA) in 1925. In 1926, he was instrumental in the establishment of the Institute of Political Sciences (Államtudományi Intézet), which reified Ratzel's organic state, an idea Teleki himself had echoed in his 1903 dissertation under Lóczy. The institute operated under the aegis of the Central Statistical Committee in Budapest, responsible for the interwar census.[74] Under Teleki's guidance, the Institute of Political Sciences and the Hungarian Statistical Society completed a massive joint project in 1926, a giant 159-page atlas, *Hungary before and after the War in Economic-Statistical Maps* (*Magyarország a háboru elött és után gazdaságstatisztikai térképekben*), with 119 maps. It was immediately sent to Isaiah Bowman and the AGS in New York City.[75] Teleki thought the institute could succeed where in 1919–20 he had failed, in propagating data and maps to the metropole of politicians and to researchers involved globally in studying Magyar rights and Hungary's minority issues.[76]

Always closer intellectually to Ratzel, Teleki engaged with the work of Karl Haushofer in a Hungary-centric vein. Settings for knowledge transfers by map men are significant: institutionally, the xenophobic aggressiveness of Teleki's revisionism was most intense in the depression of the mid-1920s. In January 1926, the count was allowed to lecture on geography at the Hungarian Institute of the University of Berlin. Teleki's arguments appeared in Haushofer's *Zeitschrift für Geopolitik* in 1926. In the original German version, he stressed the geographical unity (*Einheit*) of Hungary. He endowed the Middle Danube (*Közép-Duna*) region with natural geographic harmony, in addition to the landscapes of the Great Plains (*Alföld*). He emphasized the "power factor" (*Machtsfaktor*) in Hungary, which grew out of the settlement aspirations of tribes of foreign origin (*stammfremden Volkes*) moving into the European area, cultivating the lands "with strong geographical Energy." He urged an understanding of this *Machtsfaktor*, which "depends on whether and how a man understands the historical factor, and also fits the community into the harmony of factors pertaining to the geographical landscape."[77]

The count further resituated his older Anglo- and Francophilia inside Ostmitteleuropa, as transmitted by luminaries such as Hunfalvy and Lóczy to Hungary. Teleki emphasized the human aspects of Hungary's place in the global economy. He celebrated *Wissenschaft* in service to the state and the founding of disciplines such as linguistics, history, jurisprudence, medicine, and technics. He lauded science in Hungary, disciplines that helped the country form its real "national character." Teleki lamented the inherent weaknesses of empire and the Austro-Hungarian army. He praised Hungarians in nationalist terms for fighting bravely in World War I, yet noted how the "zeal and enthusiasm" with which many had greeted the war in 1914 soon dissipated. Turning against Béla Kun's dictatorship of 1919, Teleki protested against the "worship of alien ideologies, the economic downfall of the main pillars of support of the national intellectual world, the middle class, the penetration of our culture-bearing status with nationless and denationalizing elements, and a false, international conception of liberalism [that] partly prepared souls for the ensuing radicalism." The psychosexual metaphor of "penetration" (*Durchdringung* in German, *beszivárgás* in Magyar)—one could translate it as impregnation as well as fusion, exploration, or saturation—suggested Teleki's own racial antiurban fears and biological understanding of difference.[78]

From Budapest to Berlin, Teleki sought therefore to accentuate the *Energie* of the pre-1914 years and historical links of Hungary to Germany. As he described Trianon and the Little Entente, "The neighbors would have never penetrated so far if the armed forces would have been Hungarian." In an anti-Semitic defense of Christian Europe, he finally rejected "alien" bodies and "foreign" propaganda, liberalism, and communism, which to his mind had invaded Magyar historical and spiritual space. Appealing to conscience, he railed against Kun's appeals in Europe to peasants and workers. He criticized the League of Nations and self-determination as hypocritical, for Versailles and Locarno drained the nation's vitality. He called democracy "humbug" and Wilson's project a "degraded idealistic utopia" that led to the "anarchistic-chaotic conditions" left by the Károlyi republic. He turned away from England and France to solidify the cause of Hungary's geographers and "German" expertise. Borders became sites of anxiety and danger, points of geo-bodily entry, forces of invasion at the edge of Transylvania's heartland. He appealed to the "historic" St. Stephen crownlands; Transylvania's past; the "defense of Europe" by Hungary and Poland in 1683; expulsions of the Ottoman Turks in the eighteenth century. He paralleled Hungary's frontiers to the U.S. West, arguing that the Alps and Carpathians were akin to the Rockies and Appalachians. He celebrated the Great Hungarian Plain (*Alföld*) with allusions to the plains and prairies of North America, comparing these areas to "Ohio, Kansas, the Dakotas, Manitoba, and Saskatchewan." He thought Hungarians "fought valiantly" in World War I and referred to the communist revolution and Little Entente invasion as "two great national blows." Teleki defended a "strong Christian-national reaction" in 1919 for the creation of a Hungarian bourgeoisie, and against Bolshevism "whose radical leaders were predominantly Jewish" and "militant cosmopolitans . . . having multiplied in the last decades."[79]

Borne of frustration, Count Teleki's lecture was an America-loving rant. Like Romer, the count tried his hand at developing a neocolonial Prussian model of *Beruf*, for the new state-serving map men of Hungary. He found reliable compatriots in Budapest, such as the geographer Károly Kogutowicz (1886–1948), son of the Polish-Hungarian cartographer Manó Kogutowicz (1851–1908), to recycle his ideas and design even more compelling maps based on the "Carte Rouge" (plate 7). He continued in 1927 as a professor in the School of Economics and Public Administration, and in varied political consultations in Budapest with administrative organs of Horthy's regency. He cofounded the Hungarian Revisionist League. In *Földrajzi Kozlemények*, he reviewed a seminal book by Károly Kaán (1867–1940), published by the Hungarian Academy of Sciences, on economic geography and forest administration in the *Alföld*. Teleki described Ottoman occupation of Hungarian lands as a "sad destiny," pointing out the destruction of villages and towns in Transylvania. He objected to any foreign interference in the Great Plains. He called for greater use of illustrative maps, graphs, and statistical data to expand the "Carte Rouge" and display conditions in the counties of Debrecen, Kecskemét, Szeged, Nyírség, and Békés.[80] American economic growth was model number 1: referring back to his lectures and trip to the U.S. in 1921, Teleki sought the unity of regions and commercial integration. In his 553-page *General Economic Geography* of 1927, he again declared that the state was a biological unity, with its "organic life" and economic geography "essentially belonging to the state."[81]

Teleki claimed it was not only the nation-state, but also an expansionist overseas empire, that "gives life" to landscapes. In geography, England was a model for Germany, and Germany was the biggest boon for Hungary. Teleki grafted the vitalism and corporatism of the 1920s onto Anglo-German nineteenth-century norms. He shared in Penck's Anglophilia before 1914. In line with his reading of Ratzel, he concluded that the "unity of the British Empire is a life-unity (*életegység*) on the earth's surface precisely because it is an organism."[82] The goal of geographical study in the "old world" of East Central Europe was now to imitate the new one, to achieve full-scale global integration and state-capitalist unification of regionally diverse landscapes. *General Economic Geography* was used as a textbook for scores of Hungarian university students. By applying an evolutionary model to commercial growth, Teleki remapped post-Trianon territories into the West, but imagined it as still open in its frontier space. Lands of future Hungary belonged to an Anglo-American Europe, guided by German expertise, populated by a professional European middle class, pursuing a strong federal project across the Great Plains in an era of globalization. His hazy blueprints were laid out in collaboration with his protégé Ferenc Fodor, with whom he arranged at least twelve more maps in Budapest for the *Encyclopedia of Economics* of the early 1930s.[83]

This hodgepodge geography of the Transylvanian count, not without hackneyed ideas or prejudice draped in science, was surely characteristic of an insecure man who dabbled in studies of the natural world. He tried, like some sort of wise chieftain, to play all sides—his diplomatic small-state custom out of Karl May and East Central Europe. Like other tightroping statesmen, he leaned toward Berlin or Rome when hopes in Paris, London, and Washington fell short. After all, Germany had a wealth of knowledge to offer. In early 1928, he and Cholnoky prepared the MFT members for the Berlin Geographical Society's hundred-year jubilee.[84] In the spring, he also published an essay called "National Spirit—National Culture," in the *Budapest Review* (*Budapest Szemle*). In the text, also not terribly innovative, Teleki drew on writers from Herder to Ratzel by putting a positive spin on "neonationalism," as he called it. He favored a biological understanding of race, and the ethnogenetic idea that Hungary's "living unity" (*életegységnek*) was linked spiritually and materially to the universal progress of "humanity" (*emberiség*).[85] Teleki added the Magyar word prefix *élet*, or "life," suggesting a cosmic vitality and virtue. Vigorous nations on frontiers were borne of struggle. Nations had lives of their own. Hungary's earthen body was remapped in the 1920s and 1930s on transnational, ethnocentric, and emotional terms.

After Teleki finally took a break to recuperate in summer 1928 on the coasts of France and Greece, he returned to Budapest. In August 1929, the count participated in the English Scouting Jamboree, selected among the nine members of the international committee. He was elected a member of the Academy of St. Stephen in Budapest, and in the 1929–30 academic year he served as dean of the faculty of Economic Sciences at Budapest University. His short geography of Hungary, *La Geografia dell'Ungheria*, was published as a pamphlet in Rome with two maps—one physical and one hydrographical map of Hungary, both made at the Hungarian Geographical Institute by Károly Kogutowicz, to stress the ecological unity of Hungary's lands. He summarized his thoughts on "the dictated peace of Trianon" and "mutilated" (*mutiláto*) nation-state. In protest, Teleki offered in Italian a distillation of his thoughts on the pan-regional unity of the Danubian Basin, the Hungarian plains, and non-Magyar nationalities in mountainous regions and outside the cities, including Germans, Czechs, Slovaks, Romanians, Ruthenians, Poles, and Turks.[86] The "Carte Rouge" lived on.

Just weeks before the market crash of October 1929, Teleki gave his dean's inaugural address at the university. He spotlighted research and higher institutions in the U.S. and Germany as models for future Hungarian science. He spoke of agriculture, geology, and economic geography. Teleki reminisced fondly about his 1921 lectures and experience at Clark University. He praised departments of geography at Columbia, Berkeley, Harvard, Cornell, and the University of Pennsylvania, comparing them to German-speaking Europe—conferrals of degrees and the rise of geography-related agricultural, commercial, and economic subjects across universities in Göttingen, Halle, Kiel, Königsberg, Leipzig, Breslau (Wrocław), Jena, Giessen, Rostock, München, Köln, and Frankfurt. Following German models of research to success, the U.S. was an inspiration due to its "tremendous economic development," which ensured the country's "leadership in the world."[87]

Teleki's German-American outlook then took aim at geopolitics in a surprising way. His right-of-center ideology of revision arrived at a tense moment in 1929, for he was presented with a geopolitical choice and "spatial turn." He reviewed Haushofer's compilation of essays, *Cornerstones of Geopolitics* (*Bausteine zu Geopolitik*) in Berlin, and a volume by Richard Hennig called *Geopolitics: The Theory of the State as a Living Being* (*Geopolitik, die Lehre vom Staat als Lebenswesen*), printed in Leipzig. In fact, the German Haushofer did not impress the Transylvanian count at all. *Cornerstones* was divided into three sections, on (1) *Ostforschung* and geopolitics as applied to Poland; (2) geopolitical questions and processes in specialized monograph literature; and (3) the practical application of geopolitics to the press, commercial enterprises, schools, and civic education. Teleki found major faults with Haushofer's German book. The authors neglected "the human factor" (*az emberi faktornak*); they failed to provide a complex understanding of how geopolitics related to physical landscapes, or economic or social geography; they even lacked citations to what they had read. Teleki's unusually cranky assessment was dismissive—in his most pointed statement, "the book does not take a big step in terms of the academic science of geopolitics."[88]

Distancing himself from Germany, the count took a similar stance on the second book. "Geopolitics is in vogue" (*A geopolitika ma divat*), he announced disapprovingly. "But what the manual does is not entirely clear." Teleki preferred "more meaningful" literature in political geography. He found Hennig's compatriots guilty of "extravagant generalizations," since their work was "not scientifically robust enough" and the book offered "individual problems . . . [that] are not carefully thought through, nor thoroughly processed." They failed to account for factors pertaining to natural production, systems of land cultivation, or regional landscapes. To traffic in geopolitics, Teleki opined, "first and foremost you need to build up a system of political geography." In his judgment, fashionable discourses from Berlin and Leipzig were unhelpful to Ostmitteleuropa's science—and thus unsuitable for Hungary.[89]

This was understandable. From the mouth of an America- and frontier-loving revisionist, German geopolitics represented a heresy and a foil to geographers' lettered lives. In 1929, Teleki elected to review the fourth edition of the original *New World* of 1921 in Hungarian in *Földrajzi Kozlemények*. He praised Bowman's revised bibliography, for it featured "rich materials and periodicals." He found it to be strong on the geography of the States, the British Empire, and especially South America, where Bowman had done empirical research. This time he wrote for his Hungarian academic audience, "A geographer has written the book, but I cannot say that it is scientific (*tudomanyos-nák*). The book is not geography; it is a popular manual for politicians." For its treatment of Hungary, the diplomat Teleki criticized the U.S. "Scientific Committee" (i.e., the Inquiry, later the Committee to Negotiate Peace) and its delegation for being "extremely biased . . . even aggressive." He contrasted Bowman's outlook in the 1921 first edition, when Bowman claimed that non-Magyar nationalities suffered "thousands of years of oppression" and "hatred" under a traditional Hungarian economy, with the new fourth edition of 1929 in which he characterized the Magyars as a "small" population that lacked "from time immemorial" the same right of self-determination as the "stronger" nationalities. Bowman was emphatically wrong to claim that this was a "law of nature."[90]

Teleki thereby "mapped" Bowman. He objected to Bowman's characterization of the "small" (Masaryk and Seton-Watson's code for "uncivilized") Magyars, viewed as unable to form a strong government. He reproved Bowman for ignoring Hungary's history and reducing Magyars to a group by nationality. The European cast out "the American," considering him "barely able" to understand diplomacy or maps.[91] Teleki objected to the loss of Hungarian influence in frontier towns, cities, and areas that became part of interwar Romania (i.e., Transylvania). It also suited him to argue that territorial rights of "the Czechoslovak Empire" were not natural; no rights, claimed the conservative, were universal givens. The count from Transylvania argued from home, a European taking on the new world from the old.

ILLUSIONS

Of all our men, Isaiah Bowman by the late 1920s probably had the most illusory sense of his influence, but this made him take his *Beruf* even more seriously. Despite the fact that he saw himself as a guileless American peacemaker, adversaries and even friends easily called his credibility into question. In the eyes of the defeated, he was compromised by his work on the Polish Boundary Commission in 1919. After the publication of *The New World*, German geographers were suspicious of him, but their influence was curbed. Members of the IGU "permanently" set up the organization again in Brussels in 1922 (its first international meeting was in Antwerp in 1871), where the ninth IGC was held. At the tenth IGC in Cairo in 1925, U.S. and German delegations of geographers objected to the League of Nations' regulations, and boycotted it. Bowman together with Martonne in Paris and Romer in Lwów tenaciously worked to repair relations. In 1927, the league officially invited back German membership. Even after the Penck-Davis fallout in the early 1920s, Bowman received Penck in 1928 as a guest of the AGS. Bowman and Martonne wanted to ensure that German geographers were encouraged to attend subsequent congresses. Through the National Research Council that was established in the U.S. in 1916 as an extension of the National Academy of Sciences (it was set up in 1863 by Abraham Lincoln and a congressional charter), Bowman tried to steer U.S. membership in the IGU and among friendly European geographers, including the Germans.

The boundary expert could not resolve all tensions—and transnational did not mean non-national or anti-national in the 1920s. West of Soviet territories, Romer was involved in practically every postwar bourgeois-nationalizing Polish initiative in geography. He was head of the editorial board of Książnica-Atlas Publishing House and president of the National Council of Geographers, a position he held until 1938. As such, he kept a close eye on his friend Bowman. On 17 February 1927, Romer wrote of his busy year of work with the second Slavic Congress of Geographers and Ethnographers in Kraków, "a nice addition to my burden."[92] Bowman in turn politely sketched out their roles in his diplomatic way: "I follow all of your publications with the greatest interest, first because you are my friend and second because of the importance and value of the publications themselves. You are doing . . . wonderful work in geography and I send my best greetings and wishes for your happiness and prosperity."[93] Romer was happy to inform Bowman of his role at the congress in Kraków, held from 2–12 June, during which he "made a journey around Poland."[94] Polish geographers classified a vast number of subdisciplines of their growing field of geography into categories, a continuation of the nineteenth-century world of Humboldt and Ritter's interrelated knowledge.[95] Romer planned future research and congresses in the old tradition. When the twelfth IGC was held in London and Cambridge in July 1928, Bowman attended. He was elected the next president, with Romer his vice president.

It was a critical feature of our "German" map men to maintain their illusions by staying within the loop of men of power, and inside comfortable academic networks of thought. By the late 1920s, Penck, retired in Germany, failed to grasp how rapidly the transnational spirit of cooperation had vanished, how fast gentlemanly geography had grown captive to ambitions and personal politics. For instance, when he came to America in 1928 on a diplomatic visit to see Bowman and the AGS, he left behind a world of intrigue in Leipzig. In a fascinating case in his absence, an intense quarrel erupted between Metz and Volz, the two younger men in the Leipziger Stiftung. Metz's conservatism was closer to Penck's. Volz, the foundation's manager, reconciled himself in Germany to Stresemann and the Locarno system, and the league's minority rights policy based on precepts of international law. Without consulting Penck, Volz tried to fire Metz after objecting to Metz and Penck's favored assistant, Emil Meynen. Meynen and Metz in turn tried to get Volz fired. They criticized Volz's scholarship and accommodation to 1919, and also his resource management style as unduly excessive. The attack turned bitterly ad hominem: they next focused on the fact that Volz's wife was Jewish.[96]

By the time Penck returned to face the fallout, he was no longer in the know. He stood behind Meynen and Metz, having come to disdain Volz over Versailles and for his limited support for his own brainchild—not the League of Nations or the IGU/IGC, but his "Volks- und Kulturboden" of 1925. It was a Pyrrhic victory for Penck, Meynen, and Metz. The Reich interior minister, Carl Severing (1875–1952), a social democrat, subjected the Stiftung to more parliamentary supervision in 1929. Walter Goetz (1867–1958), a left-liberal historian and supporter of the league, was appointed to oversee its work. On 3 June 1929, the Leipziger Stiftung was brought under full control of the Interior Ministry, as per Severing's wishes and those of the liberal Reichsbank president, the wonderfully named Hjalmar Horace Greeley Schacht (1877–1970). When Volz was finally fired, he was denied future access to funds for doing research. Professional worlds of geographers were layered by emotion, once the stakes of *Beruf* and of geography as a science had been raised.

———

Geography as science is also geography as noir, and one has to delve behind the scenes of academic posturing. There is an incredible comedy-of-errors true story in 1925 that gets us into the realm of maps as politics, how systems of power really work, what makes transnational experts who they are (if they can move around). To this day, it is euphemistically called the "Franc affair" (*Frank ügy*), and it makes rounds to refresh many a dull national textbook. A nonacademic who bore into the bowels of how history is made, the Hungarian-American writer and diplomat Andor Sziklay (1912–96), who became Andor C. Clay in the United States after leaving Horthy's interwar authoritarian regency, worked for the Office of Strategic Services (OSS), predecessor to the CIA, then for the U.S. State Department after 1945. He has retold the tale brilliantly.

Count Teleki, busy with Mosul in 1924–25 and other affairs, briefly appears but apparently was not directly involved in the scandal. Many others were, however. They included men in the highest echelons of the prime minister István Bethlen's government, in office after Teleki from 1921 to 1939 and serving under Admiral Horthy.[97] The scandal broke out in The Hague on 21 December 1925, when a monocled gentleman showed up at a private bank with a French thousand-franc banknote to exchange. The regular teller had stepped out to take a call. The manager of the bank, who happened to be an expert on counterfeit money and worked with the city's police, inadvertently served the mysterious client instead. He provided the change, but sensed intuitively that something was not right. The manager alerted the bank's detective, who followed the gentleman back to his elegant hotel and then notified the police. When the police accosted the man, he produced a passport identifying himself as "Arisztid de Jankovich, special courier for the Royal Hungarian Ministry of Foreign Affairs" and claimed diplomatic immunity. Back in his luxury hotel room, Jankovich was interrogated further. One police detective noticed another thousand-franc note lying near his bed. They then found a trunk of fake money, closed but not locked, in the room's corner. The bumbling Jankovich (who used his actual name) demanded to see the Hungarian consul and was taken to the embassy. Dutch officials quickly turned him over to the Foreign Office, where he claimed hysterically to have been "assigned by the chief of Hungarian national police to engage in special activities."

On the next day, 22 December, all hell broke loose. The chief of police back in Budapest, Imre Nádosy, was questioned about his role. Jankovich turned out to be a colonel in the Hungarian army's general staff. In The Hague, two other high-ranking Hungarians were arrested, both captains of the staff, including a nephew of the former minister of the interior, a baron and acquaintance of Count Teleki who headed one of the country's main irredentist organizations. When the international press caught wind of the scandal, Hungary's highest-ranking police official, who issued the documentation to Jankovich, was forced to admit his guilt. So, too, did the former minister of supplies, a prince. Implicated more deeply were some key middlemen in the affair: the director of the National Cartographic Institute, General Sándor Kurtz; a leading technician for map production, Major László Gerő; and his boss, General Lajos Haits.

When the full story unfolded (and not every detail is known), it seems that the patriotic Gerő, Haits, and Kurtz had arranged for an initial quantity of nine thousand pounds of treated paper to be shipped from Cologne to Budapest. Freight cars brought the necessary machinery by railroad from Bavaria to Budapest. Thirty thousand thousand-franc notes, the first batch, were printed in earnest at the National Cartographic Institute, ready for shipment in September 1925. These were transported in secret to storage rooms in the huge residence of István Zadravetz, a bishop and the chief chaplain of the Hungarian army. The director general of the state's Postal Savings Institute, a former minister of finance, even served as a "consultant" for circulating the bogus bills abroad in Europe. Now Jankovich was no rube, but he was actually among twelve "special couriers" assigned initially to cover the Netherlands, Sweden, Italy, and Belgium. Each active or reserve officer of the army's general staff had instructions to have the bills changed at private banks and, once the action was completed, to transfer heaps of genuine bills to the institute through the diplomatic pouches of Hungarian embassies. Only then were ambassadors to be informed about the nature of the broader plan. The bishop took from each courier a religious and patriotic "oath" concerning the secrecy of the assignments, and he blessed them ceremonially before their missions.

Such desperate transnational lives were lives in the shadows. What's remarkable here, besides a labyrinthine conspiracy for the history buff, were the incredibly raised diplomatic stakes of *Beruf* and *Politik* by the early 1920s, given Hungary's revisionist milieu and the sheer absurdity of the manner in which so many of our men trafficked in maps. When more facts in Budapest and The Hague came to light, it was revealed that Prime Minister Bethlen got at least one intelligence report about the counterfeiting in fall 1925 from his deputy foreign minister. Bethlen, who did little, probably should have said less. He testified in 1926 at the trials of the accused that he had passed the report on to Nádosy, the head of the national police, who took the fall. Attacked by parliament and lambasted in the domestic and international press, his government through the late 1920s and 1930s survived the episode. Count Teleki emerged, his body and noncounterfeit maps neither unscathed nor untainted, while his opponents grew vocal on the interwar left and right.

CHAPTER FIVE

A LEAGUE OF THEIR OWN

Can we ever believe maps capture lives into nations? The boundary expert Bowman of Paris had grander aspirations than constructing national identity or waging wedge-tactical struggles on red-blue electoral maps. He aspired to build a league. In an address in Warsaw at the start of the fourteenth International Geographical Conference (IGC) on 23 August 1934, he trumpeted geography before a large audience that included a German delegation. Geography was science and not geopolitics, he stressed. As president of the International Geographers' Union (IGU), he spoke with aplomb in deference to Europe's geographic traditions. He paid homage to Penck and Davis, his Harvard mentor who died seven months earlier, and "in earlier years [was] an active participant and leader in international geographical excursions and congresses whose professional interests and friendships recognized no national bounds." Bowman beamed about Poland's independence after World War I, the thirteenth point of his hero Wilson. He noted the presence of Ignacy Mościcki (1867–1946), the Polish president, by profession a chemist, as a sign of the "favorable conditions we are assured at the start of a useful interchange of ideas and a substantial strengthening of the bonds of friendship." The "German" Bowman welcomed back Germans to the IGU for the first time since 1918–19.[1] With the League of Nations struggling for purpose by the mid-1930s, Bowman's diplomacy was Wilsonianism revisited, grafted onto nineteenth-century emotional lives. "Friendship," he announced, invoking the Paris of 1919, was the key to geography's enterprise.

Bowman's aims could be called naïve, a directionless project doomed to failure. Yet he always understood intuitively how much maps were loved objects. Maps were part of a "great community of interest . . . through the interchange of thought that takes place here we shall be better able to return to our several countries and do our part in community life as well as in research and education by more intelligently assisting the never-ending process of adaptation of means to end in our use of the earth's gifts." Maps were like the letters he passionately wrote. Ideally, maps started conversations. Having negotiated in Paris, Berlin, and Warsaw after the 1931 IGC with leaders of the French, German, and Polish schools of geography, he and Emmanuel de Martonne personalized the enterprise. They even urged Penck to reconcile with Romer, but the Leipziger refused. Bowman persisted with a dying idea in the global 1930s, that cooperation in science could heal political rifts. Romer's maps were "the most representative map collection that has ever been displayed in the history of geographical congresses, forty organizations from twenty-three countries participating in its development. . . . [his] contributions to cartography have played so important a part in the development of geographical science." The patriarch tried to put words to it, creepily—geographers shared "a vast reservoir out of which man dips power," he proclaimed, a kind of singular, feminized earth. He approved of Penck's dictum from Ratzel that "there is no land of unlimited resources." Bowman's league was personified by a confraternity apart from "capricious action . . . ignorance or provincialism," which pursued "those sought-for understandings of the world's peoples that are required to ease existing tensions."[2] Fifteen years after the death of Ostmitteleuropa, Bowman gave this Penckian speech with Romer by his side. The frontiersman even wore a new suit for the occasion.

WISSENSCHAFT WARS

While Bowman worked in anticommunist spaces, Rudnyts'kyi managed, with difficulty, to live professionally in Kharkov during the late 1920s and into the early 1930s. As much as possible, he kept active contact with German geographers, such as Penck in Leipzig and Eduard Brückner in Vienna. He came again to Berlin to celebrate Penck's seventieth birthday from 24–26 May 1928, as he had done in 1918, at the centennial of the Berlin Geographical Society, the second society of its kind in the world. Penck supported Rudnyts'kyi's efforts to educate citizens and develop cadres for a Ukrainian school of geography, but for his own reasons. When Stalin discontinued NEP in 1928 and imposed collectivization on 1 October 1928, announcing the First Five-Year Plan, Rudnyts'kyi's career was placed in jeopardy.[3] Fellow Galicians, men like Hrihorii Velychko, were trained in pre-1914 "bourgeois" geography rather than dialectics. Others, such as Mykhailo Ivanychuk (1894–1937), Volodymyr Herynovych (1883–1949), and Vasyl' Bab'iak (1884–1950), had taught in interwar Prague at the Ukrainian Free University.[4] They came into headlong conflict with young geographers steeped in economic determinism. Marxist men of opportunity and communists in search of careers proved eager to support and develop Soviet Ukraine's agricultural and industrial sectors through Stalin's First Five-Year Plan.

Judging from the letters Rudnyts'kyi exchanged with his sister Sofia and brother-in-law Stash in Vienna, he sensed the tension acutely. He fretted about interference and heavy workaday pressures at the Institute of Geography and Cartography in Kharkov. While the process of selecting him as an academician of the Ukrainian Academy of Sciences went on in spring 1928, his son Levko and daughter Orysia came to stay briefly with him. He warned them about coming, having understood that the selection of him as chair of the geography department of the All-Ukrainian SSR Academy of Sciences (VUAN) was a political act. Perhaps the conflict with Romer in Galicia back during the 1900s was part of his deep memory. From lack of evidence, we do not know. In any case, Rudnyts'kyi divulged his worries in family letters to Sofia, Stash, and his children Levko and Iryna in 1928–29, when Stalinists resorted to ideological wedge tactics to splinter the personnel of the old institutions.[5] By that time, Kharkov communists fabricated accusations against his directorship. They attempted to enlist the two young men Herynovych and Bab'iak, the latter of whom was briefly engaged to Rudnyts'kyi's daughter Orysia. Rudnyts'kyi was never a party member, however. In a letter of 17 March 1929, he expressed dismay with party intrigues that involved the communist Ivanychuk as well as Herynovych and Bab'iak.[6] Predictably, he took Europe's high ground. Rudnyts'kyi tried to hold to old-fashioned virtues of fairness, but by the end of 1929 Stalin's science wars took over the Kharkov institute. "Bourgeois" objectivity had become a dangerous passion.

Hopeful to the point of naïveté, Rudnyts'kyi was a modest man. Stalin's revolution in Ukraine was not done overnight. Nor did Rudnyts'kyi's expert career, in the reduction of Soviet politics to nationalities building communism, come to an abrupt end. In 1929, he was appointed the first-ever chair of the geography department of VUAN, a position he held until 1933. VUAN remained a bastion for the technical intelligentsia. But in a pessimistic letter of 26 September 1929, Rudnyts'kyi confessed privately about Ukraine's future. "I am very fearful," he ominously wrote.[7] Amid the bitterly cold winter of 1929 and reports of famine and starvation, he remained in Kharkov. Success came in 1930, when geographical societies in Berlin, Vienna, and Prague elected him an honorary member, a holdover of nineteenth-century traditions. On 28 March 1931, he wrote to his family that "the Germans . . . have not forgotten me, for a few weeks I have been getting letters, literature, and books. In a short while again one German geographer [likely Brückner, Penck's colleague] will come to visit me in Kharkov."[8] In spaces of transnational exile, the "German" Rudnyts'kyi dreamt of a vanished home, but he could not return to Galicia.

On 21 March 1933, nearly two months after Hitler became chancellor, Rudnyts'kyi was arrested in Kharkov. The Soviet OGPU (the later NKVD), on the pretext of fears of the Third Reich, passportized identity in December 1932 and began liquidating Ukrainians as spies and so-called nationalists. A phantom counterrevolutionary cell, the Ukrainian Military Organization (UVO), was used as a rhetorical pretext for terror. Warning signs appear in hindsight: Rudnyts'kyi was forced to leave his positions at the Geodesic Institute (in 1930) and at the Institute of National Education (in 1932)—as shown in a letter from July 1932, ostensibly because of bad health.[9] The former student of Professor Penck published his final scientific study on the geomorphology of the Dnieper River Basin, a reflection of his doctoral work thirty years ago in the Habsburg Empire. At the time of his arrest, he was writing a piece for the *Ukrainian Soviet Encyclopedia*, a work that was mysteriously destroyed.

The charges against Rudnyts'kyi in spring 1933 were a kind of faux life narrative. They were as follows: (1) being a bourgeois national activist of Galicia and working for the Austrian General Staff during World War I; (2) working in the UVO after the fall of the Habsburgs; (3) being an advisor in the dictatorship of Yevhen Petrushevych in 1920; (4) working in the "counterrevolutionary Ukrainian military organization" abroad until 1926; and (5) moving to Kharkov with the malicious intent of setting up the UVO in Soviet Ukraine. In addition, he was said to be guilty of (6) "being a follower of German orientation"; (7) having used the interests of the UVO to establish contacts with German nationalists and fascists; (8) having used a trip abroad to contact German cells of the UVO (supposedly, this was the trip to Berlin for Penck's *Festschrift* in 1928); and (9) having been in contact, as a spy, with the German consulate in

Kharkov. Every one of these charges was false. Each was meant to destroy both his private life and his work. Soviet geopolitics after 1933 therefore remapped his Ukrainian transnational life story into a one-dimensional tale of resistance and victimhood. The tenth (10) and last charge was the one that wrecked his professional career—Rudnyts'kyi had headed the Institute of Geography and Cartography in Kharkov from 1926 to 1929 and the geography department from 1929 to 1933. There, he supervised "hostile" work and used ethnographic maps and other materials from Berlin to support German territorial aggression. Bound by association to (Nazi) Germany, framed by Stalinist charges of deviationism in an imagined West, the NKVD arrested and expelled him from the VUAN as a propagator of fascism in geography.[10] Labeling Rudnyts'kyi a class and national enemy in the context of the Holodomor of 1932–33 and Hitler's coming to power in January 1933 further meant that his German-Galician origins were selectively emphasized. His mother's Armenian ancestry was suppressed. Thus objectified into an enemy of the revolution, or pixelated by nationality, Rudnyts'kyi was stripped of his former German/Habsburg diversity and agency. He faced the double demise of being mapped onto Stalin's earth in 1933—a "fascist" in Ukraine between the geopolitical East and West.[11]

On 23 March, just two days after his initial arrest, Rudnyts'kyi was sentenced to five years' incarceration by the Judiciary Troika of the Collegium GPU USSR. Rudnyts'kyi's case was brought to light, but not where one might expect. Penck in Berlin sought to help out the expert, his friend and former student, through scientific channels. He hoped to get him extradited from the USSR to Germany, but to no avail. In 1933 alone, practically all of Rudnyts'kyi's Ukrainian students in Kharkov were arrested. Velychko, who taught in Galicia for decades and was the author of the first major national ethnographic map of Ukraine in 1896, was accused of treason and later shot in Kharkov.[12] Rudnyts'kyi's son Levko, then just twenty-five years old, and his daughter Orysia, aged twenty-one, were declared "children of the enemy of the people," a term that applied to all family members of the accused.

ASYMMETRY

On 30 January 1933, Adolf Hitler became Germany's chancellor. By the Enabling Act of March 1933, the Führer was granted legal plenary powers and able to bypass any legislative checks. In effect, the Nazi Party (NSDAP) now ruled Germany as a one-party state. Hitler could suppress opposition, real or imagined, by paramilitary force. Slow to grasp the changes in the Europe of

1933 and 1934, Bowman invoked the spirit of Europe's geographic societies. He imparted to Romer how he wished to bring German and Polish geographers together. The American wrote to Romer in May 1933,

> If I venture a few suggestions you will, I hope, take them as suggestions only and use your judgment as to their application. Two items on your list may cause a certain amount of embarrassment at just this time: 1) the geographical problems of war, and 2) regional frontiers. The first is certainly a delicate matter while the second can of course be handled so as to be quite objective and having no relation to international questions. Since we cannot *control* the writers of papers or *edit* their manuscripts, would it not be well to avoid creating an opportunity for any emotional members who would use the occasion for the expression of disturbing views? I do not refer to Polish members only but to all members from neighboring states as well who might develop the theme in an unscientific way. . . . It is most gratifying to hear that you are working to secure both Soviet and German cooperation. I am writing today to General Secretary de Martonne on this very point and he may wish to inform you further.[13]

Bowman's heavy-handed diplomacy was less an instance of blithe Wilsonianism than the kind of pre-1914 European and Ivy League dealing he learned from his mentor Davis. It was rather ingenious. To repair German-Polish antagonisms, the geographic expert did *not* play the democracy card or insist on transparency. Instead, using Romer, his trusted friend and back channel, as leverage, he aimed to edit and control the world's political stage for the production and dissemination of maps. To deal with the "geographical problems of the war," Bowman referred to science and objectivity. By admonishing Romer of these dangers, he spoke in a code, back to their shared experiences before 1914 and in Paris. Bowman's managerial style contradicted Wilson's open diplomacy but not the emotional worlds of geographers. He may have gleaned something from practitioners of Polish émigré politics in the partition era, who had an entire century of the civilizational experience of ear-bending Western diplomacy behind them.

We therefore have to read between the lines to see how asymmetrically Romer followed his patron's demands. Editing in advance of the IGC was a censoring *out*. Ultimately, three hundred maps were not shown in Warsaw in 1934, on the grounds that they did not meet the standards of the IGU (plate 8). Bowman explicitly counseled Romer to be wary of "emotional" (i.e., unscientific) currents, an order to his subordinate to approve the return of delegations of German and Soviet geographers to Warsaw. Owing in part to his contacts after the Treaty of Riga in 1921, Romer was more in favor of bringing back the Soviet geographers than in allowing German geographers from the Third Reich. He complied, reluctantly.

Then Romer wrote back to Bowman in October 1933 (in November, the U.S. government reestablished formal diplomacy with the USSR), aware of his own personal history in conflicts over Upper Silesia and the Danzig corridor. He suggested that invitations for nonmember nations be sent out simultaneously from the IGU in Brussels, with private letters from President Bowman. This was his way of saying he did not want to befriend a German enemy. Romer hoped that Bowman's "personal intervention" would make a difference "among the Scandinavians and several others," but he expressed real doubt in "not hoping [for] much from Germany and Russia." Yet he added, "What concerns USSR we expect a large representation on the Congress at Warsaw."[14] The target of Penck's fury evidently struck the right chord with his anticommunist friend. Never one to refuse something that boosted his reputation, Bowman approved of Romer's idea for linking maps and men, with invitations he himself would write.[15]

Bowman's American pivot to Poland was far from equal or democratic, and it masked his imperious streak. Ruthlessness lay beneath polite passion. From director of the AGS starting in 1915 to 1935, president of the IGU from 1928 to 1934, and subsequent president of Johns Hopkins University starting in 1935, he had entered echelons of power. He began a new kind of elite training vis-à-vis the IGU, not just as an authoritarian administrator but also as a publicity-obsessed global manager. Starting under the new administration of FDR in March 1933, he became chairman of the U.S. National Research Council and vice chairman of the Science Advisory Board. When Bowman in December voiced his preferences to Romer about themes for the conference and languages of use, he wondered aloud about content and tone for his presidential address. He proposed to give a talk "on the subject of land settlement in the United States in connection with our national planning projects now actively in hand by the Government and of wide interest to the people." He referred to "the manner in which business is to be conducted" for meetings at the IGC, professing that he was "limited to the English language." Plainly, this was untrue. Bowman reviewed books in French and German. (Recall that he helped Penck with translations at Harvard!) Yet he ruled out German and Polish as languages of commerce, diplomacy, and *Wissenschaft*. He made it clear that the languages of the IGC were French and English, as in Wilson's project for the League of Nations. Bowman ordered Romer to find a "first class interpreter."[16]

Despite their nominally equal status as experts, the IGU vice president Romer in Poland was relegated to a client's position. Eager to please, he replied, "It is clear that . . . your proposition is for me not only an order, but is at the same time a real pleasure, for which I am very thankful to you. The problem you will develop in your address—the world salvation belongs to the highest domains of geography which, as I think in these domains is an art of living on the globe." He approved of the request for an interpreter and announced that the "exposition of official cartography . . . has an assured success." Romer noted the participation of Great Britain, France, Italy, Belgium, Sweden, Yugoslavia, and the USSR, and that he had "interesting pourparlers with v. Müller Dir.d.R.A.f. Landesaufnahme and with [Norbert] Krebs." Romer agreed to everything—well, almost everything. The Polish geographer could be pushed only so far. He asked Bowman plaintively, "Have you done something for inclining German geographers for collaboration in the Congress or even for adhesion to Union?" Halting to pursue the delicate protest further, he parted with nineteenth-century gratitude to his friend Bowman for ensuring the "prosperity of the Congress-session in Warsaw."[17]

THIRD REICH

Meanwhile back in Nazi-land, by spring 1933 Penck seemed neither interested in nor confident of Franco-German, German-Polish, or German-American rapprochement. It is hard to calculate the relative impact of the Leipziger Stiftung's pressure tactics, but Penck's influence in Saxony seems to have been substantial. He shared with many Germans, Hitler included, an indulgence for Karl May's novels. It was a powerful fantasy, but it alone does not explain causality, the practices of nationalism or biological racism, or the institutionalization of German anti-Semitism, or change/continuity in the German question in 1933.[18] Although *völkisch* factions pushed agendas very far, the champions of revision were rejected by Weimar's dying middle, social democrats like Wilhelm Volz who found themselves outnumbered and outflanked. Just as not all communists were Jews, not all conservatives in East Central Europe became Nazis or Nazi collaborators.[19] The Leipziger Stiftung lasted only until August 1931, at which point it was defunded. At the yearly all-German convention of geographers in Danzig in 1931, Metz and Volz from Leipzig kept up their personal feud.[20]

Among map men, Penck survived the Metz-Volz conflict and begat a new "son" in the reproductive world of academe when Emil Meynen completed the *Atlas der Deutschen Volkskunde* and the *Handwörterbuch des Grenz- und Aus-landsdeutschtums.*[21] Meynen, Penck's protégé and assistant, insinuated himself into the career-building network of nationalist functionary elites who became fascist. Meynen was a founding editor of the atlas project who earned his ha-bil in 1934, "Deutschland: Begriff in der Volks- und Kulturbodenforschung," inspired by Penck's 1925 map.[22] The *Handwörterbuch*, like the *Volk unter Völkern* volume edited by Loesch and Boehm, suggested open frontiers for revision.[23] The difference between *völkisch* and Nazi geographers' "Greater Germany" was a thin line between Aryan biological racism and German primordialism for set-tlement and "penetration" of nature in non-German lands. This was managed (lazily) by appeals to frontiers and *Lebensraum*; few *Ostforscher* really bothered with the culture or history of their eastern neighbors. They did not frame East Central Europe outside academe, or beyond nationality, or in human di-mensions. Epistemic grounds for *Ostforschung* were pan-European colonialism, from the Balkans to Africa. If "others" awaited reconquest, Germans ethni-cized abroad as "Aryans" were reduced to pixelated groups (not legal citizens) awaiting incorporation into the national geo-body (as in Sudetenland in 1938) or unification into Deutschland. With the League of Nations hanging by a thread, mid-1930s *Revisionspolitik* on fascist Europe's terms of aggrandizement and population politics was no longer checked by international norms.[24]

Penck's geographic strategies continued to have an impact into 1933 and 1934. Walther Geisler (1891–1945), a professor of geography at the University of Breslau, gradually took up Penck's objections to Versailles. He made the argument that Polish design techniques, though found credible by the victors at Versailles, were tremendously unsound. Geisler methodically picked apart Romer's method in his 1916 atlas, in which Romer showed population distri-bution of Poles by language in the Danzig corridor. Romer, Geisler argued, used two shades of blue to show categories of less than 1 percent Poles and 1–5 percent Poles, but for remaining categories he used shades of increasing intensities of brown. The result was the impression that Poles' majorities first existed in the corridor area on the Baltic Sea, which in his view they did not. Geisler followed Penck, accusing Romer of deliberately altering the data. He claimed (mistakenly) that Romer was using Jakob Spett's map—but this was impossible because that map was published two years later in 1918. The point here was how Geisler connected the dots to a vaster conspiracy: he lumped Poles and Jews together into the commonplace *Ostjuden* stereotype. In 1933, he published a seventy-six-page treatise on the Spett map alone, in which he ar-gued that Spett used color to misrepresent the corridor and misused statistics to make pro-Polish claims. The Third Reich rewarded such science. The Foreign

Office now sought out German scholars to bring their analytic proofs to meet the grave injustices of 1919 and geographers such as Bowman and Martonne. Geisler's second edition was printed in March 1934. Racist, anticommunist, anti-Polish, anti-Slavic, and anti-Semitic, Geisler's Penckian thesis placed blame for the outcome of World War I not merely on the "Polish" Romer (who was fluent in German), but on the Polish-Jewish Spett. He emphasized that Spett had moved back to Poland from Vienna at the end of the war, adding more innuendo.[25]

Like left-wing communists against Rudnyts'kyi in Ukraine, right-wing young fascist-friendly geographers like Geisler and Meynen instrumentalized the politics of maps, which were flexible to begin with. This applied not only to Penck's "Volks- und Kulturboden" of 1925, but in the use of Penck *himself* as a stepping stone for their own careerist goals—the global fantasies of anti-Semitism, colonial power, frontier settlement, and territorial revision. *Revision á la Penck* was supported by Nazi geographers of the party-state as early as 1933. New German map men all belonged to a generation that came of age with Hitler's rise to power, but their earlier lives and work trace back to earlier Ostmitteleuropa.[26] The nineteenth century should not escape culpability. It is more helpful to recall how complicated these lives really were: to challenge further a nostalgic pre-1914 world of yesterday, or post-1945 appeals to Stunde Null, Germans (and others) becoming civilized, or Germans (and others) working through the past in shared projects of reconciliation with postcolonial, postgenocidal memory frames of Europe's ethnically cleansed nation-states and integrative aims.[27]

KNOCKING ON EUROPE'S DOOR

At the start of the 1930s, Teleki was drawn closer to German scientific currents in geography. He lectured at a meeting of the Hungarian-German Association in Munich. He published an essay in 1930 on Lajos Lóczy, "The Man and the Professor," in honor of his Budapest mentor.[28] He nurtured an interest in climatology and his work on Hungary's economic and political geography was translated into German.[29] He served as president of the editorial board of the *Nouvelle Revue de Hongrie* from 1931 until 1938. In 1931, Teleki lectured at the Berlin National Club, and then between 1932 and 1934 as the Hungarian representative for the International Committee for Intellectual Cooperation (Szellemi Együttműködés) variously in Berlin (twice), Vienna, Paris, Munich, Cologne, Hamburg, Milan, Stettin, and Greifswald.[30] From 1932 to 1936,

Teleki chaired the National Scholarship Council. In the city of Gödöllő in July–August 1933, he was part of the International Scouting Jamboree, a favorite and lifelong outdoor passion. His *On Europe and Hungary* in 1934, a collection of major articles and speeches from early in the 1920s, was a noteworthy compilation. It was reprinted in Berlin and translated into Italian.[31]

Central to the indefatigable Teleki's mid-1930s writings and speeches as a geographer was his understanding of Europe itself.[32] It was grounded, as he saw it, not in geopolitics but in scientific expertise. Teleki's fame permitted him to communicate with academies of learning in Germany, Italy, France, Great Britain, and the United States. Nevertheless, his Europe was also revisionist. Fixated on the concept of "historic" Hungary, he challenged the artificial division of Hungary into counties and instead postulated a providential, ecological unity of Hungary's frontiers.[33] He called for full economic integration of Europe's regions, as in the United States, based on "natural" river patterns and physical geography with maps and statistics to favor Magyar dominance.[34] His Europeanization of Greater Hungary met with few objections among colleagues in Budapest. Much like in 1911, when the Francophile count won the Jomard Prize in Paris for his atlas of Japan, in 1934 the MFT awarded Teleki its highest honor for geography, the Lóczy Prize, named after his mentor. *On Europe and Hungary* went beyond conservatism, for it imagined an exceptionalist Hungary-led Europe united in an illiberal vision, assisted by state-corporate experts and anti-Semitic academics like himself, in stark opposition to Bolshevism, Versailles, Trianon, and the Little Entente.

Not least due to Hitler's remilitarization and outstanding border conflicts over national self-determination after the Paris Conference, the year 1934 is likely not the first one that comes to mind when one thinks of reconciliation. However, in fact, in January 1934 the Third Reich and Poland's Second Republic signed a nonaggression pact by which the Führer "recognized" the territorial integrity of Poland's borders. The countries pledged mutual assistance and bilateralism and agreed to avoid armed conflict for the next ten years. Probably this was just ink spilled on paper, like a map, or less important than Hitler's intentions. However, it was during this brief but significant spell of relaxation that the IGC took place during 23–31 August 1934 in Warsaw. American and Polish geographers agreed to come. So, too, did British and French geographers, who had dominated the IGU for the better part of fifteen years. The threat of a German boycott to the Polish capital was real, given that *völkisch* geographers of an older generation like Penck joined forces in the early 1930s with neocons like Meynen to protest against the loss of the Danzig corridor. Undaunted, Bowman believed he was one of few people in the world who could bridge the gap. His work paid off when forty different German geographers pledged to attend. In July 1934, at Bowman's urging, Romer took the train to Berlin and met Penck at his home, in what seemed like an honest attempt to heal old wounds. Romer kept a short travel diary entry of the encounter, but never elaborated. Romer's son Edmund reported the meeting negatively in his 1985 memoirs, mentioning only that the two men talked about history and not much else. Beyond such hearsay, we do not know what happened in the exchange, or if the two map man ever in fact "reconciled."[35]

When Bowman arrived in Warsaw, the Polish press relived 1919. Recognizing the delicacy of the German situation, Bowman this time tried to manage his contacts in public. In attendance during 23–31 August 1934 were 712 delegates from forty-three countries, including the German delegation.[36] The presence of representatives of foreign embassies of Poland attested to the importance of geography as a political science.[37] Ignacy Mościcki, the Polish president who served in Piłsudski's government after the 1926 coup, was present to receive delegates on the terrace of the former royal palace, overlooking the Vistula River. Sites of grandeur conferred meaning, for it was there that Russian governors-general, in service to Tsar Nicholas II, remained in residence until 1915, when the Russian imperial army left Warsaw for good. The IGC held its opening session in the court of the Polytechnical School in Warsaw. The presence of delegations following the German-Polish Treaty of January 1934 would signal, at least in the Poles' preferred map, the historical incarnation of Polish statehood and renewed legitimacy and integrity of its borders.[38]

Bowman's opening address of 23 August was a broad echo of Paris in 1919. He offered three binding themes: (1) the importance of the map; (2) the exhibition of maps for the conference, as arranged by Romer and the Polish geographers, on a scale and with a quality never before achieved; and (3) the role of experts on the earth's frontiers, in service to states that acted in the common cause of science. Two days later, on 25 August, Romer in Warsaw was feted with a great surprise when his peers gave him an edited volume of his essays, two years in the making, with contributions from fifty-four international geographers.[39] It could just as well have been Penck's *Festschrift* of 1918, for Romer's *Collected Works* (*Zbiór prac*) included 313 titles of his original works, with 85 cartographic productions and 228 scientific articles, of which 47 were published in foreign languages. Henryk Arctowski, the Polish-American geophysicist and friend of both Romer and Bowman who had worked under Lord and Bowman, edited the volume and gave it to the master, in honor of his forty years of scholarship. Arctowski also spoke of the importance of building friendship at home

FIGURE 5.1. Poster designed by E. Bartłomiej, one of Romer's "map men" at the Geographical Institute, advertising the International Geographical Congress (IGC) in Warsaw, 23–31 August 1934. From the conference book programme.

and abroad, along with Romer's lifelong dedication to the advancement of the sciences of geography and cartography.[40]

In this way, American-Polish new-worldists all "spoke map." Romer and the conference organizers saw Europe's cartography of 1934 as a new order, but Poland, not Germany, was for the moment at the earth's center (figure 5.1). They drew from Penck's passion, at least back to Humboldt, for empirical field-work and exploration of nature. As with the AGS 1912 Excursion across North America, visiting geographers would gain sensory experience of all of Poland's landmarks and landscapes. Arrangers of the 1934 IGC continued the style of grand colonial expeditions, in a kind of nouveau managed form. Short one-day urban trips were offered in the city of Warsaw and its environs. Each of the seven total excursions lasted a week. There were three long preconference excursions, in the plains and marshlands of Polesia and the forest of Białowieża, the Podolian plateau and eastern Carpathians, and in Kraków and vicinity, the Dunajec Valley, and High Tatra mountains. The four after the Congress were in Grodno, Wilno, Troki, and the lake district of Brasław, in Pomerelia ("transferred" at Versailles to Poland) and the Baltic coast, in the new port of Gdynia, and through the valley of the Vistula below Warsaw and to the industrial center of Łódź. There was a tour of the Świętokrzyskie (Holy Cross) Highlands, then to Polish Silesia, Cieszyn, and the western Beskidy Mountains.

Planned itineraries of geographic unity demonstrated the knowledge of Poland's experts who advanced geomorphology, glaciology, climatology, and human geography. Tours modelled on the AGS 1912 trip were not just about frontiers or science. Bowman knew the goals and codes of the men in his account for the AGS journal, to "enable the members of the congress to learn something of the country in which it is held . . . [and] to provide opportunities, lacking in the hurried atmosphere of the congress, for establishing intimate acquaintanceships."[41] Against the German developmentalist trope of a backward Polish economy, the tours celebrated Poland's progress in economic geography on its frontiers (*kresy*), and the development of infrastructure and technology, diversity of regions, and transportation by way of railway trains, buses, automobiles, and river steamers. Forming friendships in the AGS 1912 style proved elusive in practice, however, when German geographers refused to tag along.

Rather than recalling turf wars and killed loved ones, polite geographers put differences aside. Above all, they appealed to science. Bowman showed how impressed he was by Romer's arrangement of the exhibition, yet another allusion to the transnational community. Emphasis fell on cartography, geomorphology, topography, and demography. True to Romer's didacticism, IGC participants were provided with an explanatory program and catalog. Romer wanted the maps produced by Książnica-Atlas to place Poland in Europe and the "new world." He linked this world in a pedagogical schoolmasterly fashion to the priorities of Poland's history through geography and cartography. Romer proudly noted that the number of maps at the exhibition was 1,664, including sheets of the vaunted 1:1,000,000 map of the world (which Penck proposed at the 1891 IGC in Bern). Actually, something like 300 maps were *not* displayed, allegedly because of their quality (and politics). It was more likely that they did not conform to the exhibition's rules, as adumbrated by IGC regulations and the Polish executive planning that he led. Romer appealed to objectivity to defend Poland's territorial integrity. He even sent a memorandum to the IGU, calling for the next IGC (Amsterdam in 1938) to permit an analysis of the lessons of the map exhibit he had furnished to international viewers. Omitting this fact, Bowman commended the exhibit as "the most comprehensive display of modern maps ever brought together" and credited Romer personally for the "magnitude" of such a sublime cartographic display, "the first undertaking of its kind."[42]

As the conference wound down, Bowman the president finally gave an academic paper on the merits of the federal U.S. Land Program. He attested to changes in American life due to closing frontiers and its contracting agricultural base, a rehash of Frederick Jackson Turner's frontier thesis. Privileging

Davis's geomorphology, he noted more opportunities for research in a pitch for American ingenuity. In line with his mentor, he praised progress in forestry, irrigation, land use at the state and federal levels, planning agencies, regional economic studies, and water resource and power development studies. When the conference ended on 31 August, Bowman stepped down as the first American president of the IGU and became the first vice president, an honorary position, with Romer as the second vice president. He showered Romer and the Poles with compliments, "A million thanks from all of us and my own personal thanks in unlimited measure for all your kindness, thoughtfulness, enthusiasm, good organization, and for the fine recognition from the Polish government. The whole thing is a monument to yourself and a well-merited one. In so far as you depend spiritually upon the admiration and respect of your contemporaries you must be indeed a very happy man. Idealism *does* come first!" To rewrite 1919 and trumpet his Wilsonian success, Bowman called on Pawłowski, the general secretary, to write letters of thanks to the Polish ministers. Bowman wove success into Romer's family and professional life, "the extraordinary degree of vitality" he had and that "everyone spoke of." He added, "On top of all the Congress material that I brought back is the photograph of your two beautiful grandchildren. I can imagine the pride and the affection you have for them and indeed which they must have for you. There is no language adequate to convey my own personal appreciation of what I saw and experienced in Poland."[43]

The redemptive joys of summer 1934 did not last long. Soon afterward, an exhausted Romer fell ill and required special care. Even worse, doctors diagnosed his wife Jadwiga with a tumor. She underwent medical tests in Lwów. Thankfully, the results came back as a relief. In November, Romer told Bowman that she was "completely recovered and out of any danger as the microscopical studies excluded any possibility of cancer." They looked forward to "recreation among friends in the country" and their "charming time in our [Tatra] mountains" during the winter months. He thanked Bowman again, "my profound sentiment . . . [is] to be to you entirely devoted. For this sentiment I find . . . no words."[44] Romer sent more photos, posters, and maps across the Atlantic. Bowman referred back to the "most pleasant and inspiring occasion" and how glad he was to learn of Jadwiga's improving health.[45] They effused about the IGC again, but situations had already changed. In 1935, the American had his final exchange with Penck, the elephant in the room, in which he tried to bury the hatchet from the professor's dispute over Davis's erosion cycle. Penck politely congratulated Bowman on becoming the next president. He wrote to the "Herr President" but this time in German, "My dear Bowman, your election . . . fills me with great joy. At Johns Hopkins, they have placed you at a site from which you can expand your fruitful activity for our science (*fruchtbare Tätigkeit für unsere Wissenschaft*) more than was possible before. I heartily congratulate you . . . and hope that a worthy successor at the American Geographical Society will be found for you. Please pass along my best to your wife."[46] Bowman regretted that he had been unable to meet with Penck in Berlin or Warsaw in 1934. He spoke with pride of the glories of U.S. geography, announcing the procurement of funds to complete the Millionth Map of Hispanic America. He promised to send copies of sheets to Berlin. Bowman passed along good wishes to Penck's family. He ended in the vague hope that "some day" he could welcome his hero back to America, to grace the halls of the university he now led.[47] His letter of May 1935 was the last exchange the two map men had.

LIVES OF A SALESMAN

Undaunted yet unreconciled, Bowman's shining career moment came in 1935, when he left the AGS to become the sixtieth president of Johns Hopkins University. When he resigned in July, he relocated from New York City to Baltimore. Hopkins's board of trustees, men in the U.S. East Coast establishment, unanimously elected Bowman as one of their own. Bowman stood on the shoulders of giants, for President Wilson himself had enrolled at Hopkins back in 1883 and finished his Ph.D. with the dissertation "Congressional Government: A Study in American Politics," in history and political science in 1886. He was the only U.S. president to earn one. Hopkins's first president, the educator Daniel Coit Gilman (1831–1908), was originally a geographer who developed the model for university research from the University of Berlin.[48] Bowman kept up correspondence and he even wrote to Penck in Berlin in March 1939 (without receiving a reply), regarding the U.S. production of the Millionth Map.[49]

Leading lights praised the appointment. Besides Penck, Mark Jefferson, his Harvard-educated mentor in Ypsilanti, Michigan, wrote to congratulate him. Bowman replied affectionately, "your letter is one that I cherish most—for you know my failings also!"[50] In Budapest, Count Teleki was also delighted to hear the news, since he was planning to come again to the U.S. in fall 1935. An excited Bowman inquired whether "you and the Countess Teleki will be good enough to pay us a visit at Baltimore and meet friends who would be interested in knowing you."[51] Mentioning that he lived "only fifty minutes by rail from Washington," he asked, "Shall you bring lantern slides with you on the Iraq-Turkey boundary [the Mosul issue of 1924–25] question? . . . We shall plan to have a luncheon or dinner." Teleki informed Bowman that he had scheduled a stay for a few days with the Budapest-born diplomat John Pelényi (1885–1974) at the State Department in Washington, and then a visit to Colonel Lawrence Martin at the Library of Congress, to whom he would give a copy of his catalogue of all the highly valued books, maps, and atlases from his personal library.[52] He and Bowman promised to contact each other. Bowman wrote to Pelényi, hoping "to have Count Teleki here for an evening dinner, to which I would invite appropriate faculty members. . . . I hope that I can at least have him run up here [to Baltimore] for lunch, to which we would invite guests from the departments of History and Political Science."[53] Teleki never made it to Baltimore.

This timing of the Bowman-Teleki correspondence in October–November 1935 came at a significant juncture in Bowman's life, just as Hopkins's successful new public face was charged with raising millions of dollars for research after the Great Depression. Bowman grasped how Wilson's rhetoric of values was a limitless opportunity for fund-raising. The Midwesterner believed the U.S. was destined to harmonize its free markets with limited government and the world's providential good. Before an audience in Baltimore of wealthy donors at the Presbyterian Social Union in October, Bowman pontificated about "improvement in international morals" and "true diplomacy." "It is the first rule of diplomacy," he claimed, "that its business is to keep relentlessly at the task of preventing war. . . . The trouble with all diplomacy all the time is that it has to deal with an almost insoluble problem; and it has to assume that the problem is soluble. That problem is how to reconcile so-called 'national interest' with the interests of other nations." Still on the U.S. Council of Foreign Relations since 1921, Bowman played a pro-business card against the New Deal bureaucracy he had come to loathe. Elsewhere in America, he now spoke of his 1919 experience on transactional terms that "every nation has to give something away in diplomatic dealing just as one does in business."[54]

Having lost his European pals by 1935, Bowman took on a new project, making his youngest son Bob into an American geographer. The intimacy between father and son reveals much about the persistence of transnational norms, gendered frontiers, and patriarchal layers of place. For thirty years starting in the 1920s, Bob was Isaiah's pride and joy, his trusted confidant. Isaiah gave him advice on everything, getting good grades in school, whom to marry, on travel and research, what to do with money, on joining the military, on how to become a geographer—in essence, trying to make him the next Walther

Penck. Bowman's project was modeled on Ostmitteleuropa's science and map men. Their exchanges are so copious that they may explain why Bowman never penned formal memoirs, or the six to twelve autobiographical books he hoped to write. Isaiah's letters read like a Gilded Age self-help manual on how to succeed in America, or else the draft of one mightily verbose commencement address. Isaiah's letters to Bunny-Boy, Bob's childhood nickname, were "confidential," one of his favorite words. Bob's life reads like an unceasing attempt to impress the old man. He and Bob personalized Wilsonian roles—the smart hardworking father with mission-like principles, and the respectful son, overachieving and mindful of authority.[55]

Like our other transnational lives, Bowman did not shy from using an old boys' network to get his son (or academic "sons") ahead into geographers' professional worlds. In November 1935, he corresponded with Dr. Ernest Horn at the University of Iowa about cultivating Bob's ambitions. Bowman and Horn knew each other from the Commission on Direction for the American Historical Association (AHA), and the so-called Investigation of the Social Studies in American schools. Responding to Horn's inquiry about someone to fill an open position in geography in Iowa, Bowman offered "to see if we can find the man of the right sort."[56] He dropped his son's credentials: "[Bob] is taking his first year graduate work at the University of California under [Carl O.] Sauer, who has both intelligence and genius. . . . [Bob] has done extensive field work with me in the West and he persists in his desire to specialize in geography. Would you like to have him come at Christmas-time by way of your town?"[57] Bowman denigrated other "young men" in rival programs at the University of Chicago and Clark University. These elite men could be "bought," he claimed, but they cost "too much." He wrote to his ally W. L. G. Joerg (1885–1952) at the AGS in New York, "Here is a letter which I should prefer to have you keep confidential. . . . My purpose in sending it is to ask you if you have any suggestions to offer concerning younger men. . . . Generally there are a few men in their late twenties or early thirties who would be considered for such a job, but not uncommonly they are rejects from other institutions and remain second-rate men."[58] Bowman asked Joerg for "confidentiality."[59] He did not apply the same rule to his family, forwarding correspondence with Horn and Joerg to Bob in December 1935, saying that Horn was "one of the best men in the educational field." He urged his son, then just twenty-three, to be ready to strategize with him in Baltimore over the Christmas break.[60]

In the 1935–36 academic year, Bowman turned into an even more boosterish champion of U.S. exceptionalism. He delivered a rousing fund-raising speech on 6 December, "Next Steps in American Universities," at the first Annual

Meeting of the Southern University Conference in Louisville, Kentucky. He argued that universities prepared men for church and state professions, against a bloated bureaucracy. He utterly disdained the New Deal. "No man can say where the expanding agencies of government are going to stop," he remarked. "A political campaign is only a promise campaign," he observed. "The government edifice has grown to such enormous size that it has slipped out of the control not only of the people but also of the chief executive," Bowman opined, urging universities in their "first duty . . . to educate our young men to these appalling facts." He criticized FDR's redistributive welfare, on curious grounds: "No president of the United States can any longer drive the team that he creates; the team drives him. The cart is now definitely in front of the horse. This means, put into other terms, that we have a government of groups and those groups are now *within* [emphasis in original] the government, whereas the groups that we once called lobbyists were outside the government. Government is now lobbying for itself."[61] Like his grandfather-preacher Moses, the Midwesterner cast himself as a new American prophet.

In the year 1936, Bowman worked night and day at Johns Hopkins, and he called in all sorts of favors among the powerful. In January, he began an exchange with Richard Hartshorne (1899–1992), the U.S. geographer who studied at Princeton and Chicago, then taught at the University of Minnesota. During World War II, Hartshorne took on a more clandestine adaptation of Bowman's role in the U.S. Inquiry in 1918, in charge of information gathering and analysis for the OSS, the predecessor agency to the CIA. Hartshorne was quite concerned about Poland. He asked Bowman whether an article he had drafted, "Permanent Factors in Poland's Foreign Policy," was suitable for publication in *Foreign Affairs*, where Bowman was on the editorial board. When Bowman returned comments, he acted as an "objective" censor in the choice of political language. He recommended that Hartshorne not call the French troops in Poland in 1920 "troops of occupation," and that he accept Piłsudski's "firm stand" against the communists. He reminded Hartshorne that as "the saviour of Warsaw . . . [Piłsudski] organised an advance that resulted in the Bolsheviks' rout"; that "the Poles' resenting the French attitude of saving Poland . . . simmered down and could have been largely eliminated if the French had been tactful"; that it was not simply that the Allies created Poland "out of hand [but that] it was the intention of the Poles themselves to set up a new state." Bowman assiduously corrected it in a handwritten note, "Pilsudski [*sic*] and Paderewski quarreled in 1919 on the issue of dependence on Germany or on Russia. Have you looked into this?" In what he called "confidential" editing, Bowman told Hartshorne frankly that the article "just seems a little elementary and text-bookish." In the name of objectivity, Bowman called for "fuller and more balanced" statements about Poland's past.[62] The article illustrated the provincial role of U.S. geographer-experts. When it was published in 1937, it appeared as "The Polish Corridor" in the *Journal of Geography*, not in *Foreign Affairs*.[63]

Learning a lesson from 1919, Bowman became choosy about whom to support in East Central Europe. The map man's sympathies are subjective. For Bowman, they extended to wherever he had the most meaningful experiences and loyal friends—in Poland and sometimes Hungary, but not in Czechoslovakia or Ukraine. For instance, Francis Deák (1898–1972), a prominent professor of international law at Columbia University whom he did not know, contacted him. Acting in the name of the newly formed board of the *Hungarian Quarterly* (modeled on the *Yale Review*), Deák tried to namedrop "friendly" Teleki to get Bowman's attention. "On behalf of its editors," Deák wrote, "and particularly *Count Paul Teleki* [emphasis in original], I have the pleasure of inviting you to become a member of a small Advisory Committee of the *Hungarian Quarterly* in the United States. The purpose of this English language periodical is to bring before the English speaking world some of the problems with which Hungary is faced and to acquaint the outside world with the accomplishments of Hungarian science, literature and culture." This time Bowman played the objectivity card again, but to hide behind it. "Under other circumstances I should be glad to join the enterprise and I hope you will so inform count Teleki," he wrote. "It happens that I have been asked to associate myself with so many other similar movements in the interests of specific nationalities that I have declined to serve on the Advisory Board of any one of them."[64]

The double-dealing president thus cultivated his image as a U.S. exceptionalist, filling his speeches with folksy boilerplate about a special mission. In a February 1936 speech, "A Design for Scholarship," he celebrated the sixtieth anniversary of Hopkins's founding. "We labor to devise ways of putting knowledge into decisions that have public impact, of substituting reason for guess-work, of fighting vanity and greed by enlarged public understanding and courage, of rewarding society and unselfishness, and of keeping wisdom in the state in a framework of stouter timber than political cowardice, bureaucratic lechery, and communist doctrine provide," he announced. He needed "to defend civilization wisely, and toward ends intelligently conceived or selected." To accomplish this, he told the audience at Hopkins, "We must educate and control both the men who run the agencies and those who elect them. With all the lapses from grace acknowledged and all the defeats of purpose confessed, conviction is with us still that education is the most vital element in a democracy and that an uneducated democracy is a democracy in peril." He evoked a fantasy past: "Colleges were sown throughout the Middle West contemporaneously with the advance of the frontier, and the first college buildings were erected while the buckskin-clad frontiersmen still traded in the new settlements."[65] He viewed Hopkins in the East as a training ground for building geography in the U.S. West. Hopkins was like the Ivy League universities of his training but of a German sort, in lines from the two Humboldts to Penck in Berlin, for advanced learning and research. Hopkins advanced young men into geography as science, instilled patriotic virtues, and trained "sons" (no mention of daughters) for service in international affairs.

One sees this masculinist point in Bowman's revealing correspondence in 1936 with Newton D. Baker (1871–1937), the tough former mayor of Cleveland from 1912 to 1915 who served as President Wilson's secretary of war from 1916 to 1921. By the 1930s, the lawyer-politician Baker was on the boards of trustees at several American universities; at Hopkins, he was on the search committee that preferred and hired Bowman.[66] In a letter to Bowman in February, Baker wondered what would have happened if Germany had won World War I. He offered a counterfactual scenario: "Incidentally, I wonder whether you could find a spare moment . . . to put on a little map for me what Europe would have probably looked like if the Germans had won a complete victory? Germany would have retained Belgium and some parts of northern France. Austria would have annexed Serbia and a lot more of the Balkans." Baker enclosed a map for Bowman and ruminated about the "German appetite," speculating in his free time that he might draft his ideas for a choice article in *Foreign Affairs*.[67]

From his new position of authority, Bowman was intrigued. He wrote to Gladys Wrigley (1885–1975), the head secretary of the AGS in New York City, the first woman ever to earn a Ph.D. in geography from a U.S. university (at Yale in 1911), with an order to work on publicizing Baker's letter. Bowman justified U.S. intervention in 1917 by asserting, "The German objective was clearly that of world conquest." Taking the anti-Penck line and getting drawn in, he informed Baker, "Certainly Germany would have attempted to carry out the conditions of the Treaty of Brest-Litovsk"—i.e., control of Eastern Europe and Eurasia, where he made his career as an expert.[68] He enclosed material from Wrigley in the response letter back to Baker in Cleveland.[69] Baker replied and they exchanged base stereotypes about German "national psychology" as Baker prepared his piece for the October 1936 issue of *Foreign Affairs*.[70] Objectivity was selective, for he never asked Wrigley for her input. Bowman praised the draft by Baker but without comment or correction. Not one to resist hyperbole, Bowman would even compare Baker's lawyerly eloquence to that of Thucydides.[71]

That Bowman was opportunistic is beyond doubt. As an idealist he was ruthless and tireless, passionately convinced that he could revive the League of Nations in another form. Right through the 1930s, he still held to the idea that it was possible to disentangle "good" (Anglo-American) scientific geography from "bad" (German and revisionist) geopolitics. In fall 1936, he lectured on "Political Geography" at Cornell University.[72] He recycled the main points in spring 1937, into "Is There a Logic in International Situations?" for the Institute of Public Affairs at the University of Georgia. Bowman drew a clear distinction between *Geopolitik*, "a word of power in Germany since the World War," and respectable "political geography" of the West's democracies. *Geopolitik* was "neither the invention nor the dispensation of God though it is all but worshiped by Nazi elements . . . Never was science so debased as when it was consciously invoked to prove the national case. *Geopolitik*, for like reasons, is not science but rather the underprivileged child of the World War."[73] Bowman argued, as Romer had done in Poland, that diplomacy and public affairs were best handled by experts. Only these "full men," dedicated to *Beruf* in their character and objectivity in their *Wissenschaft*, solidly grasped what had to be done. Beyond the nation, Bowman invoked European virtue: "History supplies one indispensable element of training for the full man—the vanity of thinking that any human being can find a rationality that will fit more than one nation except in limited areas for limited purposes and in limited times of common emergency."[74] The geographer-diplomat played favorites among those he wished to empower, but also understood the privilege, status, and mobility that geographers gained professionally. Put another way, Bowman believed in and performed the maps of an American-led Western civilization.

BOYS TO MEN

For what was more Western than the explorer's travels, in search of new landscapes and bonds, facilitated by academic careerism? By 1936, Bowman's elder son Walter was at the Sorbonne in Paris, while Bob was at Berkeley, where he studied geography under Carl O. Sauer, the eminent cultural geographer and chorologist. Bowman and Bob first traveled together to Wyoming in 1930. Bob depended on the old man for funding his studies at Dartmouth and Berkeley.

By 1933, he decided to become a geographer and traveled a second time to Wyoming, part of a summer expedition sponsored by Columbia University. Isaiah advised to Bob in October 1936: "It seems to me that you will be lucky if you are able to finish in four years of graduate time, that is two years after this one; and if it takes five years it will be no surprise to me. . . . Knowledge continues to grow in volume in every field and no one should receive a Ph.D. who is not reasonably familiar with the problems and methods of his field." Isaiah counseled Bob to "invest" in his studies, pursue fieldwork, and graduate "with distinction." As with the 1912 AGS excursion, the study and penetration of nature ought to be done, he advised, in the company of men: "My own experience in South America was transformative and it had a great stimulating effect. It also gave me a realistic knowledge of problems in the field . . . [but] there is a good deal to be learned about man through experience with men rather than the field."[75]

When Isaiah fully convinced Bob by 1937 to follow his career path, he connected his son with Horn at the State University of Iowa. In a June letter, Isaiah assured Horn that Bob was doing serious fieldwork in becoming a geographer and an American man. Horn inquired, "When does Bob expect to receive his doctor's degree?" Bowman immediately replied by enclosing his son's résumé, writing by telegraph that Bob "will not complete his work until the end [of 1938] at earliest [and] has a teaching fellowship for [his] third year and would like to do field work in either Mexico or Colombia." Bowman told Horn that his son was "working up his German" and has "the background and training that will guarantee first-class work." He finished with a request that Bob be allowed to come to Iowa without a doctorate (!) and get time off for fieldwork, as he had done thirty years ago at Yale: "Naturally, he would like to get a job and stand on his own feet. . . . I have written you extensively because I can say some things that Bob would perhaps not wish to say himself. Perhaps I have said too much to suit him."[76] Isaiah then sent the "confidential" letter to Bob, showing off his prowess in the intervention.[77]

As Hopkins president, Bowman thus opened doors for his son's career. He adhered to nineteenth-century German models of *Bildung* (education), *Erziehung* (training or rearing), and *Wissenschaft* (science) to enable moral growth. "Keep in mind that education is only a process," Bowman wrote to Bob, "for obtaining more education later on and that the only fun in the intellectual field is in growing and in realizing that your capacities increase with time. And a lot depends on time—much more than a young man realizes."[78] In an October 1937 letter, he urged Bob to learn specific languages. This meant a strictly

European course of learning at Harvard and Yale. "Glad to see that you are working up your German. French is one of the indispensables in the life of a scholar," he wrote. "You will get more from either language in a few years than you would get in a lifetime of Spanish; though Spanish is necessary if you are going to do field work or delve into archival material."[79] Bowman admonished Bob at Berkeley not to remain one of those "linguistic cripples for life," paralleling his son's pursuits to his own with German.[80]

When Bowman learned that Bob was not coming back to Baltimore from Berkeley for the Christmas 1937 holiday, he chided his son for want of discipline. He convinced Bob to cultivate himself adequately *as a white man*: "I am disturbed about the lack of preparation for the [doctorate] exams. . . . I strongly advise against the type of cramming that requires work until midnight or after. Is there a nigger in the woodpile somewhere? Have you been spending too much time on teaching, International House, car, etc? Sauer has dealt with young men before . . . your present job is to do preparatory work for a doctor's degree in such a manner that you will be rated as an outstanding man and not just one more ordinary Ph.D."[81] In a February 1938 letter, the Hopkins president said, "I have just written to Uncle Walter and Aunt Edith to thank them for a Cape Cod barometer which is great fun. When you return in May or June I must tell you of the difficulties I had in trying to explain the physical principles to my women folk. God did not make girls for physical science—but for a much higher purpose!"[82] Bowman therefore insinuated that a woman's place in society was in the domestic sphere. He admitted to all this overbearing micromanagement: "I sign myself in all humility, but full determination, and with complete control of my well-known supervisory instinct."[83]

After Bob took his father's advice and passed his exams at Berkeley, he traveled to San Juan, Puerto Rico, for field research. He and Bowman kept up regular contact. In a letter to his son in May 1939, Bowman confessed to a problem he was experiencing with "a Negro undergraduate" who applied for admission to Hopkins. He opposed the boy's admission: "While a few colored persons have been admitted to the Graduate School in times past, there has never been a Negro undergraduate. Moreover, there is no need for such admission in view of the existence of Morgan College for Negroes, which is located here in Baltimore. The State of Maryland has recently taken over Morgan College and will improve it greatly. It would therefore be advisable to let us know if the boy is prominently 'colored.'"[84] Bowman's words reveal his nineteenth-century life and outlook defined by his biological understanding of the nation, his white Christian nationalism and racial and patriarchal privilege, and all the realities of inequality behind the façade of American democracy.

Exiled out of the Ukrainian SSR, Stepan Rudnyts'kyi's three children enjoyed no opportunity for career advancement such as the Bowmans had. Following the ideologically motivated arrest of the geographer in 1933, they were assigned passports and forced to move from Kharkov/Kharkiv to Voronezh, and into the Soviet hinterland. The forest steppes were kind of a Eurasian frontier, but not quite the American West of Karl May's lore. On 4 January 1934, by verdict of the Presidium of All-Ukrainian SSR Academy of Sciences (VUAN), Rudnyts'kyi was expelled for having participated in the chimerical UVO. Stripped of his status as an academician of VUAN, he was again counted as a Ukrainian bourgeois nationalist, a fascist, and a counterrevolutionary. The Kharkov/Kharkiv Institute was dismantled, and a new Institute of Geography was founded at Kiev/Kyiv University (it lasted until 1952), set up for Soviet Russocentric power in Stalin's new capital as of 1934. New communist maps were published including a map of Kiev/Kyiv oblast', made by the Kiev/Kyiv University director and professor Oleksii Dibrova (1904–73), who was long acquainted with Rudnyts'kyi's work. Under Stalin's helm, no Soviet scientist could manifest public memory of Rudnyts'kyi's maps or his achievements in geography.[85] Materials of the library and cartographic holdings and equipment were moved from Kharkov/Kharkiv to Kiev/Kyiv. Personal collections of books by Rudnyts'kyi also were confiscated. The Academy of Sciences expelled him and stripped him of his professorship and all titles and positions.

Since the collapse of the Soviet Union, more today is known about Rudnyts'kyi's life in prison after his initial arrest in Kharkov/Kharkiv in March 1933. According to the criminal records, Rudnyts'kyi had been working on a 342-page monograph called "Heonomiia." He completed a massive 1200-page book on the endogenous climate dynamics of the Western Hemisphere.[86] Both were seized from him, and never released. Only after 1991 did it become known that Rudnyts'kyi was sentenced by a Soviet tribunal on 23 September and sentenced to five years in the White Sea and Baltic Canal area. Rudnyts'kyi became a prisoner (zek) in a forced labor camp (Svirlag) on the Solovki Islands. Scores of Ukrainian poets, teachers, agronomists, geologists, doctors, and engineers were brought there as laborers, essentially to be worked until their deaths. By the tribunal of the UNKVS of Leningrad oblast' on 9 October 1937, the execution of Ukraine's most accomplished geographer was ordered. Stepan Rudnyts'kyi was killed by an NKVD (formerly GPU and then OGPU) firing squad on 3 November 1937. In total, 266 persons were shot in a mass grave with him. Principal victims were Ukrainian "bourgeois" scholars, writers, theater actors, politicians, and social activists. Rudnyts'kyi's family held out hope that their father remained alive, but two of the children, Levko and Orysia, disappeared around Voronezh, probably in 1940, where they had been forced to relocate before the Nazi invasion of the USSR in June 1941.[87] In life or in death, no peace or reconciliation for the family of educated Ukrainians from Galicia was near.

———

Historians are in the business of explaining events, but they have their hands full with East Central Europe's treasure of maps. In box after box of these curious artifacts, maps expressed concealed moods and illusory bonds. But all moods are fleeting, and all bonds dissolve. Like maps, leagues were temporary projects that gave an illusion of permanence. Were those on the twentieth-century winning sides, Bowman in America or Romer in Poland, any less full of illusions than colonizers before 1914, any less emotionally burdened than Teleki in Hungary, Penck in Germany, or Rudnyts'kyi across Ukraine's borders? Is it historically fair to view maps as rational plans or an abstractly ordered modernity, as if every place was a finished building, a gridded city, or a planned bureaucracy? Should one always ascribe unsavory behavior to the malcontented, the "losers" on the wrong side of war, as opposed to those who appealed to objectivity, democracy, science, or peacemaking? Stalinist authorities after all appealed to science and peace, but they ascribed malicious intent in geopolitically remapped spaces of Ostmitteleuropa and the USSR. In the terror campaign of 1937, it finally cost Rudnyts'kyi his life. Into the difficult years of World War II, our maps raise more questions about the integrity and rationality of lives and inner worlds than they can answer.

CHAPTER SIX

EX-HOMES

On 3 September 1939, two days after the Nazi invasion of Poland from the west, the Romer family of four—Eugeniusz, Jadwiga, Witold, and Edmund—shared the last supper of their lives as a family. The Sunday gathering took place in the Romer home on ul. Długosza 25 in Lwów, which they bought in 1911 from the geographer Antoni Rehman, Romer's predecessor and mentor at the university. Witold and Edmund worked round the clock to build barricades on the city's outskirts, but leaving Galicia proved nearly impossible. On 9–10 September in the middle of the night, Edmund headed toward Romania with a group of men, and from there to Ankara and Palestine and eventually to the Polish government-in-exile in London. Witold, the younger son, headed for the Romanian border on the next night, making it from there to Italy and on to join the antifascist resistance in France in 1940. He also became part of the Polish exile community in London. Their aging parents sheltered and fed some 159 persons they took in as refugees in September–October 1939, plus 23 Polish pilots fighting against the Luftwaffe. Edmund's wife Krystyna and children Marysia and Tomek came to stay in the house full of strangers. The parents did not know the whereabouts of Witold and Edmund after they left.

Jadwiga's health took a turn for the worst, while Eugeniusz struggled to work as a geographer and keep up his morale.[1] On this the eve of World War II, we will take look at "home" and its textures of meaning among those who died or were displaced by the conflict.[2]

OLD WORLDS

Bowman and Romer continued to look back to the IGC 1934 in Warsaw and the Paris Peace Conference in 1919 with a heavy dose of nostalgia. Romer sent Bowman a huge bibliography of all the works produced by the *Polish Cartographic Review*, since its inception in 1923.[3] Bowman thanked Romer for "the card with the pipe and pretty girl [figure 6.1]. . . . [It] is always nice to hear from you and especially so when the good news is conveyed that your health is restored. We are all proud of your part in the Congress, and, indeed, of the part of all Polish geographers who contributed to its unqualified success."[4] When he had recovered from his own poor health and Jadwiga from her cancer treatment, Romer gave lectures in Lwów and Warsaw (similar to Teleki's in America in 1921) on Poland's economic geography. He revisited the atlases of 1916 and 1921 by turning toward a geographical analysis of agriculture and manufacturing products. Propaganda was mixed in. He trumpeted Poland's raised standard of living, density of roads, railways, banks, post offices, culture, communications, trade, and conveniences of modern life. Romer sought to overcome Poland's nineteenth-century and integrate regions of imperial Germany such as Pomorze, Wielkopolska, and Silesia with Poland's center and east. He stressed the progress of a free Poland. The country became unified, he argued, not because of Germans, or the Habsburgs, or Piłsudski's centralism, but because of its civil society and academies of learning, which fostered growth. In the distant homes of memory, these were the values of his family from Galicia before 1914.[5]

Although Romer had approved of Poland's reconquest of Lwów in the Polish–Ukrainian War of 1918–19, he was very critical of Piłsudski (who died in May 1935) and the Polish military's intrusions into geography and cartography. He argued that Piłsudski's coup of 1926 and authoritarian campaign of *sanacja* (a Polish word for cleansing or purge) against a system of checks and balances promised a false regeneration. It impeded Poland's civil society, permitted assassinations and arrests, and invited the unlawful torture and humiliation of prisoners. Resisting this, Romer supported academic geographers of Kraków, Warsaw, Lwów, Poznań, and Wilno by advocating, after his retirement in

"The Congress dances"

The International Congress of Geography
in Warsaw. 1934.

Cartoon of E. Romer.

FIGURE 6.1. Caricature of Romer, M. Gra[m]ski, "The Congress Dances," the International Congress of Geography in Warsaw, 1934. Memories not of Vienna in 1815, or even Paris in 1919, but of the IGC in Warsaw in 1934. Courtesy of Special Collections, The Milton S. Eisenhower Library, The Johns Hopkins University, Baltimore, Maryland. Isaiah Bowman Papers (IB-P Ms. 58), Series I, Box 3, Item 10b.

1931, the autonomy of science and university life against intrusive "statism" (*etatyzm*). He defended civil liberties and supported Polish schools in the republic. He insisted on legal norms of citizenship and thought that any constitution that failed in its social contract threatened the nation's survival. His project of Enlightenment was in line with Hugo Kołłątaj, the Kraków geographer and reformer, and the ideas of Locke, Burke, and Rousseau in the Polish tradition of republicanism and its 3 May 1791 constitution. Yet he was also a close reader of Ratzel and German geography. Romer was imperfectly modern and national. He refashioned Galicia into a neocolonial frontier space where a state met its limits, particularly on issues of rights and representation facing Ukrainians and other minorities in Polish borderland spaces.[6]

Romer nevertheless sympathized with the anti-*sanacja* bloc in the Sejm in 1936 and 1937. He was no less an angry creature of Europe's empires as a Versailles "victor" than revisionists who dreamt of restored national unity. When Romer resorted to sub rosa pressure politics on the grounds of nationality, it paralleled Teleki's revisionism in Budapest and Berlin, Penck's *völkisch* tactics in Leipzig, and Bowman's bending of U.S. presidents' ears. At no point in his life did Romer consider Ukrainians a historic people. In fact, he called them "Ruthenians" or "anarchists" long after the Polish-Ukrainian War, and he grouped them on Poland's frontiers with Hutsuls, Lemkos, and Rusyns.[7] He assigned value to regions as parts of a whole. After Poland's 1921 census, he classified minorities mainly in the modern sense, reduced to language or confession. He hoped to strengthen a multiparty state and forge an assimilationist society of Polish citizens who had greater mobility, but this hardly made him a democrat.[8] Like Penck and our other patriarchs, the former Habsburg subject felt that his children should always have opportunities at home and abroad for scientific pursuits.

In 1938, the Polish authorities subjected Romer to intensified harassment. They aimed to besmirch his reputation and intimidate him to the point of expulsion from public life. Romer was forced to respond to the charge that he had disclosed state secrets, through maps, to an internal enemy. What happened to Penck in England in 1914 now happened in Poland, when military authorities in Warsaw confiscated maps and items printed by Książnica-Atlas. Apparently, the government was after the publication of "The World in Numbers" (*Świat w cyfrach*) because it contained data about the Polish armed forces and an urban plan of Lwów. These, Romer noted, were items that had been in circulation for several years, the latter for ten years.[9] Red tape became the state's weapon of choice, as Romer grew mired in paperwork that was needed to combat the charges. Although he attended the IGC in summer 1938 in Amsterdam, he had to turn down its presidency when his peers offered it to him. That same hectic year, he viewed the destruction of Czechoslovakia and annexation of the Sudetenland as political disasters. On the way back to Lwów through Berlin, he bought a copy of Hitler's *Mein Kampf* and began to contemplate an even greater catastrophe.

As Romer was being ostracized by his government, after the Munich Conference of 28–29 September 1938 Bowman was called by the U.S. president for his expertise. Roosevelt first came into contact with Bowman as governor of New York in December 1930, when a solicitous Bowman sent him one of his scholarly papers on settlement. FDR was pondering how to extend U.S. economic interests south of the border. On 15 October, Bowman thanked FDR for referencing his book, *The Limits of Land Settlement* (1937).[10] Despite his deep antipathy toward the New Deal, the pan-American careerist replied to FDR with a four-page missive and a map of the South American cattle industry. Again, Bowman had a knack for opportunity. FDR thanked Bowman for his service in a personalized letter sent to Hopkins in October 1938: "Many thanks for taking the trouble to send me that interesting letter about the Llanos of Venezuela—even though your description is not encouraging to the possibility of colonization." Two days later, on 2 November, he had a follow-up request:

> Frankly, what I am rather looking for is the possibility of uninhabited or sparsely inhabited good agricultural lands to which Jewish colonies might be sent. Such colonies need not be large but, in all probability, should be large enough for mutual cooperation and assistance—say fifty to one hundred thousand people in a given area. I take it from your letter that the territories in Venezuela that you described are north of the Orinoco River. Is there any information about the country south of the Orinoco but on a good deal higher level—say three to five thousand feet? Also, do you think there is any possibility in Western Venezuelan country on the Eastern slope of the Andes? Or do you think there are any possibilities in Colombia itself? All of this is merely for my own information because there are no specific plans on foot—but I am grateful to you for the interest you have taken.[11]

This certainly enticed Bowman. In November, he ordered the production of maps delivered to Henry Morgenthau, Jr. (1891–1967), the secretary of the treasury. FDR wrote to Bowman that the maps had arrived, and Bowman, like his friend Romer, eagerly sent more. In a fascinating (unpublished) response, the U.S. president sent a detailed letter on 25 November, connecting "Latin" America to the "Old World" and "Central Europe,"

I have not spoken of the political difficulties which a large foreign immigrant group would create if planted in this small Latin American country, assuming that Costa Rica is willing to receive them. The effect of such a group upon the state, and the possibility that through the presence of the group we might become seriously involved in European quarrels, are matters upon which reflection is needed. My own feeling is that we keep our position uncompromised in the Western Hemisphere only so long as we do not interest ourselves directly in the importation of European population elements. The moment we do so we are likely to be charged with the importation of a European quarrel into America. Even if we are right about such importation from the humanitarian standpoint, we thereby give the other fellow a chance to claim that we are wrong. Do we want to run that risk? Do we wish to confuse our position and dilute our argument respecting the Monroe Doctrine? Why not keep the European elements within the framework of the Old World? Even if we do not favor migration to Latin America, but allow it, difficulties will arise. If we both favor it and allow it, we commit ourselves to the consequences. These consequences will surely involve us in the rightness or wrongness of acts of the governments of the states of Central Europe.[12]

Reading between these lines, Bowman gave his expert opinion about "pioneer" settlement as he called it, but he also warned about the need to "keep the European elements" in Europe. He laid out options for the U.S. to avoid taking sides, what Wilson advocated up to April 1917. Bowman argued from a muddle of Wilsonianism, anticommunism (he assuredly had that), racial anti-Semitism (he had that too), and 1930s-style appeasement, not simply the geopolitics of Ratzel or Mackinder. To stay out of "Central Europe" was the main point, factoring in the nineteenth-century past, World War I, and the geographer's anti-urbanist xenophobia.[13] Bowman counseled a foreign policy and Latin America policy in sync with the (Theodore) Roosevelt Corollary to the Monroe Doctrine, a hemispheric world that was also a product of William Morris Davis's professionalization of U.S. geography. Like other Anglocentric nativists in the history of labor immigration, he wanted to keep Jewish populations from coming to the States. On 10 December, he sent to Morgenthau, to give to FDR, the African materials that the president had requested—in his own words, dealing "with the possibilities of white settlement in that continent."[14]

In this manner, Europe's persistent tensions of colonial empire factored strongly into the Bowman-Romer friendship. The men of literal letters kept on exchanging maps and articles. In December 1938, Romer told Bowman of an article he had prepared, "The Partition of Czechoslovakia as a World Disaster," which his secretary translated into English. Romer relied on their friendship to "give . . . a feeble idea upon my fury resulting on the October-World history." He asked Bowman, who was on the board of *Foreign Affairs*, for it to be "well placed." The misguided assumption was that Bowman, a past supporter of Poland in 1918–19, would surely allow a Pole in 1938 to speak for himself. To make the appeal work, he enclosed maps and lovely postcards of Polish landscapes. Romer was in crisis, on all fronts. In his inner life, he turned to religion, sending to the Episcopalian Bowman an ecumenical card. He enclosed a "Christmas souvenir . . . especially executed for Rev. Lord," the former U.S. Inquiry expert who was now a Catholic priest in the Diocese of Boston. The photo for Bowman, taken by his son Witold, was a copy of a seventeenth-century Spanish sculpture from a Benedictine cloister in Lwów. Facing the Third Reich, Romer identified Poland with a favored image of the romantic poets, and he confessed to his friend, "I look for Christ—a pain [*sic*] vision."[15]

Bowman's response showed how very American he had become. Recall his efforts back in 1936 to encourage his "Thucydides," Newton D. Baker, to publish in the American journal. Baker, who only knew English, was no expert on Eastern Europe. Now in this case, Bowman sent a letter not to Romer but back-channeled it to "Ham"—his pal Hamilton Fish Armstrong (1893–1973), the editor-in-chief of *Foreign Affairs*. "Here is a paper from Romer which I promised to send to you. Of course, you cannot print it. He must think that we live in the wilderness and have just heard of Czechoslovakia. Pathetic. But if you return it, as I am supposing you must, delay the return for two or three months and by that time he will feel less keenly the rejection of the article." In effect, Bowman damned the article in advance of a fair review. He wrote back to Romer on the same day, "I am sending [your article] to Hamilton Fish Armstrong for publication . . . Of course, it is he who must decide publication . . . I have no power in this respect."[16] The relationship between the two men therefore had clear-cut limits. It illustrated just how self-serving the Wilsonian's cartographic placement of a "democratic" Poland was.

CALLING DR. LOVE

In summer 1939, Dr. James Lee Love (1880–1954) came back into Bowman's world. They had met over thirty-five years earlier at Harvard. Dr. Love was the Secretary of the Lawrence Scientific School (later Harvard's Graduate School of Engineering), a graduate of Johns Hopkins, and now a professor of mathematics at the University of North Carolina. He politely requested a story of the geographer's life and success, and Bowman, unbothered like his European counterparts by war and occupation, took the time to oblige. Writing from the family's cottage in Wolfesboro, New Hampshire, Bowman floated the title of "Free and Twenty-One," alluding to 1899 when he became a U.S. citizen. He began the saga by placing his boyhood and education on the frontier, and focused on his early character formation. On the issue of diversity in the Midwest, he scribbled, "Jew & Catholic were rare curiosities. Everyone was poor—or all but a few Glaziers and Deans [who were] called rich. Yet each at hardest work period and worked in fields." Bowman was sure to emphasize his love of reading, despite the few books in his father's house. He thought himself a true American, a true man. He loved military exercises, studied hard, and became the first in his family to go to college.

Bowman then went on to map a family fantasy in this utopia of golden opportunity. "Both my grandparents were well-to-do people in Canada," he wrote, "but they had large families and their estates were necessarily divided rather finely. My father was caught in the hard times of [Grover] Cleveland's second administration, and there were eight children. I was the third. Only a country school education was available. Judging by present-day standards, it was very poor schooling. But it had one great quality: discipline. I use the word in the good sense, meaning not discipline as to physical behavior but rather discipline of the mind." He chalked his upbringing up to Protestant discipline and Midwestern grass-roots virtues, without ever emphasizing his specific German background. Writing to Dr. Love, he summed up the qualities as in family lore:

> I was extremely poor and conscious of my poverty. During my senior year at Harvard I had a single suit of clothes the entire year and I declined all social invitations, even those from my own class (I have never attended a class meeting) because I was painfully conscious of the fact that I was ill dressed and conspicuously so. But it never occurred to me to be bitter about it or to want to upset society because my father had eight children and was unable to help me. I had to earn every dollar of my education. This is the literal fact, for I returned to my parents, with interest, the sums which I had to borrow from them. My marriage was delayed until I was thirty-one. I suppose that if I had been more sophisticated, I might have felt more strongly and sought grounds for complaint. As matters stood, I had no feeling of animosity toward others who were more fortunate than I . . . All my life I have worked hard but if I had been a rich man I should have done the same.[17]

Dr. Love was deeply impressed. Here, in a nutshell, was Bowman's packaged path to advancement. He told of a recipe for humanity's success, part of a prescription for the future of their shared country in a global world. Like Professor Penck's pupils, Bowman's heirs could follow it.

On 31 July, Bowman had another heavy-handed exchange with his son Bob, to whom he also wrote from the summer home. Bob was forlorn after a breakup with his girlfriend. His father placed all the blame on her for refusing to play the role of a (male) scientist's wife. "One big source of difficulty no doubt," he warned, "was the lack of understanding on her side of the nature of your work. . . . There is only one piece of advice . . . that further relations with the family would be highly inadvisable." Then he went on:

> [A] man in academic life has a great deal to offer a girl in the way of good associates, interesting occasions, and a relatively high degree of security of living. You have a chance to meet a good many nice girls in the future and they will not be unresponsive to the advantages of your position. You may think this is a very unromantic way of looking at things. It is. But when a girl marries she knows that babies and their care and cost are items she can not overlook. If she is the right sort she will want to be in love first; but she will also wish to know that the future is not going to be slavery, hard as she may work willingly to help both her husband and the children. . . . You will not know how much until later in life when you realize what you are aiming to do for children.[18]

Bowman's "life story" to Dr. Love and meddling marriage advice to Bob came in this fictionalized and gendered space, to which he attached all kinds of strings. He expected not only that his son would have a heterosexual relationship that ended in a lifelong marriage, but also that Bob's future wife would be like his wife Cora, glad to trail her scientist-husband in the world. She would tend to domestic affairs, raise children, and enter civil society in this way. In becoming a bourgeois geographer and a map man, Bowman's prejudices came out in epistolary form.

Five days after the Molotov-Ribbentrop Pact between Stalin and Hitler carved up Poland and Eastern Europe on 23 August, Bowman wrapped for Dr. Love another packaged tale about the U.S. Inquiry at the end of World War I. He spoke of Walter Lippmann (1889–1974), assistant to Wilson's secretary of war; Robert Lansing (1864–1928), who initially called for benevolent neutrality; David Hunter Miller (1875–1961), Wilson's legal adviser; and Sidney

Mezes (1863–1931), the director of the U.S. Inquiry and president of the City College of New York. Bowman emphasized the role of experts like himself: "In July [1918], when I left for a vacation in New Hampshire, Lippmann and Mezes entered the drafting room and looked at each of the maps in turn. They discontinued work on the 1:3 M. map of Europe, the one which, subsequently at Paris, was most in demand as a base map for Photostatting newly drawn lines and solutions. They did not understand the technique but were willing to make a decision about it!" Mezes was Jewish and Bowman abhorred him, regarding him as a schemer and an infighter. He was also the brother-in-law of Edward House, Wilson's campaign manager in 1916 and negotiator in Paris. Mezes, through House, once had the inside track Bowman coveted to Wilson's inner circle.

Then Bowman fondly recalled the U.S. Inquiry: "Shortly thereafter Lippmann joined the military intelligence division and went to France. This put him into action again. I understand that he prepared material for distribution by airplane behind the German lines. Mezes let things drift. There had been constant threats of resignation from members who had no confidence in him. I begged each complainant not to stop [working on the Inquiry]. The work was important and a change would come in due time." In Bowman's fable, he alone saved the day: "David Hunter Miller broke the ice. He talked for forty minutes (so he told me) with Colonel House . . . [He] let me know that he told Colonel House that Lippmann was no good for the job and Mezes had a paralyzing effect. He suggested to House that he consult me." After Bowman's lunch with House convinced him of his merits, "House asked at once, 'What's the matter with Lippmann?' I told him that it was hardly for me to say and that in any event Lippmann had entered the army and appeared out of it. I answered that I thought administration bored him." Bowman spun the last yarn: "Then he [House] asked, 'What's the matter with Mezes?' We both laughed. It wouldn't do to criticize his brother-in-law. He then said, 'Will you take charge of the Inquiry if I give you complete charge of men, money, and plans?' I replied that these three terms included everything and agreed to serve."[19]

By September 1939, Bowman's mythic world tied up his experience in Paris and East Central Europe with a personal American story of "men, money, and plans."[20] He held to the Wilsonian ideal of a man of intellect and integrity, despite the failures of the League of Nations, the havoc of the Great Depression, and a Europe now at war. To sideline Mezes, he performed the role of a Penck, a stateside scientist, the *homo geographicus*. Never a proponent of the New Deal or of ethnic diversity, the Hopkins president now gave lurid speeches to both Democrats and Republicans to peddle geography for the government. The prophet "spoke map" in a common post-1930s vocabulary of America's civilization against fascism and communism. Yet his stories to Dr. Love were also nineteenth-century fictions of region- and class-traveling mobility, out of the transnational world of frontiers. In America or Ostmitteleuropa, it was never quite a liberal education. Bourgeois progress for Bowman thus became a gendered tale of success from the U.S. West to the Eastern Seaboard, symbolized by his New Hampshire summer cottage and letters to Romer, Bob, Dr. Love, and anyone else who would listen.

Romer deeply shared in Bowman's dreams for geography, but by mid-1939, the obstacles he faced were far more formidable. For one thing, the Polish state subjected him to greater hassles. On findings by their own experts, Poland's Ministry of Religious Confessions and Public Education (WriOP) in Warsaw ordered a trial after they had determined which maps should be used and in what way as scientific supplements for pupils in Poland's schools. They elected to censor and prohibit the atlases and maps made by Romer's commercial Książnica-Atlas firm. The government ordered a judicial process to take place in Lwów, after Romer and a few others were accused of disclosing military secrets to the public. For the "crime" of map production, the accused would have to defend themselves; no participation of counsel was allowed; no evidence of any offence committed by Romer was provided. Romer appealed directly to the minister, Wojciech Świętosławski (1881–1968), a professor who called a conference, after negative publicity, in which Romer was allowed to participate. Somehow, the ministry's earlier decision was reversed, and charges dropped. Because the main purpose was intimidation, the memory of humiliation lingered.[21] Clinging desperately to his Galician home and maps, the Polish geographer published his *Advice and Warnings* (*Rady i przestrogi*) just before the war, a series of commentaries about the fate of Europe, akin to Teleki's work in 1934. Romer supported Józef Beck (1894–1944), the minister of foreign affairs, whose speech before the Sejm on 5 May 1939 asserted Polish "honor" and rejected the Führer's bellicose demands.[22] On 30 June, he wrote to Świętosławski with his plans for the "parceling out" (*rozparcelowania*) of lands to account for "ethnographic" Ukrainians in union with Poland.[23] The new geopolitics had real limits.

In August 1939, an aging Eugeniusz and Jadwiga Romer celebrated their fortieth wedding anniversary in Limanowa with their children and grandchildren. On the summer holiday, their respite was interrupted by the death of Janina, Eugeniusz's only sister, who lived in nearby Rymanów. She had been widowed seventeen years earlier. In the last week of August, the Romers left to attend her burial.[24] When the Wehrmacht invaded Poland on 1 September

1939, Romer anticipated the end of a former life. He was not optimistic. He reread Hitler's *Mein Kampf* and Hermann Rauschning's *Revolution of Nihilism*. The Galician geographer and veteran of the border settlements in Paris in 1919 and Riga in 1921 penned in his memoirs, "I had utterly no illusions . . . for what awaited all of humanity during this total war, and above all us . . . it seems for the first time in my life I felt utter powerlessness."[25]

YOU CAN'T GO HEIMAT AGAIN

Nazi propaganda maps typically depicted colonial living space (*Lebensraum*) in the east by the mid-to-late 1930s, but Penck's relation to the Third Reich is more complex than it would seem. Sometime late in the summer of 1943, an eighty-five-year-old Albrecht Penck began drafting his memoirs from his home in Nazi-occupied Berlin, then under siege by Allied forces. His wife Ida was in failing health and nearly deaf. Albrecht got shellfish poisoning, after which he was bedridden for three months. He reflected on a long career, how he found his *Beruf* and learned his profession. Penck remembered his youth in Leipzig, his father and grandfather, Fräulein Auguste de Wilde, who gave him a stipend, and his early mentors the paleontologist Karl Zittel (1839–1904) along with Alfred Kirchhoff and Ferdinand von Richthofen. He saw his adventuresome spirit and love of nature as German qualities he inherited from his father and mother. He reminisced about Munich, and how he had been the youngest professor of geography in Vienna during the 1880s. He professed a deep love of family above all else, his daughter Ilse (1886–1952) and her husband Armin (see below), the grandchildren who stayed with him in 1941–42, and his eighty-two-year-old sister Hanny, whom he called the "devoted nurse" (*hingebende Pflegerin*). Penck recalled their Alpine summer home, in the Bavarian forests of Mittenwald. He spoke of Ida's depression (but not his own) after their son Walther died of cancer in September 1923. He celebrated their fifty-year marriage as a profound love of landscapes, "And yet she still loves Mittenwald even more than me. How often she says, 'We still would have remained there!' "[26] As winter approached, the reflective explorer pined away for a place that was accesssible only in memory.

Here maps of home get murky. Penck's can be unclouded by the detailed exchanges he had between 1938 and 1944 with Sven Hedin, the pro-German Swedish geographer. Hedin, who backed the kaiser in World War I, also praised the Führer on the eve of World War II as part of his admiration for *Kultur* and support for Nazi expansion. These manly explorers pictured Europe's civilization out of a peaceful and prosperous North. Penck was proud of the fact that his *Besinnliche Rheinreise* (*Contemplative Journey along the Rhine*) was being translated into Swedish. He even tried (and failed) to read Hedin's travel works in the original.[27] Grateful for Hedin's friendship, Penck also clung to dreams of settlement in Germany's East and across Eurasia. He connected the "future of humanity" (*Zukunft der Menschheit*) to the "future of the German people" (*Zukunft der Deutschen Volkes*) in a colonial space.[28] Hedin was elated, for instance, when Hitler announced "reunion" with Austria. He wrote to Penck, "The last days' news is very encouraging. For the leaders and for Germany, it is again a great victory. *Anschluss* is only right and natural. Thus the power of Germany grows into some new armies." Hedin concluded, "It is amazing to see this labor of love: the devil of Versailles is being destroyed before our eyes. There's a justice in heaven, and the history of the world has a conscience!"[29]

Penck's supposed "return" to Mitteleuropa was a long-held German fiction, and a German historical fixation. He revisited the fictional unity of a homogeneous, premodern ethnolinguistic kin-state—the Second Reich as a kind of economically prosperous small town, and a racial village. He remembered the mystique of Bavaria: "Mittenwald is like a *Heimat* to me, and the city market recognized this in 1933 by making me an honorary citizen." The whole family came from Stuttgart and Prague in summer 1938. They gathered on the northern foothills of the Bavarian Alps, to celebrate the professor's eightieth birthday. Penck was feted, too, by the Berlin Geographical Society when it presented him with a bust in his honor. This time, the getaway among his German family and compatriots was the real, transnational-to-national home he treasured—in the summer retreat, he rewrote Mittenwald since 1914 as a site for his excursions as a geomorphologist. It was a refuge from the outside world. "All were present . . . and we all remember the joy within us," Penck recalled, as he compared the festivities to his sixtieth (in 1918) and seventieth (in 1928) birthdays. Coming to Mittenwald was a family tradition since 1934. He thought his grandsons Wolfgang and Helmut (without mention of his granddaughters) really loved being there, out in nature.[30]

In war and conflict, all homelands become borderlands, real or imagined sites of imminent danger. By now, Penck's frontier America was a distant memory. Penck referred to the "coming home" (*Heimfall*) of Austria (he taught at University of Vienna from 1886 to 1906) back to the Reich and the "return" (*Rückkehr*) of the Sudetenland as a result of the Munich Conference of 28–29 September 1938. In the fall, his daughter Ilse and her husband Armin Tschermak von Seysenegg (1870–1952), a doctor and professor of physiology, were living in Prague. Of mixed noble origin, Armin descended from a prominent

Bohemian family; his father Gustav Tschermak von Seysenegg (1836–1927) was a professor of mineralogy, a colleague of Penck's from his time in Vienna. In late Habsburg imperial academe, it is a fair conjecture that this was how Armin and Ilse were introduced. He and Ilse gave their five children very Wagnerian-sounding names—Elfriede, Hildegard (she died at age four), Hildegund, Inge, and Wolfgang. At the "hour of uncertainty" as Albrecht called it, Ilse and two of the daughters had come with Armin to stay with him and Ida in the safety of Mittenwald. When the German-Czechoslovak border railway was shut down, the family had to fly back to Prague. Penck associated foreign policy with his family's plight, how good the news was that Germany was "again strong and powerful." Returning to geography, he explained: "Despite the years . . . I feel more fresh enough to be able to continue to work scientifically. Work to me is the great joy of life, and I hope I can still accomplish a lot, which is still under preparation." Penck thought Hitler resolved the Sudeten question for German home seekers: "Hitler is sincere in his desire for peace, but he has also created a readiness for war. He had impressed this on the neighbors, and made it clear that the responsibility falls on them if they would make a sacrifice of ten million people because of three-and-a-half million Sudeten Germans. The meeting of the four statesmen in Munich in my opinion was therefore a major turning point in European policy. The old game of intrigue (*Ränkespiel*) is ending."[31]

Penck's fascist apologetics coupled with German anti-Slav and anti-democratic colonialism were evident when he recalled a root-searching trip he took in 1939 as a sentimental quest for self-definition. He spatially adapted his family's history into a generational saga of social uplift, "I went with the Pencks of Darmstadt to the land of our ancestors. In the Odenwald in 1939, I met with my sister Hanny in Freiburg in Breisgau. Together, we stayed one week in the area and another week in the Odenwald, sharing family memories. We did the same in Nordhausen. There I discovered the home of our Penck grandparents. We visited Ilfeld, where our great-grandfather had owned the paper mill."[32]

The trip also meant rejuvenation. He was unable to go to Mittenwald because he fell seriously ill in July 1939, and had to check into the Martin-Luther-Krankenhaus. For the first time in his life, he was hospitalized for an extended time. A remarakbly vigorous man, he now needed a prostate operation and had to stay in Berlin for eight weeks. When our map men confided in each other, Penck too began with the travails of his own body—he told of how he thought of the men fighting in the "first operation" of September 1939 while he was being nursed back to health by doctors and nurses, including his own sister.[33]

In a December 1939 letter to Hedin, he proclaimed the Germans' "quicker and greater success" as "essential to my recovery." Penck informed his friend in Stockholm, "Oh, the year has not been good for me. Twice I was seized by the flu. And then a bladder disease set in, for which an operation was needed. It was successful, but for nearly five months I have not been out of the house." He fretted about the Soviet "appetite" for land and thanked Hedin again for the "loyalty and friendship that you have always shown us Germans."[34]

Judging from available evidence, however, Penck's *völkisch* closeness to the NSDAP is harder to interpret. He was not a typical fascist or conservative, insofar as "typical" can be described. For one thing, he did not find his friend Hedin's nationalist support for Hitler's policy at all objectionable. Yet he noted that since 1930 he had been watching "the awakening of the Germans" as a "silent observer." He referred to a 1933 speech by Bernhard Rust (1885–1945), the Nazi minister of science and education (*Reichserziehungsminister*), how it caused him *not* to join the party. Penck by his orientation remained provincially bourgeois, both Prussian and enduringly Saxon.[35] At the Berlin Geographical Society in 1933, he lectured on geography and military cartography in the era of Frederick the Great, whom he praised. Penck noted that his "harmless" talks with German teachers and telephone conversations were monitored by Nazis "for some time" in the mid-1930s. Living in Berlin, the conservative academic even took Sunday walks with Friedrich Meinecke (1862–1954), the liberal professor of history and later author of *The German Catastrophe* (1946), which famously criticized the Reich (after the war). Penck did work stateside for the Nazis in 1940 on "questions of the destiny of European peoples," but, as he tried to explain away, "I remained outside the party."[36] What linked Penck to Germany was not the Führerprinzip but an ordinary academic service, the *Beruf* of a duty-bound, professionally mobile Germanophone expert, harking back to his Protestant roots and the nationalization of geography in Ostmit-teleuropa. In 1940, he collaborated on so-called Landesaufnahme projects in toponymy, as he had done with Hans von Beseler's commission in World War I. Similar to Rudnyts'kyi and Romer, Penck held to the notion that expertise could remain objective and independent of party politics. The German nation was transnational, too.

In history and memory, Penck's two world wars collapsed into one. "We want to finally eliminate the damages of this thirty years' war," he wrote to Hedin, "and to unify the nation together in appreciation of a Greater United Germany (*Grossdeutschland*)." He closed with family greetings from his wife Ida and sister Hanny, who were taking care of him during his prolonged illness.[37] He and Ida physically retreated, while they still could, to the idyll of

Mittenwald as the war raged on in the winter of 1940–41. As the geographer put it, they were "in the middle of the forest, enjoying the mountain winter."[38] Summer 1942 was the last time they saw their beloved Mittenwald. They were unaware that their grandson Martin, the late Walther and Aenni Penck's son, perished on the Eastern front. He was twenty years old.[39] By the time Albrecht reemerged in early 1943 and resumed regular exchanges with Hedin, Allied bombs fell near his home in Berlin. On 1 March, "three houses of the nearest neighbors" were burned to the ground. Penck told Hedin in April that he had just finished reading his book, *Amerika im Kampf der Kontinente*.[40] He was glad he no longer had to lie in bed all day. Though his eyesight was failing and he walked around with a cane, the globetrotter said he felt vigorous. He lamented the fate of the "German middle class" (*das deutsche Bürgerthum*) and compared attacks by the Red Army to the Spartacus Uprising (of early January 1919). He always personalized World War II through the places that were changed by World War I. Albrecht referred to the Tyrolean-Bavarian frontier life and fate of another relative named Penghofer, part of a regiment in Munich. Penghofer was forced to flee to Tegernsee, today a small spa town in the Bavarian Alps, which was freed by American forces in May 1945. Like the young Martin Penck, he was killed in battle, never able to return home.[41]

REVENGE

The years from 1938 into the early 1940s were also pivotal for Count Teleki's revisionist diplomacy in fascist-communist East Central Europe, where prejudice was inured with race- or class-based "science" and geography became geopolitics. On 12 March 1938, the Wehrmacht crossed the Austrian border. The Führer rode across the border at Braunau, the place of his birth in 1889, rendering the map's bold line obsolete. Three days later, on 15 March, Hitler officially proclaimed the Anschluss in the Square of Heroes (Heldenplatz) in Vienna. On 14 May, Horthy in Hungary handpicked Béla Imrédy (1891–1946) as the new prime minister. Imrédy, a mainstay of the Party of Hungarian Life, espoused the right-wing Szegedist idea of counterrevolution and called for national renewal; he had been appointed minister of finance in 1932 by Gyula Gömbös (1886–1936), the fascist leader and previous prime minister. Teleki tried to stay aloof from politics, but by summer 1938 this no longer seemed possible.[42]

In July 1938, Teleki accepted the post of minister of education and confessions. He embraced Imrédy's Catholic ethnonationalism and pro-British orientation under Horthy's regency. Still wary of Nazi designs, Teleki went along with Horthy and Imrédy, but Hungary could not afford to alienate the Führer. By the time the Third Reich annexed Czechoslovakia on the pretext of Aryan reunion with Sudeten Germans in September 1938, Teleki jumped at the chance for territorial revision—a testament to the depth of his emotional ties to homes in Budapest-Transylvania and experience of 1918–19. Even still, the Czechoslovaks, Romanians, and Yugoslavs of the Little Entente were deemed adversarial. Teleki supported the Hungarian delegation in Czech-Hungarian border talks with the Czechs at Komárno by providing maps and cartographic data. Hungary's Ministry of Foreign Affairs authorized him to negotiate with the Axis powers for the so-called transfer of Southern Slovakia to Hungary.

In Hitler's fascist Europe, Teleki supported the reversals of Versailles and Trianon. He failed, however, to consider that the delegation's aims were circumscribed by the Rome-Berlin axis (of October 1936) and the Berlin-Tokyo Anti-Comintern Pact (of November 1936), not to mention Hitler's anti-Semitic planetary aims. With the First Vienna Award of 2 November 1938, Hitler granted Hungary the "historic" St. Stephen lands in Southern Slovakia and Carpathian Rus'. In trying to keep Ferenc Szálasi (1891–1946) and the fascists of the Arrow Cross at bay, Teleki placed trust in the Euro-Magyar ethnocentrism of 1938, which followed again from the logic of his "Carte Rouge" and Paris in 1919, to develop a professional middle class that had been absent in Hungary's history. In this "between states" moment over the fate of Hungarian Jewry and the long history of Hungarian-Romanian relations in Transylvania, as Holly Case has pointed out, Hungary's Europhile elites became opportunistic collaborationists with proto-fascist leanings or fascist motives. Small-power elites had to balance out grand illusions all at once with priorities of economic development, religious conviction, and political survival. In his diplomatic quests for patronage, Teleki tried (and failed) to keep every channel open for territorial revision. His anti-Semitic balancing act in Hungary did not rule out an alignment either with "democracies" or with the fascist Axis when the moment arrived.[43]

On 16 February 1939, when Imrédy was forced to resign after his enemies learned that his great-grandfather was Jewish, Teleki accepted his second prime ministry. In March, his government under Horthy took advantage of Nazi territorial gains by laying Hungary's claim to the Carpatho-Ukraine region. Teleki's calculations this time were in line with a "historic" Polish-Hungarian fraternity as a bulwark of Christendom, the Polish-Hungarian border as Europe's moral frontier against communism. Catholic friendship with Poland, which reflected the tropes of nineteenth-century romantic nationalism, was instrumentalized for purposes of border revision.[44] Operation Barbarossa, targeting Jews and Slavs as inferiors in a colonial "Aryan" space, was no less deep and fanciful. (It was named for the twelfth-century German Holy Roman Emperor, Frederick I Barbarossa.) The Führer combined Teutonism with Ratzel's *Lebensraum* for Germany's colonial expansion and exploitation of the East. Yet the Führer mapped the Hungarian-Polish frontier differently, and racially, as an integral part of an anti-Semitic German civilization. Hitler approved of Hungary's revisionist aim to occupy the rest of Carpathian Rus', but he warned against the occupation of the rest of Slovakia since he planned to use it as a transit point. He looked to the Organization of Ukrainian Nationalists (OUN) in Western Ukraine and Southeastern Poland and collaborationists based in Uzhhorod; the OUN worked with fascists on the ground against Poland.

One might conjecture that the Führer himself probably reasoned back to 1914–21, that by using "inferior" Slavic stock as a means to an end and for the Third Reich in this way, ethnocentric Ukrainians would demand their own nation-state against the Poles. In effect, this cogrievance was shared with many German revisionists after World War I. Hitler's men would exploit resources in the newly built German empire to the east, while never allowing non-Germans nationalities to have their own territory, or history of East Central Europe. Teleki was too blind to grasp that he was being mapped *colonially* from an imperial Berlin in 1938–39, but in his defense, he could not read Hitler's mind. (Neither could Stalin.) The count was not privy to Hitler's surrealist map of Ukraine as Europe's breadbasket, integral to the Third Reich, or a fictive North American frontier space in which masculinized German colonials of a superior race "penetrated" fertile nature like a woman's body and settled across Eurasia's soils.

As late as spring 1939, Teleki appeared to believe that Hungary's elites could deal with the Führer and achieve some of their post–World War I goals. Advantages seemed to lie ahead. On 14 April, Hungary withdrew from the League of Nations, which it joined in 1922. When the Molotov-Ribbentrop Pact was signed on 23 August, Hungary stood to gain. But when Europe's interwar architecture collapsed, it fell quickly. Teleki watched in horror as Poland's fate was sealed by plans for a Nazi-Soviet two-front invasion. Representing Horthy's government, he averred Hungary's neutrality from the start of the Blitzkrieg, but by that point in the summer neutrality meant accepting Hitler's gains to the east. Teleki hoped to make gains against Czechoslovakia, Romania, and Yugoslavia. In so doing, he was left relating to the Führer on the terms of the "Carte Rouge" and St. Stephen space, or Trianon victimhood, refusing to allow Germany the use of its territory to launch an invasion of Poland. Teleki held desperately to a dying fraternity of Poland and Hungary through 1939.

Europe was a home that no longer existed. When the sole remaining option for an "honorable" geographer of the past century was to regain national losses of 1920, he no longer belonged to the transnational confraternity of map men.

SUICIDE

While Prime Minister Teleki pursued neutrality for a homeland in Hungary, he continued to aim for Trianon revision. The lifelong bibliophile of Transylvania regarded all of his maps, atlases, and books as almost sacred artifacts. Times were very hard: in correspondence with his friend Bowman, Teleki entertained the idea of selling or storing all of the maps he had collected in the wartime safety of a distant United States. An exchange ensued between Teleki in early 1940 and John Pelényi of the Royal Hungarian Legation in Washington, Bowman at Hopkins, the AGS office in New York, and Colonel Martin, head of the map division of the Library of Congress.[45] On 27 January 1940, Pelényi sent Bowman a letter from Washington, in which he spoke of Teleki's spectacular 1935 catalogue. Since Hungary was a unity, Teleki would only sell the library as a whole. When Pelényi sought Bowman's advice based on Bowman's thirty-year friendship with Teleki, Bowman was very flattered.[46] In a reply of 5 February, he expressed his "sorrow that he [Teleki] is compelled to dispose of it." He continued, "Without doubt there is a market for it but it is not the kind of market that existed in this country following the World War. On account of my long friendship with Count Teleki, I want to do everything in my power to help him in the manner proposed."[47] Pelényi told Bowman that Teleki had put the "fair price" at six thousand dollars—the previous price was six thousand Dutch guilders, the offer Teleki had refused from Holland.[48]

In a fascinating unpublished letter, Teleki made his pitch. On 12 February, he said that it was not just the money he was interested in. He included various geographic materials and described the country's travails as a deep spiritual struggle:

> A French friend of mine informed me that the French are earnestly at work to investigate and plan out the after-war situation and proposed to me to provide them with a general outline and other materials of information. Considering the matter of great importance, I dictated the outline personally. Now that this outline is ready I believe that reading it over would interest you since you studied so much the problems of Europe right after the first world-war. . . .

These world-war-times are quite terrible. I am nailed to Hungary or better said 23 hours out of 24 to this prison called the Prime Minister's Palace. It is still quite good that world-wars, economic difficulties and the planning-out of territorial readjustments are not only political but geographical matters and so I remain a little inside my own business. But it is very sad not to be able to meet old friends as usual and not to be able to get personally a glimpse of the outside world. . . .

I would like to see this whole mess finished soon. Not for my sake but for a small land it is hard to stand the pressures, especially economic ones for a long time. I have fears that Europe shall have a bad crop this next summer and that would aggravate the situation of the neutrals who are in a bad geographical position. I would like very much [that] the big ones would hurry up.[49]

Teleki equated life with his precious maps—"nailed to Hungary." He also appealed on the basis of a shared home, a shared past, and the common code of geographers. He signed off to his American friend with a solemn message, "Well, good-bye now, with very best regards from house to house and give me news."[50] Reading "house to house" back from English, literal Magyar would be "házról házra," suggesting the homes he had lost. It was an old-fashioned gesture of intimacy on the count's part.

In the first months of 1940, Teleki sent to Bowman and the AGS in New York a vast treasure of artifacts from his "home," with maps and texts in English and German. He expected Bowman to emerge as the leading postwar boundary expert in the U.S., able to influence the president. This may well have been a ploy to gain Bowman's attention, but it worked. Teleki took advantage of the moment to rectify the silences countered by Hungarian activists, to do what Masaryk in Czechoslovakia and Romer in Poland were doing for decades—spamming the West with maps. He sent articles from the *Hungarian Quarterly*; reports and works by Hungarians on the global dispersal of Magyars; "proofs" that stressed Magyar presence in Romania and Transylvania; and maps of Transylvania that he and András Rónai (1906–91), another of his protégés, had prepared. Teleki and Rónai in 1937 affirmed Magyar population density and its "different types of ethnic mixture of population," along with various other documents offering historico-geographic proof of Hungary's enduring economic vitality, political-ecological unity, and need for the rectification of its borders vis-à-vis the Little Entente.[51]

In March 1940, when Bowman acknowledged receipt of all the memoranda, he turned impassive. He vowed to read them with "the greatest care . . . and discuss [them] discreetly with others," but this was another self-serving truth.

In April, he took up the issue of Teleki's library and maps with John Kirtland Wright (1891–1969) at the AGS. He turned also to Colonel Martin at the Library of Congress.[52] Bowman wrote to Teleki that "we have a widespread desire to stay out of the war. . . . If England and France were to be in real danger we might come in again but the danger would have to be brought home very directly to our people. This does not mean that we shall stay out of the peace settlement. There is a strong inclination to do so but our economic interests are involved and a betterment of world trade is indispensable to our well-being."[53] In addition, Bowman wished it were possible "to sit down together and talk over the whole European situation . . . [to understand] the shortcomings of 1919 on the part of all parties involved in the settlement." He mused about losses on frontiers and the fears of a bad harvest, with concerns about the production of wheat and rye, and of soil moistures for the coming summer of 1940. He passed along good wishes to Countess Teleki, who had been ill. He ended by alluding to their families' "old life" and special bonds: "My wife joins me in sending best wishes to you and Countess Teleki. . . . May we hope that the storms of the moment will leave something of our old life still intact and that in a happier day we may once again exchange visits. I think of you almost daily—the terribly hard task in which you are engaged, the plight of your country, the hardships imposed upon neutrals by the big powers, and the uncertain outcomes that make planning of any sort a precarious business."[54]

The modes of male bonding did not stop there. Having gained the ear of FDR in the late 1930s, Bowman probed for more information. On 9 April 1940, he suggestively prodded Teleki for intelligence in the name of a scientific peace, as in 1919. He informed Teleki that Armstrong, the *Foreign Affairs* editor and "one of my most valued friends," hoped to meet him personally. He professed to identify with Teleki's bearings in Europe as a home: "We are intensely interested in all that goes on in Europe. Participation is another thing. For those of us to whom European affairs have been a preoccupation for the past twenty-five years the growing intensity is felt almost as keenly as if we were living in Europe. We cannot free ourselves from a sense of responsibility nor can we suppose that in any world that is to follow the war we shall not help to pay the costs in one way or another." Bowman played his "confidence" card again, assuring Teleki that Armstrong was a friend to Hungary. So, too, was the U.S. Council on Foreign Relations. "I have discussed with him the memorandum which you sent me, and I can assure you that you can talk as freely with him as you could with me. He will respect your confidences and upon his return I hope to hear that you have had a good talk with him," he wrote.[55]

Teleki profusely thanked Bowman and offered his opinion on Europe in two key letters sent from Budapest on 12 April and 20 April. Not until 3 June did Bowman receive the second letter in Baltimore with an accompanying note from the U.S. State Department (but no explanation). In the first letter, Teleki informed Bowman of the Hungarian Ministry of Foreign Affairs' intense gathering and publication of a vast and comprehensive collection of documents on Hungary's foreign relations since 1919, being distributed in the United States through Columbia University Press.[56] Neither Teleki nor Bowman mentioned Hitler directly.

The men zeroed in on the maps. On 16 April, Bowman asked his secretary to forward Teleki's correspondence and the 1935 catalog of his books, maps, and atlases, to Colonel Martin at the Library of Congress.[57] In the second letter, that of 20 April, Teleki aggressively chided the Little Entente and Romania's aims in Transylvania: "In South-Eastern Europe the League of Nations and the powers let time pass without trying to find and carry through a plan of sound settlement based on good-will. . . . Both the League and powers were deaf and blind to all oppression and imperialistic policy from the part of the enlarged small states."[58] As pressure mounted, Pelényi informed Bowman on 25 April that "owing to the present unfortunate conditions in Europe [the prime minister] does not feel that it is the proper time to attempt to dispose of his collection."[59] Teleki backed out, but not before he lashed out. Bowman received the letter six weeks late and passed it immediately to the State Department, ensuring that his government had an idea of what the prime minister was doing. Bowman then sent the catalog to Martin, who annotated it carefully and stored it at the Library of Congress (where it is held today). When Teleki decided not to sell his maps in mid-1940, both the issue and the friendship with Bowman were closed.[60] Bowman and Teleki had no further contact.

Individuals hid their emotions behind the overlay of maps, for maps were bonds that reinforced "friendship" and a threatened set of objective norms. In his second term as prime minister in the early years of World War II, Count Teleki clung to his maps as a mark of international diplomacy, right up to a fateful and tragic end. On 30 August 1940, Hitler granted the so-called Second Vienna Award to Hungary, by which the country "recovered" parts of Transylvania in interwar Romania. Lands included Szatmár (Satu Mare) County in which the Teleki family's estates were located before 1914. Hungary's gains represented a reversal of the Trianon treaty and a partial fulfillment of Teleki's revisionism. The victory was Pyrrhic. In December 1940, the prime minister signed a so-called friendship treaty with Yugoslavia, but by March 1941, Hitler insisted that Hungarian territory be used for Germany's war aims. The Führer reshuffled fascist Europe's map by promising to "transfer" parts of Northern Yugoslavia to Hungary, but Teleki's breaking of the Yugoslav treaty would put an end to Hungary's brief neutrality. Crucially, this breach of trust portended the demise of the count's credibility in the West, since Hungary would become a military outpost for the Third Reich's aggrandizement. Teleki's faith in a code of civility met its end against Hitler's outlandishness and his own nationalist anger directed at Trianon. His European revisionism, anti-Semitism, and Christian anticommunism combined in 1940–41 to doom him as a statesman of the interwar period.

On the early morning of 3 April 1941, Count Pál Teleki ended his life in Budapest. Many details are unknowable, but powerful twentieth-century myths and illiberal tropes of victimhood live on in today's Hungary.[61] It seems evident that he died with a profound sense of shame. The suicide took place shortly after Hitler directed the use and occupation of Hungary's "neutral" territory for the invasion of Yugoslavia. Having failed to chart out an independent foreign policy, the count's final note expressed the failures of the country ("nailed to Hungary," as he put it to Bowman in 1940) he embodied. It read both as a letter to Horthy and as a political confession with religious undertones:

> We have broken our word out of cowardice—in contradiction with the eternal agreement based on the Mohács peace [with Yugoslavia]. The nation feels it, and we have thrown away its honor. We stand on the side of villains because of trumped-up atrocities. No word is true, neither against the Hungarians, nor against the Germans! We will be graverobbers of bodies. We are the nation's trashiest. I did not hold you back. I am guilty.[62]

On 6 April, as strikes by the Luftwaffe began on Belgrade, the Nazis and their collaborating forces crossed from Austria and Romania in transit through a country of over nine million people. The most famous geographer of modern Hungary did not live to see the rest of World War II.

MANPOWER

Outside of Europe, Bowman's years in America from 1938 to the early 1940s read like the account of a man on a different planet. Prior to the Nazi invasion of Poland, Bowman served as an adviser in FDR's administration for the top-secret M Project, a position he kept until 1943, in which government-funded U.S. scientists starting in 1938 researched issues concerning migration and population resettlement from continental Europe.[63] Despite the efforts of Romer and Teleki, by 1939 neither Poland nor Hungary was at all high on the Anglophile Bowman's list of concerns for the United States. Privately in 1939–40 and before the Japanese attack on Pearl Harbor in December 1941, the Johns Hopkins president's binding concern was local and provincial. In private letters, he fretted day and night over his son Bob's career as a geographer. Bob Bowman increasingly came to represent his map-loving ex-Canadian father's dreams of U.S. success and power. In his first letter on 23 September 1939, Bowman recommended an elite path: "I doubt if we shall be drawn into the European war. If we do it will not be at once and just like that! I share your distaste for a front-line trench. I didn't raise my boy to be cannon fodder!"[64]

Bowman was an opportunist. He thought of the war in business terms, as evidenced by his missives in defense of U.S. interests. He wrote, "Whether the war be long or short, the cost of the preparations already made, and the treasure yet to be expended are so vast that taxes are bound to be yet higher. Whoever wins, in the long run America will help pay the cost of the war indirectly no matter what advantages in trade we may and I believe shall 'enjoy' for the time being. To that cost will be added the immense armament bill that we quite properly run up on our account to make sure that the war will not spread to the Western Hemisphere or that we shall not be drawn into it unprepared."[65] Bowman felt that academics across the sciences and humanities should be smart enough to invest in free trade. If wars made long-term profits for America's market empire, so be it. He wrote in a follow-up letter of 16 October, "The account that you read in the paper to the effect that a California man had left a million dollars to Johns Hopkins is correct. He was a professor of history for twenty-five years. Put his money into Dow Chemical Company when he was young and just watched it grow."[66] For the sensitive class- and region-hopping Bowman, maps of the past were about an enriched future.

Wars brought Bowman not just money, but a sense of purpose. At Hopkins, he used his connections to devise plans for advanced research in geography, a goal he shared with old transnational counterparts. He wrote to Bob at Berkeley on 9 November that "a department of geography of a new type . . . [would] involve appointments here of men like Joerg and Martin and one or two of the members [probably John Kirtland Wright and Gladys Wrigley] of the AGS staff. . . . There would be no classroom instruction, only individual guidance. . . . There would be strong support for the solicitation of money for geography . . . I still have hopes!" Sitting on the U.S. Council on Foreign Relations, Bowman's "new world" at Hopkins would shift the base for geography away from the AGS, the old European colonial institution, fully into the council with its government-related personnel and probusiness interests. He wrote to Bob and described these plans for academic geography as integral to an American war economy: "The AGS has very little money and the prospects are not particularly good, though I say this confidentially. I am helping them all I can through the Council to keep the Council interested at a high level and to encourage donations. The Council is interested and active and I am hoping for the best. . . . All this should be kept confidential of course."[67]

Just as East Central Europe crumbled, Bowman extended his vacation. On 31 July 1940 in New Hampshire, the patriarch-frontiersman wrote his annual circular to his two sons (daughter Olive was again excluded) from the family's summer home. He said, "I have just finished repairs on the old shack and I want to tell about it! Nothing much—screens renewed, shingles tightened. Enough to start the sweat and I mean sweat! Hot as blazes here the past week." He worried for Walt and Bob about the possibility of a draft, but managed some levity: "If there is a war we will all be swept in. Even old codgers of 61 like myself will register and defend home areas." He read through a "history of the Revolutionary War period." He even had a new suit to talk about: "An idea has just gathered headway in my brain-pan: to take my despised suit to Baltimore and there get a dress-maker to patch it thoroughly, each patch a different color. If it is to be a museum piece this will enhance its value." He signed off to his sons as the "chief tinkerer and supervisor [and] also the bringer-home of the bacon and the bathing suiter."[68] When Bob, following his father's advice, took on field research in Australia and New Zealand, he wrote of manly virtues from Christchurch, New Zealand. In a September 1940 letter, he spoke in a macho way of a conscripted friend named Andy Clark: "One understands under such circumstances what war means to the progress of a people. My only philosophy at present may be summed up . . . [as] 'Be tougher than the other guy.' Saying which, I spit in the waste basket."[69]

Back in the U.S., Bowman peddled Bob's career, his son's Walther Penck–like global travels as a star researcher, imagining him as vital to America's destiny. In October 1940, George L. Warren, the executive secretary of Roosevelt's Advisory Committee on Political Refugees, wrote a letter at Bowman's request. He introduced Bob and his credentials: "This note will serve to introduce Mr. Robert G. Bowman, geographer, . . . the son of Dr. Isaiah Bowman, President of Johns Hopkins University of Baltimore, U.S.A. Dr. Bowman is a world famous geographer and has organized and completed many important studies of the possibilities of refugee settlement in different parts of the world. . . . Mr. Robert G. Bowman desires to continue certain inquiries under the direction and guidance of his father, Dr. Bowman, particularly in New Zealand and

Australia."[70] To secure his son's career back in the States, Bowman even resumed writing to Ernest Horn in Iowa. In November 1940, Horn's colleague A. C. Trowbridge in the department of geology wrote to the Hopkins president, whereupon he name-dropped Horn and then asked Bowman if Bob could come back from his research by summer 1941 to teach geography courses at the university. Essentially, the job was Bob's without so much as an interview.[71] Bowman wrote to Trowbridge from Hopkins that his son was still engaged in fieldwork, having just finished his first year of teaching at Canterbury College at Christchurch. He could not resist self-promotion: Bob had studied under Sauer at Berkeley; he served as his father's assistant for two summers "on settlement problems in the western part of the United States and in the Prairie Provinces of Canada"; and he traveled widely and had been to Mexico and Puerto Rico (for only eight months). Bowman wrote, "[Y]ou can assume that he would accept the appointment unless he is called into military service."[72] Trowbridge wrote back to Hopkins with an immediate offer.[73]

As it turned out, in the name of the Bowman family Isaiah accepted the position for Bob before he knew anything about it. Only afterward did the father cable the son in New Zealand. "It is your decision and not mine," he dissembled. "Nor had I anything to do with securing the offer"—again patently untrue, given his correspondence with Horn back to 1938.[74] In December 1940, the father wrote to Horn and sealed the deal, "I am sure that Bob's decision to teach in the Iowa summer school was influenced very largely by your kind reception of him several years ago and your continued interest in him. I have just received his first field report and it sounds strangely mature! Youngsters in their twenties do move fast!"[75]After Bob defended his thesis at Berkeley in the last week of May 1941, his overjoyed father wrote, "Three cheers for thesis at Berkeley! Yes, that's the way—earlier work always looks and seems juvenile. But *diversity of experience built in* [emphasis in original] is what counts in one's twenties. You've had it!" He signed it "Love 'Pa' and Grandpa," in with reference to his first grandchild, whom Walter and his wife had just had.[76] Bob's career as an Anglophile German-American geographer, in the explorer's world of the great outdoors, was launched in this way. He would move to Iowa in summer 1941. When Bob published his very first article based on field research, it was in 1942 in the *Geographical Review*, where his father served on the editorial board.[77]

Back across the Atlantic, the Romer family was in desperate shape. The photographer Witold, who was working in 1940 at the Royal Geographical Society in London after having served in the antifascist Polish army in France, wrote in despair to his father's famous American friend. In a letter of 4 August, he told Bowman how he had been forced to leave the entire Romer family behind in

September 1939—his father, his mother (he did not know of Jadwiga's death), his wife and sister-in-law, his two children, and his brother Edmund's two children as well. "They are all in good health," wrote Witold, though he had not seen any of them in nearly a year. "But their situation is very bad. The information found in the newspapers, concerning the persecution of Polish people, and especially the educated class, are confirmed by the persons who get [out] from there and in other ways. The Polish population under the Russian occupation are deported into Russia, sometimes children are separated from their parents and everybody is living in constant uncertainty [of] what the next day will bring. My greatest trouble is now to get them out." He described Eugeniusz as "very depressed." He hoped for "diplomatic" repatriation, that "if an American university would like to invite my father (to offer him a chair) an inquiry could be directed to the Russian government concerning the permission for my father and his family to leave the country." In his father's name, Witold appealed to Bowman's old-fashioned sense of honor. He knew that Henryk Arctowski had been fortunate to get out of Lwów in fall 1939, after which he got a position at the Smithsonian (until his death in 1950).[78]

Bowman's reply was terse. "I am indeed grateful to you for word about your father," he wrote to London. "If you ever have an opportunity to communicate with him safely, please send him my most affectionate greetings. As for getting him out of Poland, the difficulties are enormous. Yet we shall not be dismayed by them. To help an individual leave the country has been found to be impossible after many attempts. What may prove possible in time is group transportation—scholars and their families—several private agencies are considering the problem. You may depend upon me to do everything possible when, and if, an opportunity presents itself or can be created."[79] Witold certainly could not have been pleased. True, his father was a renowned geographer, so any attempt to get him out of Soviet territory could not happen quietly. Bowman connected science with family, as Penck had done in 1914. His response showed the limits of "friendships" among the map men of Ostmitteleuropa. Friendships could not move oceans.

Bowman soon learned after April 1941 of the suicide of his friend Count Teleki in Budapest. When he wrote about it to Professor Domokos G. Kosáry (1913–2007) of Eötvös College, a modern European historian and one of Teleki's antifascist and anticommunist confidantes in Budapest,[80] Bowman chose his words carefully. Bowman described the count as "a very dear friend" whom he had known well, all the way back to 1912. Bowman sent a beautiful condolence letter to Countess Teleki, who had been ill after her husband's suicide for several months. "I do want you to know how deeply I feel about the events that must be closest to your heart and to tell you that Paul's friends will always keep in mind so long as they live a warm remembrance of his generous personality. You would have been delighted . . . to know how deep is our feeling for his character, what he stood for and what he did." His farewell was a twentieth-century prediction: "They are the things by which our spirits live. . . . One day Europe will be changed over into something better, and in the history of that time there will be a high place for Paul Teleki."[81]

After Japan's attack on Pearl Harbor in December 1941, Bowman became a kind of court geographer to FDR, a trusted adviser to the U.S. State Department. He shuttled up and down the Eastern Seaboard, as the White House invited him for consultation. In his scarce free time, he continued to pour his energies into Bob's future as an American geographer in the world. He wrote in January 1942, "Mother and I . . . both feel that putting your knowledge of Australasia to use is the important thing at this time. Very few people have such intimate knowledge based on field work and Australasia is now very much on the map. I am saying to the students here that the easiest thing to do is to enroll in combat service. The hard thing to do is to appraise one's own skills and make a real effort to put them to the best possible use. This is a war of intelligences."[82] Father and son exchanged ideas about the "Jewish question," as they both called it. Bob wrote in March 1942 that he liked the Aussies. He thought Australia with its "flies, heat, mosquitos and geographical position alone" could be a "grand place for settling the Israelites," whom he termed a "clan." Isaiah replied to his son that "to invite them into a country (as was once proposed for Alaska) and confine them in a given region, not permitting them to come into the cities later on, is to establish a modern form of Pale. . . . Neither Jews nor modern democratic governments would stand for that."[83]

One of the reasons Bob Bowman is so important to this story is how he brought the map man's intensely goal-oriented emotions to life. By spring 1942, the protégé secured a place in the American heartland, in the geography department at the University of Iowa. Thanks to Bowman's channels, he was slated to start on 8 September teaching courses in geography.[84] In summer 1942, however, he and his brother Walter were both drafted into the U.S. armed forces. Bob reported to the local draft board at Berkeley. He was designated 1A. He noted that the married men were being summoned in a big call-up. In August 1942, Bob passed his physical and was inducted officially, as Private Robert G. Bowman, first-class, out of Monterey, California. He was sent to Army Training Headquarters at Camp Ritchie (C-17) in Cascade, Maryland, in preparation for intelligence training and the European theater of war.[85] In the military, he made his father proud. By January 1943, he was already promoted to sergeant.[86]

In addition, by the start of 1943, Bob was only twenty-three. He and his father trusted each other with all their ambitions and strategies. Isaiah reported a "confidential" announcement to his son that "I have been asked to serve (and have accepted) on a committee of ten, nominated by the American Council on Education and appointed by the Army, to give over-all examination, correction, or approval to the curriculum of the students referred to several hundred colleges for training under the new Manpower Commission and Army-Navy regulations. We are to assist in the setting up of the 'Screening' process by means of which students will be selected for assignment. . . . So all this as one Army man to another!"[87] Isaiah's perspective was not without racism or anti-Semitism: when he sent Bob an "article on Geopolitics" in February 1943, he referred to the French geographer Jean Gottmann (1915–94) as "a Frenchman and a Jew." (The Frenchman was actually born in Kharkov during World War I, then part of the Russian Empire.) Gottmann was "an agreeable and a quiet person." Isaiah told Bob "confidentially" that since he needed 500 students in the Army Specialized Training Program by June 1943 and was in "need of a couple of additional geographers," he would include Gottmann as the sole Jewish member and not include another one, the future UCLA geographer Henry Bruman (1913–2005). (Born in Berlin, Bruman moved with his mother to Los Angeles in the harsh years after World War I.) Isaiah wrote to Bob, "All this I write for your information and for any contents if you have time to write me and have any light to shed on Bruman. I understood you to say that Bruman is not a Jew. One of the new men could be, as for example Gottmann, but I do not want two of them in the same department."[88]

Isaiah in spring 1943 responded with more harping after his son complained, as all our map men did, that his expertise was being ignored by the state or not utilized properly. He aimed to get Bob into a position in wartime intelligence or higher education, or both. Bob wrote, "Lately I have given a good deal of thought to the problem of my immediate future. Camp Ritchie, I have at last realized, is no place for me. Despite earlier suggestions, I have not yet been given an opportunity to reach . . . I know more about mapping than most of them, but I don't have those bars on my shoulders!" He mapped his father's vices as virtues, in the hope that "you can pull the necessary political strings to get me into something better."[89] Isaiah pressed his son up to another level. In an exchange in May 1943, General Milton A. Reckford of the Third Service Command in Baltimore told Isaiah, "Give me his number and record briefly and I will get him out of there."[90] Astoundingly, the old man's clout worked. The military transferred Bob in September 1943 to Fort Belvoir, Virginia, into the Officer Candidate Regiment. Bob thanked his father and was delighted to have a "job" for the first time in his life (outside academe).[91] Bowman contacted Richard Hartshorne, the OSS director, and even got him to talk with his son. When Hartshorne replied, politely at first, that this was against regulations, Isaiah found a loophole. Hartshorne wrote in October 1943, "It will be impossible for me to call Bob, because it is against the orders of the O.S.S. for one of its members to contact a member of the armed forces with regard to a transfer from the service to this organization. However, if he could call me . . . it will be satisfactory."[92] "Pappy" got results and boasted, "Here is Hartshorne's response to my suggestion that he call you . . . You see we remain very much of a family, and that's what makes life worth living."[93]

Though geopolitics is real and had global effect, all geopolitics is local. The end to the telling father-son story was that the Polonophile-judeophobic Bowman, averse to helping out Witold Romer or Poland's citizens or the government-in-exile, went out of his way for his kin. In November 1943, Bob from Virginia reported to his father happily that he and Hartshorne carried on a fifteen-minute conversation. He told his father that Hartshorne "hopes to get me to Australia, and later, points north. If not there, then New Delhi, Chungking [Chongqing] or possibly Europe . . . He promises action, plenty of interesting work, opportunities, and possibly danger. Which just fits the mood of RGB, who aims to crowd as much into life as his famous pappy, if possible." To impress his father, Bob covered the entire earth as a site for U.S. power: "If I go to Australia, he said I'd probably be attached directly to MacArthur's Headquarters. Maps and photos would enter into the work, and field excursions would be likely. Sounds as if he is thinking mainly of reconnaissance, and liaison between Army HG and the O.S.S. . . . I'll be delighted to get into something really useful and interesting when I leave here. Army instructing is dull and to a great extent childish because most recruits are childish—intellectually if not physically."[94] Bob informed Isaiah that "the big cut came yesterday . . . [and] nearly half the men were dropped . . . they lacked 'punch', 'drive', 'speed', 'experience', or something"—all self-evident Bowman qualities.[95] While Professor Penck's other pupils were arrested, deported, repatriated, or killed, the American father and son generated twentieth-century norms and myths that were still complemented by the codes of a confraternity of geographers.

CONTEMPLATION

Places and maps are layered with meaning. It is hard to ignore the coincidence of the ethnicized and disentangled Penck and Romer, having severed their relationship in 1934, penning parallel memoirs from Berlin and Nazi-occupied Lwów in 1943–44. Romer hid in a Catholic monastery and was lucky to survive the war. He told his life story as a Polish struggle and referred to "victims of Poland" (*ofiary Polski*). By this, he principally meant a lived mid-twentieth-century collective experience—the failures of Versailles and Locarno, the destruction of Europe, or Poland as a proxy for Europe, between Hitler and Stalin, the German annexations of Austria and Czechoslovakia in 1938, Poland destroyed in 1939.[96] In Romer's micronarrative, the geographer focused on a single street near his house on ul. Długosza 25. His loss under foreign occupation represented an end to Poland's civil society. In fall 1939, he kept up a regular schedule of secret meetings in one of the old university buildings. His "quartet" of experts regularly met, comprising Kazimierz Bartel (1882–1941), Franciszek Bujak (1875–1951), Stanisław Grabski (1871–1949), and himself. On 22 September 1939, the whole city of Lwów was lost to Soviet occupation. In spite of Grabski's arrest by the NKVD on 27 September, Romer continued the meetings at the same house from which his two sons had just fled. More Polish geographers came to see Romer, including Julian Czyżewski (1890–1968), Józef Wąsowicz (1900–1964), and August Zierhoffer (1893–1969). There was also Jan Treter of the Książnica-Atlas firm, now confiscated by the Soviets. These men listened to BBC Radio at 1700 and news from Ankara at 1900, during which they held anxious and lively discussions. Romer fretted that the war could last many more years. Poland's future was unclear.[97]

Romer's life on ul. Długosza 25 and in Soviet-occupied Lvov in 1939–41 bore all the shock and stigma of clinging to home in a city under occupation.[98] After the Nazi-Soviet Pact of 1939, he lost his academic pension. For this comfortable middle-class Polish family to avert ruin in Galicia, it had to sell off works of art, jewelry, and furniture. Jadwiga, who underwent her first operation during treatment for thyroid cancer in 1934, had a relapse first in 1937 and again in early September 1939, right when Witold and Edmund escaped. Jan Karol Glatzel (1888–1954), a well-respected Polish-Jewish oncologist in Lwów, made house calls for her chemotherapy. Romer's once cordial relations with Soviet scientists and geographers came under suspicion in Stalin's Europe. By 1940, he was made into the figurehead of the Cartographic Institute in Soviet-occupied Lvov, the position Rudnyts'kyi in Soviet Ukraine after 1933 was not allowed to retain. Romer worked on a new socioeconomic atlas of Stalinist Poland with nearly 100 maps on agriculture, forestry, mining, manufacturing, horticulture, communications, and demography on behalf of the Soviet war effort. His role was to draw, seamlessly, the newly incorporated Soviet political territories. He cooperated with the Ukrainian Academy of Sciences in Kiev to determine economic regions (*raionirovannie*) and bring Polish geography into conformity with Marxist economic geography. He opposed measures to use Russian phonetics and toponyms, instead preserving Polish place names. It was prepared, but destroyed in Operation Barbarossa. Into the spring of 1940, even while learning of Soviet deportations of Poles into the Russian interior and Siberia, Romer had to devote at least three hours per day to Soviet cartographic labors. In most of his other hours, he cared for his wife at the family home.[99]

Sometime in spring 1940, Jadwiga Romer suffered the final relapse of her cancer, and her husband had to admit that the chemotherapy no longer worked. His beloved Matusia, as he called her in the dear way, died on the night between 30 September and 1 October 1940. Romer fell into a period of mourning and deep reflection. He lost interest in work for the rest of the year and cut off all remaining contacts, even with his friend Bowman. He turned inward and re-treated to his library of classical antiquity he had accumulated in the 1930s. He reread Aristotle's *Politics*, Plato's *Republic*, and the writings of Xenophon and Demosthenes. He found solace in St. Paul's epistles and Augustine's *Confessions*, Christian mystics, and other patristic literature. He who grew up agnostic in a positivist era of liberalism and rationalism in Austria-Hungary now sought out a priest at St. Michael's (sw. Mikołaj) in occupied Lvov/Lwów, Franciszek Janicki, through his personal secretary Róza Skrochowska, a longtime friend of his mother in the family's multigenerational household. He also met with Father Jan Stepa (1892–1959), the last dean of the department of theology at Jan Kazimierz University, and Father Jan Nowicki (1894–1973), a professor of canon law, to undertake regular theological studies. To offer an analysis of Romer's transference of energy (psychiatrists call it cathexis), one might imagine that the geographer "remarried" by taking Catholic pieties into the troubled, intimate space of his old Lwów home. In a sense, the church became his new Polish bride. This "ex-home" was one of the precious few things he had left.[100]

When the Nazis commenced Operation Barbarossa on 22 June 1941, Romer was forced into hiding. The city was occupied by the Wehrmacht on 30 June. The Nazis redrew Poland's sovereign territory into the General Government and "Distrikt Galizien." Romer narrowly escaped being among the first victims of Nazi terror against civilians, when a group of twenty-five Polish professors were arrested and shot on the city's outskirts, in Wzgórze Wuleckie. Edmund seriously thought his father was dead. In the son's memoirs of 1985, he did not hesitate to associate Penck with the NSDAP and referred to the German by name as the "master and denunciator (*mistrz i denunciant*) of my father since 1916."[101] For his father's asylum, it was Jan Ciemniewski (1866–1947), another priest-activist in the city, who first suggested the Resurrectionists' Monastery (OO Zmartwychstanców) on ul. Piekarska 57 as the venue of this desperate choice. Eugeniusz Romer entered the monastery as a "postulant" (or layman) with the assistance of yet another priest named Kowalski, who agreed to hide him. He kept this secret even from his own family until after the war. The timing was fortunate, because the Wehrmacht started to bombard Lvov on that very afternoon.[102]

Under Nazi occupation in December 1941, Eugeniusz wrote to Edmund, informing his son that he was still alive and had delved into religious studies.[103] In the city's monastery, these studies were also of maps. They entailed the kind of disciplined work he did in advance of Paris in 1919 and Riga in 1921. Romer also prepared geographic reports for the London government-in-exile. In one written in 1942, "The Ruthenian Question in Poland's Past," Romer argued that *sanacja* in the Second Republic, premised on the idea of attracting individuals from different regions of Poland to foster strength, had the opposite of the intended effect and led to attenuation of affinities to the Polish nation and language. Romer's ethnocentric target, then as before in World War I, was diversity and the Ukrainians: he referred to the Polish-Ukrainian cooperative experiment of Henryk Józewski (1892–1981) in Volhynia as a "holocaust" (*hekatomba*) and a "denial of state authority" orchestrated by the Ukrainians against the Poles.[104] He revisited the Polish state census as Penck once returned to the Prussian census of 1910, to prove the Polishness of the minorities of the southeastern provinces. He accused his *own* state of falsifying economic statistics in 1931 and hiding results from the public. Poland's government, he concluded, had "no courage" to listen to its geographers, or address the most pressing issues of land reform vis-à-vis Ukraine. He even compared *sanacja* to Prussia's anti-Polish policies and Germany's colonization in the East, in partitioned Poland before 1914.[105]

In this manner, Romer in the monastery went back to Ostmitteleuropa geography to turn modern Polish politics around. He criticized Nazis and Soviets, but without once using the word totalitarian. He called Hitler's foreign policy "reckless" (*karkolomna*) and compared it to the land invasions of Russia by Charles XII of Sweden and Napoléon. He commented on the deportations of Poles in February 1940 into the Soviet interior, the Central Asian steppes, and the Siberian hinterland—as is now known, including the killing of some 22,000 Poles at Katyń.[106] In a subsection called "No Negotiations with the Bolsheviks," he declared that despite Stalin's propaganda, the Soviet dictator was neither prepared for an "offensive" in contested Lemberg/Lwów/Lviv/ Lvov in September 1939 nor war with the Wehrmacht in June 1941. Though he criticized the Second Republic and blamed Nazis and Bolsheviks for the end of Poland, Romer reserved the greatest antipathy for Ukrainians in his home city. "It does not matter with whom they ally, either the Germans or the Bolsheviks," he wrote, "because their only goal is to leave Poland in ruins." He demonized "Rusini-Ukraincy" as a mass of opportunists who executed Poles, in collaboration with *either* Nazi or Soviet rule.[107] Romer's frames for World War II were shaped profoundly by the memory of World War I and the Polish-Ukrainian War.[108]

Romer intended the long report of 1943 on the Eastern situation to get to the Polish government-in-exile in London, via the geographer Józef Wąsowicz. By then, his sons Edmund and Witold were living in London. They arranged for their father to be compensated with a small salary of 600 złoty per month, barely enough to pay for costs of living. Bujak visited Romer secretly in the monastery in September and October 1943. The Catholic geographers coauthored a study called "Poland and Western Europe" for a London audience. Romer argued that Poland was integral to *Western* Europe by its history, physical geography, and culture, while the anticommunist Bujak pointed out the contradictions of an atheistic USSR and cast it fully out of the West of Christendom. The geomorphologist Czyżewski also visited Romer, urging the former professor and his friend to take on the role of scientific expert again.

But this time in crisis, Romer turned less to science than to religion. Romer in the Nazi-occupied city penned his autobiography as a conversion tale and a struggle for Poland's future. Having read his Rauschning, he saw fascism as a nihilistic revolution. This, reasoned the Pole, came out of the peculiarities of the "German soul" (for he knew something about that). He announced that the war epitomized "the conflict of man with God" and called for a "spiritual revolution."[109] He labeled "the Moscow Bolsheviks . . . a monstrous giant standing on feet of clay," who left Lwów in ruins.[110] The Poles, Romer claimed, had fallen victim to all the traps of modern European politics. The geographer referred *back* to the high Polish romantic culture in the nineteenth-century, Adam Mickiewicz (1798–1855) and Zygmunt Krasiński (1812–59).[111] He associated Jadwiga's death in 1940 with Poland's resurrection. Using poetic language such as *grom* (thunder) and *srom* (a double entendre meaning ignominy, also slang in Polish for a woman's vulva), he appealed to Krasiński's 1846 "Psalm of Repentance" ("Psalm Żalu") part of "Psalms of the Future" from the depths of the monastery. Romer conjured these geosophical lines: "When Polish history befalls / Murder and ignominy! / Better the thunder! / It shall be resurrected from the thunder! / Let it be reborn from ignominy (the vulva)!" (Kiedy w

polskie spaść ma dzieje. / Mord i srom! / . . . Lepszy grom! / Zmartwychwstaje
się spod gromu. / Nie zmartwychwstaje spod sromu!)[112]

Romer now turned inward—the man and his maps, the man into his maps.
He blamed the Nazis, Soviets, Allied powers, *and* the Poles themselves for Poland's demise. He returned in March 1944 to his major themes of 1938–39 in the
lengthy pieces, "The Pacification of the World" and "Everyone Is Responsible," now as a Catholic convert.[113] He appealed to a shared Christian humanism
by pointing out the mistakes of science and that all the belligerents had been
blinded by self-interest. He argued the war's "main culprit . . . was rampant
individualism, materialism and sensualism of a philosophical and vitalistic
(*życiowy*) kind," and that modern politics lacked a "relationship of man to God,
man to man, and nation to nation." Romer referred to the pacifist Friedrich
Wilhelm Foerster (1869–1966), a philosopher and Nazi opponent who fled the
Third Reich in 1933. He adapted Foerster's criticisms of German policy from
Bismarck to Hitler, for the "Teutonic-Prussian deformation of the German
nation . . . the mission not as a nation/people (*narod*) but as a nation of super-
men (*nadnarod*)."[114] He protested against Germanophilia in England and the rise
of anti-Semitism in Poland (more on this in chapter 7), though he apparently
never thought of the Final Solution as a central, or even a principal, concern.
Finally, the widowed Romer in 1944 concluded that Poland belonged in its
spiritual, physical, and historical geography to the West and a Christian Europe.[115] These emotions—his prejudices and convictions—were all his own.
Inside the Resurrectionists' Monastery, trying to stave off arrest and certain
death in Soviet, then Nazi, then Soviet-occupied Galicia again during World
War II, Romer in Lwów thought on the spatial terms of stateless Poland's lost
nineteenth-century Atlantis and twentieth-century Atlantic world.

———

Homeless in his own city, Romer's monastic geography in 1943–44 transposed
his life of maps into a new Ostmitteleuropa key—not modernity but a kind
of medieval science. He eventually came to believe, as Teleki did after Trianon, in his special mission and Polish Catholic conversion. Eugeniusz Romer's
"ex-home" tale is a mission in maps, but where Bowman's story collects
dust in an archive, Romer's persists in an illiberal Poland. "My mind and heart
turned to the knowledge of God," Romer reflected, and "my love did not fade
for the nation . . . [or] the old Polish emblem, 'God, Honor and Fatherland'"
(Bóg, Honor i Ojczyzna).[116] He appealed to dogma and prayed to the Holy
Spirit as his "helper . . . light and power."[117] He blamed Nazi-Soviet geopolitics

abstractly on a hidden enemy, the secular world, and he faulted those who failed to heed his late 1930s warnings. The god-seeking "postulant" Romer, at the age of seventy-two being sickly, penitent, confident in his rectitude, twelve years after his formal retirement as a professor, came back to his life and expertise. He worked and prayed over his beloved maps. Habits die hard. He threw himself back into geography and cartography, just as he had done during the First World War.

CHAPTER SEVEN

TWILIGHT

The twilight of our geographers' lives and careers lays bare their passions, prejudices, and struggles. During the Second World War, families were torn, friendships permanently severed. Many of our men, despite fraternal codes of honor, never spoke to each other again. They wrote for the drawer, but they also wanted to lead a civic life for others. By the early 1940s, Penck's epistolary confidante was Sven Hedin, while Romer without Bowman was left with no confidante at all, first under Soviet, then Nazi, and then Soviet occupation. Modern wars sorted individuals into hierarchies by race and nationality, reducing characters to pixels and lives to national histories. Geographers' long-held universalized bourgeois aspirations collapsed into emotional maps—Penck's "Volks- und Kulturboden," Teleki's "Carte Rouge," the Galicia of Romer and Rudnyts'kyi, the Bowmans' new-worldism. Displacement bred frustration and isolation, which fostered solipsistic fantasies and searches for life's meaning on a global chessboard of civilizational clashes and geopolitics. Subplots included the writers' own defenses of their life choices, survival tales of war, losses of homelands, gendered mappings of their families into legacy narratives. Trans-Atlantic dreams emerged from symbolic places of East Central Europe, layered sites of deep continuity. These were not just catastrophic spaces, battle fronts or borders, or lines drawn on a grid, but the symbolic landscapes of mountain retreats, familiar streets, monasteries, tourist spots, and summer homes.

A DRIVE TO THE EAST

In late November 1943, Penck left his home in Berlin for good. He worried about his family and collection of scientific works. He kept all his notes, notebooks, photographs, and letters in boxes in the Berlin apartment at Maierottostrasse no. 5. During the Allied air raids, the windows and doors of the apartment in Berlin-Wilmersdorf were damaged beyond repair. On the weekend of 28–29 November, with neighbors huddling into basements and bomb shelters, Albrecht and Ida and his sister Hanny were aided by their relative, Gastel Ganghofer, whom Penck called their "guardian angel." He would help transport them to the hospital at Hindenburg-Stadt in Upper Silesia, then from there to his daughter Ilse's house in Prague. Writing two weeks after the escape, Penck graphically detailed his ride across an apocalyptic urban landscape. He lapsed into the amateur style of the *Heimat* writer and his World War I writings. "It was horrible" (*greulich*), he said. They passed the burning Wilmersdorfer Stadthaus, and then over the Rankestrasse. Everywhere there was debris. Thus the traveler's lament, "Hardly a house was intact, most had burned down, many were windowless as on Maierottostrasse . . . on Kantstrasse here and there were intact houses, but also many that were completely destroyed. . . . So we carried on to Adolf-Hitler-Platz in the west end of Berlin, where the most beautiful large houses were burned down." They turned into Thüringer Allee and arrived at the Hildegard-Krankenhaus, a hospital, also bombed out. Penck found a room with three beds and a small table. No washbasin was there. The window siding had a hole in it. Then Gastel rushed off for his own safety and headed home. "This became our refuge," Penck recalled. Penck tried to make the best of it. There was at last some time for a meal. "A quiet Sunday followed," he remembered, "and we were fed very well."[1]

On 1 December 1943, the Pencks began their final journey to the East. Transports were arranged for men and women, separating Albrecht from Ida. A stoic Albrecht wrote a week later to Ilse, explaining that Ida was being cared for well but that she was "often apathetic" (*oft apatisch*) given the harsh strains.[2] Penck further wrote from the Hindenburg-Stadt hospital on 10 December. "I could not read the station name from my berth," he recalled. "Where we passed I heard Kottbus . . . in the early morning we were in Breslau, then in Oppeln."

He mostly preferred to give German names for cities, but definitely he knew where he was. "In the Upper Silesian coal fields there were long delays . . . ambulances . . . individual travelers. We stayed for three hours . . . in Gleiwitz. We arrived in Zabrze and I found a special room for Ida and me at the hospital. For three hours I waited in vain for Ida, when at last she was brought in." Penck clearly intended to leave this legacy of his life for his children and grandchildren. The geographer ended his tale of danger with a family reunion, as he had done in *Detained by England*. Germany was nature, the center of his map and the world, where his wife Ida remained.[3]

In 1944, Penck penned three final letters to Hedin in Stockholm. These unpublished sources and another written to Prof. Dr. Eduard Spranger (1882–1963) in Berlin shed valuable light on the last full year of his life. Penck recounted the family's dramatic escape from Allied bombs and the ruins of Berlin. Home was a settled, longed-for space. He informed Hedin that if he came to Berlin, he could no longer stay in the Kaiserhof because it was completely burned down. The family's flat on Maierottostrasse no. 5, though it still stood, had its windows blown out. Having packed only essentials in their suitcases, Albrecht and Ida were transported to the hospital in Hindenburg-Stadt with about two hundred others. He spoke of "our people" who have "lost everything" in the war. He praised Nazi efforts for "peace." In his mind, German militarism was never at fault. "We are all looking forward to the new weapons and [Joseph] Goebbels holds out the prospect that we will go far afield to the last blow, until our opponents finally can be defeated," he wrote. "We do not lack iron resolve, but the number of our enemies is very large. I would like to see our victory, as this sustains me in my isolation!" He thanked Hedin again for his "love and loyalty" and passed along greetings to Sven's sister Alma.[4]

By the time Penck next wrote to Hedin in April 1944, from Ilse's house in Prague at Budweiserstrasse 80 in the 14th district of Ober-Reuth, Ida had passed away. As Eugeniusz Romer had described Jadwiga's cancer relapse, Albrecht Penck recounted in 1944 how Ida became sickly right after being forced to leave their home in Berlin. She had been in a state of shock, in fact. Albrecht's sister Hanny—"my Alma [Hedin], so to speak"—was now caring for him. He received some good news about his other grandchildren and grandchildren-in-law on the Eastern front. (Recall that Walther's son Martin, his grandson, had died in 1942.) Son-in-law Armin helped to arrange for care in Prague, ensuring that Penck was close to the university. The geographer reminisced about the *Wissenschaft* he and Hedin had shared. He wrote about Richthofen, whom he had succeeded in 1906 at the University of Berlin, and Krebs, his successor in 1926. He lamented the end of the Geographical Institute and the Oceanographic Institute in Berlin. He ended the letter there: "These are perhaps the last lines that I can ever address to you. Let me therefore say how much I appreciate you as an explorer and as a friend! The era of Sven Hedin will never be forgotten in Berlin."[5] In May 1944, Hedin wrote back in sympathy but drifted into geopolitical litanies: the defense of "Grossdeutschland"; the "meanness and brutality" of air raids carried out by "Anglo-Saxons"; Soviets behaving "like Mongols" at Katyń and Vinnitsa; the "gangsterism" of the Americans, always in the guise of democracy. Penck's last letter to Hedin in July 1944 described the Eastern front through a multigenerational lens. "We fight on three fronts now," he wrote. "My grandson Helmut is at Normandy. Wolfgang fights in northern Italy with his malaria. One grandson by marriage was wounded on the front in Southern Russia, and he spent a week in the hospital. My brother-in-law is in a navy squadron on the Baltic Sea. My daughter is caring for me, and for that I feel very grateful."[6] This was how the two explorers' parting exchange ended.

In Penck's 10 August last letter to Prof. Dr. Spranger, his friend and former colleague in Berlin, he stressed the importance of geography in German education. He was worried especially about the fate of his collected scientific work. Spranger was Penck's own "Dr. Love," a professor of philosophy and education who could be counted among his admirers. On the value of topographic maps for the study of land forms, he emphasized how geography was "the study of the surface of the earth (*Erdoberfläche*) in the sense of Ritter and especially given the influence of Richthofen." Any geographer worth his salt had to develop empirical observations from excursions and professional training. A German cosmos emerged again: "For a long time, I have juxtaposed Geography as the *Wissenschaft* of the present to History, but what we see and describe belongs to the past. The current vanishes too quickly from sight. Today I am rather of the opinion that everything is in flux in space and time, which are not sharply separated." He finished with a flourish, his vision of a colonized German earth, "the *Heimat* of mankind."[7]

Then sometime in December 1944, Penck came down with a violent fever. His daughter Ilse wrote that he retained his vigor and interest in the war to the very end. On 7 March 1945 at the age of eighty-seven, Albrecht Penck died of natural causes in Prague-Reuth (Praga-Krč). Four days later, on 11 March, she wrote in confidence to Sven and confessed how often her father spoke of him. He "always felt a great honor and pleasure in your friendship," Ilse wrote. Penck's daughter bid a farewell to her father and the family's German romance in its transnational life and solemn death: "Until the last hour he was full of love and concern for the Fatherland, and he had the honor of serving it with his greatest happiness and pride. He is to be laid to rest in Stuttgart, beside my mother and brother."[8]

"BEFORE DEATH PLUCKS MY EAR"

From his office in Baltimore at Johns Hopkins University in 1943, Bowman threw himself fully behind the American war effort. He mostly left the everyday management of the university to his provost, P. Stuart McAuley. The veteran of Wilson's U.S. Inquiry consulted regularly with FDR's State Department. He was elected president of the American Association for the Advancement of Science. This was another career irony, for Penck in July 1914 received an award from the British Association for the Advancement of Science. (The BAAS, modeled on the Gesellschaft Deutscher Naturforscher und Ärzte, renamed the British Science Association as of 2009, had been set up in Leipzig in 1822, near Penck's birthplace.) Isaiah half-joked in a letter to Bob that "the Army feeds [us] well."[9] From August to October 1944, Bowman advised the U.S. delegation at Dumbarton Oaks on drawing blueprints for the creation of a new United Nations in place of a defunct league. He kept reminding Bob, who by then was serving in the Air Corps on the Pacific, that the "job [in Iowa] is waiting for you."[10] Bob, who addressed his father as "Pappy," "Papacito," "Paterfamilias," and the "Fixer-Extraordinary of Boundaries," was glad to play along.[11] When Isaiah fell ill with influenza in January 1945, it prevented him from going to Yalta in February. "If I were a free man I'd write 12 books before death plucks my ear," he proclaimed.[12] In March 1945, Edward J. Stettinius, FDR's secretary of state since December 1944, sent Bowman a letter personally inviting him to be part of the U.S. delegation for the founding United Nations conference in San Francisco from April to June 1945.[13]

Behind Bowman's string-pulling were the frontiersman's anxieties about power and kinship, race and sexuality, reflecting both his love for his sons and the multigenerational prejudices of his transnational life.[14] Bob, now promoted to captain, wrote in April 1945 that he was not only becoming a man but also the "General of the Bowman Armies." He informed Isaiah, "Congratulations on your San Francisco assignment, SIR! Great Stuff. Give 'em hell, and don't surrender an acre of land that we have fought for and won from the little yellow illegitimates. We shouldn't be too sensitive about hurting their economy or just who owned this land eight-six years ago. I write this to you instead of my Congressman because I think you will carry the ball farther."[15] In July 1945, the patriarch boasted to both his sons, "During the last three weeks at

San Francisco I was responsible for preparing a report on the Conference for the President. This and other matters brought me into touch with the President [Truman], first at a reception and then for a consultation in his suite. . . . I was obliged to organize a team of 27 writers, and an editorial committee of seven that met for an hour and a half every morning."[16] On 4 August, two days before the U.S. dropped the atomic bomb on Nagasaki and five days before Hiroshima, Isaiah wrote to Walt at the headquarters of the Fifth Air Force. He asked facetiously of his elder son, "Are you lost in the jungle? Or are you assembling the armada that is to invade Japan?"[17] Upon Bob's arrival in Yokohama, Bob sent a chilling note back to Baltimore. Describing Tokyo on 8 September, he doubted that two million of the original seven million people were left as survivors in the city. Isaiah shrugged it off, entirely. He deflected his feelings back, as his mentor Davis had done, to the old Penckian assignment of the Ivy League, to mobilize the geographic sciences in fall 1945 on behalf of President Truman's postwar government. Bowman went on figuring out the flow of commerce and populations transnationally, assuming from the Eastern Seaboard that the U.S. military and American business interests harmonized globally with the rest of the world.[18]

REPATRIATION, IN PLACE

By early 1944, Witold and Edmund Romer in London were trying desperately to rescue their father. They simultaneously prepared to ground the family's place in a changed postwar world. Irena, Witold's wife, and their children Andrzej and Jan were brought to Jodłownik near Limanowa. Krystyna, Edmund's wife, took their children Maria and Tomek to stay in Żerisławic, in the Myślenicki district. In late March 1944, Romer committed to fly to England, a move that required a false identity document. He prepared for the journey and even made maps. One was based on an idea suggested by a report of his old friend Arctowski, the Polish-American scientist and U.S. Inquiry member. Henryk thought in 1919 to generate a "who's who" map of 100,000 famous Poles from the *kresy*, based on birthplace, in contested territories such as Prussia and Galicia which the Central Powers had claimed. Romer derived the data from a small encyclopedia by Erazm Piltz, and he passed it to Robert Lord and then Isaiah Bowman, to show the Polish names of some 100,000 inhabitants in imperial Russian provinces, in Habsburg Galicia and lands of the "Prussian-Polish administration." To emphasize continuity before 1914, he presented "outstanding Poles" to "prove" not only that Poland belonged in Europe, but also that Poland's landscapes were unique to the continent.[19] In a sense, this memory project of *German* emotional politics became a *Polish* bourgeois passion: he emplotted Poles by place of birth, province, village or town, or "abroad." He divided their professional careers into sixteen categories: (1) humanistic sciences, (2) religion and philosophy, (3) natural sciences, (4) technical and applied sciences, (5) fine arts, (6) music, dance, and theatre, (7) sport, (8) industry and trade, (9) military, (10) politics, (11) jurisprudence, (12) finance, (12) civil administration and local government, (14) ecclesiastical affairs, (15) agriculture, and (16) forestry. He wrote up the text as "The Spiritual Structure of the Polish People and Regional Differences." Romer saw it as his personal legacy.[20]

Romer left the monastery on 3 April 1944, bound by train for Warsaw under the assumed name of Edmund Piotrowski. He brought some belongings, reports, and "spiritual" work for a pilgrimage to the West.[21] In Romer's few remarks on Jewish mass murder and the Holocaust, he continued to think obsessively of Poland's restoration and place in the postwar order. He referred to the "bestiality" of German and Ukrainian murderers of Poles and Jews; they, in his mind, were mainly to blame for Polish and Jewish deaths. He viewed the extermination of Polish Jewry as a weakness of will and moral character, of Poland's enemies *and*, more controversially, by the Poles themselves. He emphasized that the aims of Hitler's army were not only to kill Jews, but to export "murderous sadism" from Warsaw to Volhynia, Podolia, and Lwów, where "Ruthenians" were willing collaborators. He thought as a Polish Catholic that there should be no place in any postwar order for any manifestation of anti-Semitism, Polish or otherwise.[22] By summer 1944, the man called Piotrowski was moved at the urging of the Polish Home Army. He survived the journey, made it to the vanquished country's capital, and witnessed part of the Warsaw Uprising.

On 10 July, Romer's maps and diagrams were transmitted by courier to produce multiple copies. These were lost or destroyed during the war, or simply never returned to him. Romer's "cultural" maps never saw the light of day.[23] Yet the geographer's value in Warsaw and in London as a scientific expert did not diminish. Piotrowski lived on the Powiśle, on ul. Leszczyńskiej 71 in an apartment on the first floor. Soviet defeat of the uprising made life more dangerous.[24] By late September 1944, he tried to move under the alias again, this time to Kraków. The Red Army captured the expert and detained him at the camp for political prisoners in Pruszków, on the outskirts of Warsaw. Somehow, he got passed back eastward. Details here get very fuzzy: in January–February 1945, his daughters-in-law Irena and Krystyna left Jodłownika for Kraków with their possessions.

The first place they stayed was with their aunt, Olga Bukowska. Krystyna was able to find a three-room apartment on ul. Konarskiego 31, and the daughters-in-law and their children lived all together in one room. With the Yalta Conference going on from 4–11 February, the family home on ul. Długosza 25 in Lwów was finally lost to the Red Army's occupation. It was a crushing blow, but Romer was able to have his correspondence, his manuscripts, and some furniture (the items not sold off) transported from Soviet Lvov/Lviv to Kraków, with a working desk, beds, bookshelves, and part of his library. It seems that the new home on ul. Konarskiego 31 was furnished especially for him. Romer used the large room all at once for his work space, library, and sleeping quarters. The third of the three rooms was occupied by Wiktoria Kudła, Romer's housemaid, and her daughter Stefa. In the Polish gendered spatial romance of a settled postwar, posttraumatic home, the two women nursed and cared for the widowed and displaced professor until the end of his life.[25]

When Edmund Romer recalled his father's account and plight, the story that emerged was the scientific, political, and religious struggle of a patriotic man, on a messianic journey to save Poland's future by sanctifying its past. Like Teleki's Budapest disciples, Edmund too was the product of a certain national-religious age. He valorized his father's suffering, marked the family's professional dedication to organic work, and associated Eugeniusz's life and work in a hagiographic sense with geo-bodily struggle. Physically, it was partly apt. By early 1945, the septuagenarian had such a severe stomach ache, possibly an ulcer, that he began vomiting heavily and had to be hospitalized. After the stressful years of hiding, doctors now recommended that he stay permanently in the new Polish People's Republic (PRL). None of this prevented Romer from going back to work. In summer 1945, he consulted with Grabski, the interwar minister of confessions and public education, who pursued Polonizing policies in the 1920s. Grabski, a cofounder of Piłsudski's PPS in 1892 who had switched political allies to Dmowski's National Democrats (Endecja) before World War I, had been the architect of the so-called Lex Grabski of 1924. It was he who had aimed to eliminate the teaching of Ukrainian language in schools of the Second Republic. He represented yet another connection back to Romer's "anti-Ukrainian" years and pre-1914 Ostmitteleuropa orientation.

Like many in the transplanted Polish Galician intelligentsia, Romer never reconciled himself to the geopolitical loss of Lwów in Stalin's USSR. In July 1945, he agreed to take part in the fifty-six-person scientific council for the so-called Ministry of Recovered Territories. Grabski, arrested by the NKVD in 1939 and released, since 1944 was lobbying from London together with Stanisław Mikołajczyk (1901–66), Poland's de facto prime minister, for Stalin to "return" the city to the Poles. When they ran up against Stalin's aims at Yalta in February 1945, Grabski shifted again to work with the communist Wanda Wasilewska (1905–64), in support of an ethnonational Poland. Such nationalism was easily instrumentalized by and for communist Poles in the more homogeneous Poland of 1944–45, to take advantage of Stalin's policies to achieve the ends of Polonization and create a Poland for the Poles.[26] Romer, like Rudnyts'kyi, was neither communist nor Leninist nor Stalinist, but Grabski saw the value of experts and encouraged expat Polish experts, who now could freely apply his anti-German (or "antifascist") and anti-Ukrainian prejudices to resettle permanently out of Lvov/Lviv. Thus Romer and his sons Witold and Edmund would be prized by Stalinist science, never objective *Wissenschaft*, if they headed West into the new communist Poland (PRL).[27]

So the geographers' trajectory followed a familiar logic. The seventy-four-year-old Romer "repatriated" to Kraków in 1945. In true Habsburg fashion, unlike when Rudnyts'kyi left for Soviet Kharkov/Kharkiv in 1925, Romer's repatriation was an act done in place, from a country he had never left. At Jagiellonian University, he returned to the position of full professor from which he had retired in Lwów in 1931. He directed the geography department there, reintroduced seminars, and aimed to train graduate students as new experts in the many subdisciplines of geography.[28] Romer researched safer subjects such as geomorphology and climatology, getting back to his training. He took rejuvenation trips to Rabka, the health resort town between Kraków and Zakopane. He could no longer write about political geography as he had done in the Second Republic. Witold and Edmund, also having lost their Lwów permanently in 1944–45, likewise decided to repatriate from London, to work as scientists and academics in Poland's new frontier West. All three transnational Romer men thereby came home to an "ex-home" with their changed lives, families, and careers in post-1945 Europe.[29] Their new nation-state was anything but free of tension. Their time and place was never a Stunde Null.

A MULTIGENERATIONAL AFFAIR

Given that most of Teleki's correspondence has not survived, we are left with a lot of guesswork about his interior family life. The prime minister's suicide in Budapest in 1941 left behind a corps of disciple "sons," but no discernible American sympathizer or West European ally in professional geography. The count's most prominent followers were left in Budapest, colleagues and former students such as Ferenc Fodor and András Rónai. Bowman conveyed intelligence he gathered from Teleki to the U.S. secretaries of state, as in a "Confidential Memo of the Council on Foreign Relations: Memo on Hungarian Claims and Policies" prepared by Philip E. Mosely, based on a summary of the revisionist aims offered by Teleki.[30] Teleki's Transylvania legacy and relationship with Bowman were carried on after 1941 by his only son, Géza (1911–83). Like Walther Penck, Géza Teleki became a geologist in the Ostmitteleuropa tradition. With the Second Vienna Award in 1940, he was able to find a position as a professor of geology at the University of Kolozsvár (Cluj), in the very same place from which his father's dear geographer-friend Cholnoky, an expert in hydrology, fled from the Romanian army in 1919. Not too much is known about the inner dynamics of their father-son relationship, but Bowman and the younger Teleki wrote in summer 1945 after Hungary's authorities appointed Géza the new minister of culture and education (in 1944). Right away, Bowman invoked the AGS excursion of 1912 and spoke of it fondly: "Your father, Count Paul Teleki, was one of my earlier European friends. We traveled together for six weeks in the United States in 1912. Our correspondence continued until the end. I treasure especially his last letters." He passed along regards to Géza's mother and offered to help personally with Géza's "career and success."[31] For his part, Géza all but begged Bowman for help from the United States, mentioning "all that I have heard of you during my childhood from my father." He signed his letter in December 1945 "as my father's true son and follower in thought and mind, to turn to you as my father's friend and ask your support in the name of our country."[32]

In spring 1946, after the communists dismissed him, confiscated his estates, and stripped him of his "count" title, Géza asked Bowman for help to repatriate from Hungary. He reasoned he had little choice but to leave. Teleki the Younger called him a "good friend" of his father Pál. He appealed "to get an invitation to one of the United States University [sic] in the degree of an assistant beside a professor of economic geography or economic geology . . . I would be glad to work on any scientifical institute where my knowledge of Central Europe can be used and, in meantime, I could study for several years the scientifical progress and the life of the U.S.A., which I am deeply interested in."[33] Now that the transnational Transylvanian expert was a landless aristocrat without a homeland, the early Cold War geopolitics of 1946 dictated what happened to him next. The subsequent Paris Peace Conference from 29 July to 15 October 1946 bore some similarities to that of 1919. The U.S., USSR, Britain, and France reached territorial settlements with Bulgaria, Finland, Hungary, Italy, and Romania. Treaties by the victors carved up postwar Hungary, established war reparations, ended Italy's colonial claims, allowed for Hungary's place in the UN, and provided for the rights of nationalities (a change from 1919). The victors also drew lines for the borders of Italy with France and Yugoslavia, of Romania with Hungary, Bulgaria, and the USSR, of Finland with the USSR, and of Hungary with Czechoslovakia.[34]

The conquests of Hitler and his collaborators in executing the Vienna Awards of 1938 and 1940 were nullified, but shifts in ideology meant new opportunity. "I am not proud of my professorship in Hungary, nor to have been minister of public education," Géza wrote, sharply disavowing communism. "Both were duties I performed with my best will and knowledge. I never wanted to be a politician." These could have been his father's exact words. Géza promised to work hard in the U.S., to improve his English and "prove . . . [my] ability in a society before one gets to merits." He enclosed a short bio and résumé of his scientific work as a geologist. He referred to the Second Vienna Award of 1940, which "reattached" Northern Transylvania as "my homeland" and "brought a change to my life." Yet he was silent both about Trianon in 1920 and the Führer and his father's foreign and domestic policies. Absolving himself of anti-Semitism and his father's tightrope diplomacy twice as prime minister in 1920–21 and 1939–41, Géza fashioned a nineteenth-century story of geographic science directly to a fellow map man of Ostmitteleuropa. It was a Europe-to-America tale of progress Bowman could relate to.[35]

Flattered in the usual way, Bowman spoke with Philip M. Hayden, secretary of the board of trustees at Columbia University, to find for Géza an honorary (i.e., unsalaried) visiting lectureship that would begin immediately, on 1 July 1946.[36] Bowman vouched for Géza's character, what he could provide for the United States. In fact, however, he and Géza had never actually met. The younger Teleki was "said to be a man very much like his father, a man of wide knowledge and good personal character."[37] Bowman described to a skeptical Charles H. Behre, Jr. (1896–1986), professor and head of the department of economic geology at Columbia, how he and Pál Teleki had been "intimate friends."[38] Then he covered his tracks by asking Géza afterward for personal information about his character and family life, to "let me know when you would be free to come and whether you are married and would bring your wife with you."[39] Géza divulged that his "beloved mother . . . died June 9, 1942 [from] shrinking of the kidneys. She could not support the death of my father. So did my grandmother on the father's side . . . who died October 28, 1941. My only sister is married in The Hague to a Dutchman."[40]

What was came next was a bombshell. Géza confessed a secret, "I am married now for the second time and have four childrens [sic]. My divorced wife and three children live in the country, where they have an estate of 300 acres left, which means very much today in Hungary, as it is the maximum a family can retain after the land reform. They will stay on their estate in Hungary in any case. My second wife and my boy of three years, out of this marriage, would accompany me to the States in the case I can manage to get a job there. But until I can settle this question they would remain in The Hague at my sister's house." Teleki went on that Columbia "would give me the possibility to enter the States." He then admitted, "I have lost everything in Transylvania and my wife too, I could not get on without a salary." Géza anxiously and very politely asked "for a fund or endowment as a research fellow (perhaps through Carnegie or Rockefeller Foundation)." He hoped to settle his private affairs so that he could come to the United States between February and April 1947, and he thanked Bowman for his "benevolent support and advice."[41]

It was this red flag that made Bowman lose his patience and back away. He handed off Géza, suggesting that he contact Dr. Joseph H. Willits (1889–1979) at the Rockefeller Foundation.[42] By November 1946, Behre wrote back that "we do not really need him on our staffs [at Columbia]."[43] For Bowman, it came as a relief. He wrote to Willits's secretary at the Rockefeller Foundation in November, enclosing Behre's letter. He stated, "Since I brought Count Teleki's name to the attention of Dr. Willits it appears that Count Teleki's family arrangements are complicated by a divorce and re-marriage, but his desire to bring his family to the United States and by other circumstances that make it impossible for those of us who knew his father and wish to help the son, to continue to do so."[44] He wrote to Behre in agreement at Columbia on the same day, with emphasis, "I would drop the whole business."[45] The Hopkins president even wrote further to Dr. Paul Kerr in the Columbia geology department, "I am afraid that if we encourage Count Teleki to come to this country he will be very much on our hands. I have therefore written to The Rockefeller Foundation that the complications seem too many and too great to warrant our further encouragement."[46] From America to communist Hungary, Bowman raised, and then disappointed, the hopes of another Transylvanian in search of an open society and frontiers. It was a doubly lost tale of Ostmitteleuropa as an oceanic saga, in which transnational geographers' entangled personal lives, science, and professional service persisted into the early Cold War and Europe's post-1945 reconstructive years.

FREUNDE UND FEINDE

By early 1946, Bowman at Hopkins dreamt of setting up a cutting-edge international institute for scientific research in geography. He declined an appointment to the U.S. Atomic Energy Commission, in part to pursue such aims. The dream never came to pass. America's early Cold War years were full of scandal, confusion, and self-serving ambition, with oversized expert personalities in politics and diplomacy, civilizing missions, fears of communism (again), and accusations in academe such as those at Hopkins directed against Owen Lattimore (1900–1989), the liberal history professor and China hand who was later a target of Senator Joseph McCarthy.[47] Bowman really did nothing to defend Lattimore, or other faculty. Another major affair was the shutdown of the geography department at Harvard, Bowman's alma mater, in 1948, which he was glad to assist (for his own reasons). President James Conant (1893–1978), an accomplished chemist, argued that geography was insufficiently scientific and not a valid university subject. Scant resources ought to go elsewhere. The human geographer Derwent Whittlesey (1890–1956) chaired the department, and he got on the wrong side of Conant (and many others) by hiring his partner, Harold S. Kemp, who hardly published a thing, to be a lecturer in the department.

The grand explorer tradition returned again. Conant also had to deal with Alexander Hamilton Rice, Jr. (1875–1956), a geologist of this type who married the heiress Eleanor Elkins Widener. She was formerly wed to George Dunton Widener (Widener Library is named for their son, Harry), who died aboard the Titanic in 1912. Rice, the founding director of the Institute for Geographical Exploration at Harvard, essentially bought himself a professorship in geography. Then he refused to teach courses so that he could go off on expeditions. Resentful of Bowman's role in the AGS, once upon a time Rice rejected one of Bowman's books, *The Pioneer Fringe*, for use as a textbook. In defense of a discipline ruled in a not-too-distant past by the likes of his father-mentor Davis, Bowman the class-traveler was glad to be on the external advisory committee that got rid of the entire Harvard department. He described the Harvardites to Jean Gottmann, the French geographer whom he had fired the previous year at Hopkins (Gottmann became the director of research at the United Nations in 1946–47), as "a bad bunch of men."[48] Gottmann, the same person whom Isaiah called "a Frenchman and a Jew" in an earlier snide letter to Bob, was now being used by Bowman in order to trash talk other people. Beginning in May 1948 at Hopkins, a search began, conducted by the board of trustees, for Bowman's presidential successor. A lonely figure in the end, he finally stepped down in December 1948, succeeded by the biophysicist Detlev Bronk (1897–1975).

Bowman's career ended with mixed results. At a moment when geography met its catastrophic disciplinary ends (at least temporarily), he took consolation in the U.S. with his family and epistolary European bonds. He claimed publicly the same transcendent objectivity in the service of presidents and the business of academe as he claimed in Paris in 1918–19, to be above party politics and bickering. In 1947–48, Bob Bowman and his newlywed wife Jean just had their first child, a baby girl named Barbara. Barbara was Isaiah and Cora Bowman's fourth grandchild. (Walter and his wife Erna had three children already.) Isaiah talked Bob out of taking a position with the Air Force, at Air University in Montgomery, Alabama, for he wanted his son to stay in academe and develop his "passion for research and independent field work."[49] The proud man was disappointed by how his professional life ended. "The University," he swore, "was more important than anything I could gain by arguing in politics." In his mind, he acquired this unique trait from his experience with FDR, when his "social acquaintances were against [the president] and the New Deal almost to a man."[50]

Isaiah's last and biggest private project was Bob, whom he counseled to stay objective publicly, but also to keep confidences. While Bob served as chair of an internal committee in Iowa, he addressed his "Pappy" jovially with a barrage of "titles falling my way in future: Director, Governor, President, Generalissimo, Fuehrer, Emperor, etc. etc., paralleling those long since acquired by my illustrious ancestor, Isaiah, Prophet of Baltimore."[51] Bowman warned Bob about the American Association of University Professors (AAUP), "a place where professors can blow off steam," and coached his son deviously, "The surest way to quiet a group of professors is to give them something to do. They do not like it. They complain of committee work because they are not interested in it and often do it badly. The secret of administration of such a group is to invite their opinions and little by little have them feel that the policy that you desire to establish sprang out of their own brains. This is another way of saying that the job of administering such a group is political in the good sense of the word. . . . You keep them happy while recognizing that left to themselves policy disappears and each man works for himself. One of the jobs of administration is to listen to professors belittling each other."[52]

If Bowman's sad twilight arrived in 1948–49, it was the frontier myth that sustained his world map and remained most powerful for his imagined sense of authority.[53] Isaiah looked forward to his retirement and seeing Bob, Jean, and his granddaughter Barbara more often. In a letter to his son, he alluded to the imminent excursion: "In another week I start West! Westward Ho! Westward to Barbara! . . . Tell her that I am an explorer and show her a picture to prove it. Here it is! . . . No, this is not Don Quixote, nor is it the nag Rosinante! It is Isaiah as he would be now if he tried to ride a mule in the high Andes. I have other pictures, taken in my youth when I was erect, had a flat belly, was svelte, and wore a beard like the conquistadores of old!"[54] (See figure 7.1.) Hoping to become a "free man," he would travel to Iowa City with Cora for the family reunion in October

FIGURE 7.1. Photo of Bowman and his "nag" in the high Andes before 1914, one of his favorite collectibles. Courtesy of Special Collections, The Milton S. Eisenhower Library, The Johns Hopkins University, Baltimore, Maryland. Isaiah Bowman Papers (IB-P Ms. 58), Series I, Box 3, Item 10b.

1948.[55] When he retired, it was much to the relief of the Hopkins faculty. On New Year's Eve, the board of trustees threw Bowman a fancy gala, with his family and the Baltimore business community in attendance.

Once his comfortable retirement began in America in 1949, Bowman grew reflective. He assembled his private papers and remembered his cherished

Polish friend Romer. On 3 June, he sent a letter to Kraków in customary style. He thanked Romer for an article on climatology in Poland and reminisced about the geographer's outdoor life: "I read your paper at once and now understand better the different climatic zones of Poland. This, however, is of less importance than your own good self. I trust that time and events have dealt not too badly with you. . . . Presumably you have a house full of grandchildren. At least I have four. . . . How pleasant it would be to sit down and have a long chat with you. I still hope that this pleasure may be ours one day." Romer was moved by the gesture. He wrote about Jadwiga's death in 1940, sufferings during the war, his theological studies and conversion, and his scientific pursuits. What endured above all was his "childish nature, filled with reveries." He hoped to send "larger dissertations with many maps."[56] It was one of the last letters Isaiah Bowman received, and across the Cold War divide, this was the last contact the two friends would ever have.[57]

On 5 January 1950, after one of his customary fourteen-hour workdays, Isaiah Bowman had a massive heart attack. He died at 8:20 AM the next morning, 6 January, at Johns Hopkins University Hospital in Baltimore, with his wife Cora by his side. Eugeniusz Romer (figure 7.2) survived all our other men, a hard thing to fathom given Poland's fate and occupation in the course of World War II. The party-state in communist Poland never revived Romer's Książnica-Atlas firm. The Geographical Institute that had borne his name was relocated to Wrocław. In 1950, during the late Stalinist years, the firm was changed to the State Cartographical Publishing Company (PPWK, or Państwowe Przedsiębiórstwo Wydawnictw Kartograficznych). Romer's topics pertaining to Poland's borders no longer could be political, so he reverted to his training in climatology and geomorphology. After his eightieth birthday in 1951, he received the Commander's Cross and Star of Order of Poland's Rebirth. In 1952, he became a member of the newfangled Polish Academy of Sciences. He even outlasted Joseph Stalin by one year. On 28 January 1954, Eugeniusz Romer died in Kraków of natural causes, survived by Witold and Edmund, who came "home" from Wrocław and Gliwice to pay respects and settle affairs. Unable to return to the old Galician Lwów, now redrawn into Soviet Ukraine, Romer's body was interred at Salwator Cemetery in Kraków, where it remains today.

AFTERLIVES

How can we write history as transnational biography, to investigate subliminal texts, peel back the layers of places, detect feelings, bodily pains, family

FIGURE 7.2. Caricature of Romer by an unknown artist c. 1919, sent to Bowman and saved privately with his Paris Peace Conference papers. Courtesy of Special Collections, The Milton S. Eisenhower Library, The Johns Hopkins University, Baltimore, Maryland. Isaiah Bowman Papers (IB-P Ms. 58), Series XIII, Box 10, Item 142a.

dynamics, musical moods, ineffable prejudices, blurred emotional worlds? Stepan Rudnyts'kyi in Ukraine of the 1990s and 2000s came to be regarded as a hero, a founder and scientist who struggled for geography as a discipline, all for the glory of a very fragile nation and territorial state. It is hard to deny that narratives of grand struggle remain important. In ex-communist spaces, they can be genuine and consoling, but there are all sorts of holes in them. Then there are the families, archives of lived experience and untold secrets, socially performed roles, repositories of oral history and myth. Emilia Rudnyts'ka, the only surviving daughter of Stepan Rudnyts'kyi through World War II, settled after the war in Lvov/Lviv. She was married, widowed, and married a second time. She lived well into her nineties. From the thaw in the 1950s through glasnost' in the 1980s, and well beyond in independent Ukraine in the 1990s, she wrote to Soviet authorities for information in patriotic attempts to rehabilitate the patriarch. The heroic impression that emerges from Stalinism is that she too was tireless in her quest to learn about his disappearance and the whereabouts of his remains.[58] Soviet authorities partly rehabilitated Rudnyts'kyi in 1955, fully in 1965. Despite the family's efforts, not until October 1991 was it known precisely when the Ukrainian geographer was shot by the NKVD, or where his body was buried in 1937.[59] Where discipleships get to take place, whether inside countries or expat communities, in "reproductive" institutions or academic climes that push research, achievements of scholars supplant tensions of spatial biography. The case of Rudnyts'kyi is hardly uncommon.

Eugeniusz Romer's two sets of memoirs were censored until the late 1980s, when they were published by the independent Catholic firm, Znak. If Polish scholars make the error of lining up the German Romer into a Polish canon, the geographer has been seriously neglected in the English-speaking world. In the first detailed history of Wilson's U.S. Inquiry, published in 1963 by Lawrence Gelfand, Romer's work was not mentioned, despite the fact that his base maps for East Central Europe's postdynastic reconstruction were among the most circulated ones in Paris in 1919.[60] In his life, Romer never saw his memoirs in print. He thought to call them "Felix Culpa," or Fortunate Fall, in the spirit of a theodicy.[61]

Romer's sons and grandsons kept up the Galician traditions of Polish organic work from between the three nineteenth-century empires into the twentieth and early twenty-first centuries. Witold graduated with a degree in chemistry from Lwów Polytechnical University in 1923; by the early 1930s, he was a leader in Poland in the fields of photochemistry and phototechnical engineering. He studied lithography in Paris and introduced new methods for map reproduction at his father's firm Książnica-Atlas in Lwów and Warsaw, years before Walter Benjamin's "The Work of Art in the Age of Mechanical Reproduction" famously appeared in the Frankfurt School's journal. By 1935, he earned his doctorate in physical chemistry, and he headed the new department of photography and photomechanics at Jan Kazimierz University in Lwów. With his talents in image reproduction, he worked for the Kodak Company and Royal Air Force in Britain during World War II. In bumpy continuity from Galicia to London to Silesia in 1946, when he emigrated from

London to Wrocław, he set up Wrocław University's department of technical chemistry, and the newer Wrocław Polytechnical University's department of phototechnics. As a professor, he received the Prize of the City of Wrocław in 1957 and Commander's Cross Order of Polish Rebirth in 1965. A lifelong agnostic unlike his father, Witold died in Wrocław in 1967. His granddaughter Barbara, a designer and respected stylist in the city, found and catalogued the photographer-professor's old crate of nearly 1,500 negatives and 3,000 celluloid negatives. Between 2014 and 2016, she arranged many public exhibits of these spectacular photos, including of mountain landscapes and architecture from the family's archive. "Wrocław through the Lens of Witold Romer" and "Witold Romer—Scientist and Artist" were maps of sorts. They were dedicated in Poland to rare and unknown aspects of the life and work of the Romer family in their lost Galician home and the Tatra spaces they deeply loved.[62]

The second son Edmund had a no less transnational twentieth-century life. Born in 1904, he studied engineering at Gdańsk Polytechnical University, specializing in electromechanics, industrial surveying, and metrology. Before making it to England in 1943, Edmund worked in the wartime aerospace industry in Ankara, and then in the field of automechanical repair in Palestine. He also returned to "Silesian" Poland in 1946. He lived in Bytom until 1948, where he worked in factories and on industrial mechanics, also teaching physics in secondary schools. From 1949 to 1961, he was the chief organizer (and dean from 1958–60) of the department of optics and precision mechanics at the Silesian Polytechnical University (now the Silesian University of Technology) in the city of Gliwice/Gleiwitz. Edmund was a professor in the new departments of electronics from 1949 to 1964, automechanics from 1964 to 1970, and automatics and informatics starting in 1971. He retired in 1974, wrote a biography of his father in 1985, and died in Gliwice in 1988. A public exhibit was first held in January 1989, at the Jagiellonian University Library in Kraków, dedicated to Romer's cartography.[63] In 2004, when the Polish National Library in Warsaw, which lost its map collection when the Nazis destroyed it in 1944, assembled a conference in tribute to Romer on the fiftieth anniversary of his death, the library produced a volume called "the geographer of three eras," inspired by Edmund's mostly flattering account of his father.[64]

As for the landless Count Géza Teleki, only son of Pál, his cross-border entanglements were complex. He had to wait until 1948 to come to the U.S., without any help from Bowman. He left communist Hungary with his second wife Hanna and their children. Initially, he secured a small stipend from the Carnegie Endowment to settle in Washington, D.C. Starting in 1955, Géza worked as a professor of geology at George Washington University until his retirement in 1978. Géza had been married twice previously in Hungary, first to Jolán Darányi in Budapest in 1935, and then to Hanna Mikes in 1943. He had three children from his first marriage—Pál or Paul (b. 1937), Ilona (b. 1939), and Fruzsina (b. 1942), and one from his second marriage—Géza-Pál or Geza-Paul (1943–2014). They, too, joined mobile communities of experts: Paul as a geologist at the University of Florida, Geza-Paul as an anthropologist. Grand-daughter Ilona, daughter of Count Géza Teleki, fled from the 1956 Hungarian Revolution to Australia, where Fruzsina and Ilona also lived. She started work in 1965 as a librarian in Sydney. Géza's only sister, Countess Maria (b. 1910), was twice married to European nobility and died in exile in Estoril, Portugal, in 1962. Géza's third wife, Zsuzsanna (Suzanne) Gilbert, born to a Jewish family in Hungary in 1926, converted to Catholicism. After a prolonged illness, Suzanne and Géza Teleki committed suicide together in their home near Washington, D.C., on 5 January 1983.[65]

The Teleki family's extended genealogical saga of maps runs even deeper. Countess Ilona (b. 1939), daughter of Géza, was not to be confused with Countess Ilona Teleki DeVito di Porriasa (b. 1940), daughter of Béla, who fled from Soviet Hungary and Romania. Cousin Count Béla, whose properties including the family's multigenerational castle were seized by communists, was imprisoned for twenty years. He also sought asylum in the States, but was only able to make it in 1964 after bribing Romanian authorities to leave. Ilona, who spoke no English and arrived penniless in New York from Transylvania in 1945 with just a few suitcases, took a series of jobs including at a hosiery factory in the Bronx and as a teletypist at Merrill Lynch. Throughout her life, she held steadfastly, if quietly, to passed-along national and family myths that Count Pál Teleki was not an anti-Semite. She also refused to believe that he had committed suicide in 1941. Countess Ilona the Second learned English fluently, climbed her way up through the financial firm, married an Italian nobleman, and worked for decades as a Wall Street banker. She died of breast cancer in 2013.[66]

Maps are like buildings, and they have layers of history. In 1983 just outside Budapest, in the suburb of Érd, the Hungarian Geographical Museum (Magyar Földrajzi Muzeúm) was founded. It may be even less frequented than the baroque-style Teleki-Degenfeld Castle (built in 1748, reopened as a hotel in 1985) in the tiny picturesque village of Szirák in Northern Hungary, not far from the Slovak border, or the Teleki Library (Bibliotheca Telekiana, built in 1802) in Târgu-Mureş, Romania, but it is no less significant. The museum is surely one of Teleki's other distant progeny, the fulfillment of a dream from before 1914 to build an institution for geography modeled on the 1906 Museum for Regional Geography in Leipzig (today the Leibniz-Institut für Landeskunde, a major research center).

Critical attention to the Teleki family's past has been shaped and reduced by Hungary's memory wars involving the politics of anti-Semitism, EU enlargement, and the return to (and exit from) Europe. Controversial proposals were made in 2004 to erect a statue to Count Pál Teleki in Budapest, mainly by activists on Hungary's center-right and with support from the Teleki Memorial Committee and, at least initially, Gábor Demszky (b. 1952), the long-serving Hungarian liberal mayor of the city from 1990 to 2010. It drew attention from the Simon Wiesenthal Center (in Los Angeles), and met many objections due to documented anti-Semitic policies of Teleki as prime minister. Under international pressure by Hungarian liberals and politicians even in his own party, Demszky changed his mind. He decided that the Teleki statue had to be moved outside the city to the resort town of Balatonboglár, near Lake Balaton, in the courtyard of a Catholic church. After twenty years in office, Demszky was voted out of office, and today his SZDSZ party, once a powerful strategic alliance of liberals and social democrats, has all but disappeared with the rise of Viktor Orbán (b. 1963), the Fidesz and Jobbik parties, and Hungary's aggressive far right. Tensions of illiberalism are not resolved: Teleki's "Carte Rouge" still hangs prominently on display in the Hungarian National Museum in Budapest, and the Hungarian Geographical Museum in Érd.

For the Leipzig-born Albrecht Penck, family loves and displacements persisted even as Anglo-German cooperative notions of fraternity were shredded by two world wars. The May 1939 *Newsletter of the Association of American Geographers* noted that a special session on "Walther Penck's Contribution to Geomorphology" was scheduled for the annual meeting, at the University of Chicago in December. The planned session never happened. Not until 1953 was Walther's seminal book of 1924, published posthumously, rendered into English.[67] Pieces by Norbert Krebs and other acolytes congratulated Penck in the nineteenth-century tradition as a "master of earth science" (*Meister der Erdkunde*) for his eighty-fifth birthday in 1943.[68] At the family's Berlin flat on Maierottostrasse 5, Albrecht had all of his cherished scientific notebooks, photographs, and letters put in boxes.[69] Hanny, his sister, died in 1948 at the age of eighty-six. Anna Maria (Ännie) Lampert Penck, Albrecht's daughter-in-law and Walther's widow, died in Stuttgart in 1982, at the age of ninety-two. Aside from his wife Ida and Sven Hedin, Penck was closest to his daughter Ilse and son-in-law Armin. Ilse died in 1951, Armin in 1952. Their surviving son Wolfgang and three daughters, Elfriede, Hildegund, and Inge, lived on after the war. In 2011, the local newspaper *Leipziger Volkszeitung* reported the creation of Penckstrasse in the northeastern Sellerhausen area of the city. The small street was named in honor of "the geographer and geologist Albrecht Penck [who] was born in Leipzig . . . the most important German geographer of the first half of the twentieth century."[70]

Today at Johns Hopkins University in Baltimore, Bowman Avenue is totally defunct. A single portrait of the ex-president hangs today in the Hutzler Reading Room in Gilman Hall, with other past presidents. A bust stands outside of Shriver Hall, on the lower quad.[71] As for the family, Cora Olive Goldthwait Bowman died after a stroke on 11 May 1952 in Chevy Chase, Maryland, having survived Isaiah by two years. Both sons, both outdoorsmen, lived into their nineties. Daughter Olive, born in 1915, married Walter H. Gerwig, Jr., a lieutenant colonel in the army during World War II. The children inherited their parents' love of the natural world, as epitomized by the family's summer cottage in Wolfesboro, New Hampshire. Walter, born in 1910, earned his doctorate and also became an educator. He died at the age of ninety in 2001, in Rochester, New York. Erna, his wife, was a marriage and family counselor who received her master's of social work from Columbia University in 1942, became prominent, and was active in her Presbyterian church in the city for over sixty years. Erna Bowman survived Walter and died in 2011, at the age of 102.[72] Eldest son Walter treasured and cared for the family cottage that his parents bought in 1911. One of his great passions was in writing a book on nearby Lake Wentworth, first published in 1956, reissued in 1996. He and his wife were active in the Lake Wentworth Association, in which he served as president and an honorary life member. By Walter's last wishes, his remains were interred at Wolfesboro's Lakeview Cemetery.

––––––––––

The harshest American generational story comes last. Robert Goldthwait (Bob) Bowman, born in 1912, by his education followed Isaiah's trodden path into Ostmitteleuropa geographic science. Bob earned his bachelor's degree in 1935 from Dartmouth College, and his doctorate in 1941 from the University of California-Berkeley under Carl O. Sauer. Bob served with the Army Air Corps in the South Pacific for three years during World War II, after which he taught for three years at the University of Iowa. Starting in 1949–50, he worked as a professor of geography at the University of Nebraska-Lincoln. Isaiah, after the shutdown at Harvard in 1948, had hopes for him to set up a high-performing department there.[73] Bob and his wife Jean had three daughters and tried to settle in.

For decades as an academic and particularly following Jean's death in 1982, the army veteran battled alcoholism and depression. A living embodiment of solitary manhood and the frontier life, he never became a productive academic. He took his father's death hard and spent many years estranged from his family. Bob acted as the caretaker and co-pruner of his father's archive until it was brought to Johns Hopkins. He had psychological fits and habitually arranged bonfires in his backyard, during which he burned original copies of his letters to his father. Some indeed are missing from the Hopkins files. He remained on a modest pension and lived in Nebraska until 2000, until his eldest daughter Barbara convinced him to move to Visalia, California, to be cared for. Cherished by the Prophet Isaiah as his and America's hope, the Walther Penck successor for the lived transnational dream of geography into the twentieth and twenty-first centuries, Bob Bowman died in obscurity in 2007, at the age of 94.[74]

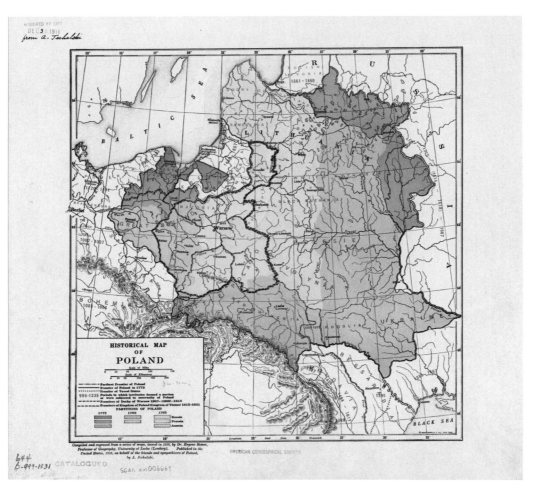

PLATE I *Historical Map of Poland, Compiled and Engraved from a Series of Maps, Issued in 1916, by Dr. Eugene Romer, Professor of Geography, University of Lwów (Lemberg), Published in the United States, 1918, on behalf of the friends and sympathizers of Poland, by A. Jechalski* (New York, 1918). Activists distributed this separately from Romer's 1916 *Geographical-Statistical Atlas of Poland*. Note the appeal to friendship in graphic form, the map's trans-Atlantic reach, and the credentializing of Romer. The map's "fantasy East" orientation bears a frontier focus in red and pink on a Greater Poland, which incorporates the Duchy of Courland, Samogitia, Lithuania, White, Black, and Red Ruthenia (with the city of Lwów), Polesie, Volhynia, Podolia, and Ukraine, along with pre-partitioned lands before 1772, 1793, and 1795. Contested portions in blue absorb the Baltic corridor, including Gdańsk/Danzig and parts of West Prussia. Courtesy of the American Geographical Society Library, University of Wisconsin-Milwaukee Libraries, AGSL 644 B-999-1831.

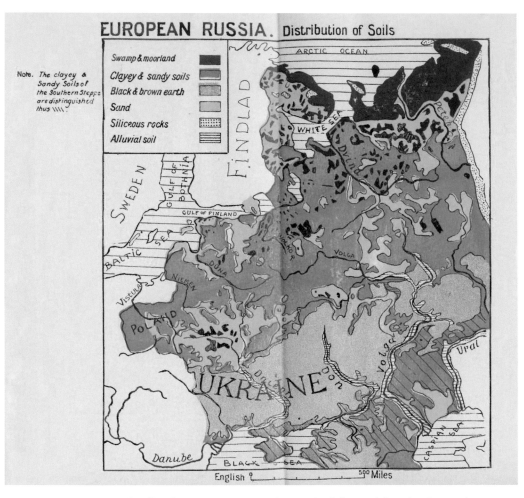

PLATE 2 Stepan Rudnyts'kyi, "European Russia: Distribution of Soils," appended to *The Ukraine and the Ukrainians* (Jersey City: Ukrainian National Council, 1915), drawn from *Ukraina und die Ukrainer* (Vienna: Verlag des Allgemeinen Ukrainischen Nationalrates, 1914). In this rare hand-drawn map of "European Russia," note the size in print of "Ukraine" as opposed to "Poland," and across Galicia. ("Findlad" is misspelled.) Rudnyts'kyi was steeped in the German scientific tradition of Penck and Richthofen.

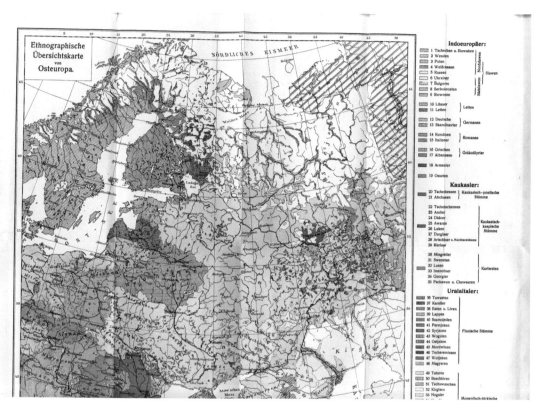

PLATE 3 Stepan Rudnyts'kyi, *Ethnographic Survey Map of Eastern Europe* (*Ethnographische Übersichtskarte von Osteuropa*), appended to *Ukraina: Land und Volk eine gemeinfassliche Landeskunde* (Wien: Bund zur Befreiung der Ukraina, 1916). Note the organic coloration and blending of green and blue Slavic populations on frontiers to the East, the imperial shades of pink (as in the British Empire) for "Deutsche" and "Skandinavier" blurred together as Germans, and the classification of "Ukrainer" in unison by a single ethnonym, with no Habsburg designation of "Ruthenen."

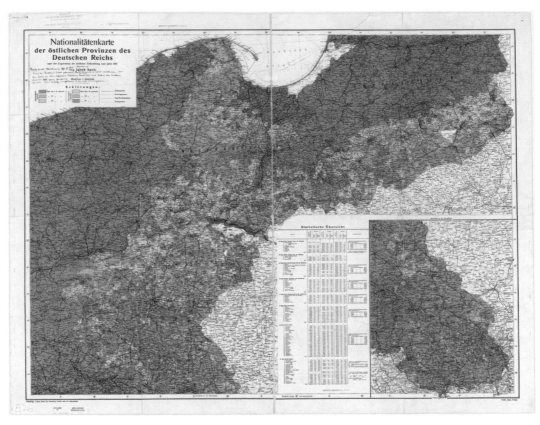

PLATE 4 Jakob Spett. *Nationalities Map of the Eastern Provinces of the German Empire after the Results of the 1910 State Census* (Gotha: Justus Perthes, 1918). Scale of 1:500,000. Likely passed by Romer in Paris to Bowman and the Americans. On the map itself, Colonel Lawrence Martin handwrote in German a note that transcribed an exchange by Penck on 9 April 1919 with his friend Eduard Brückner, in which Penck claimed the map was "a refined forgery" (ein raffinierte Fälschung). In other words, the Berlin professor feared that it was working all too well. Courtesy of the American Geographical Society Library, University of Wisconsin-Milwaukee Libraries, AGSL 64-C E2 C-1918.

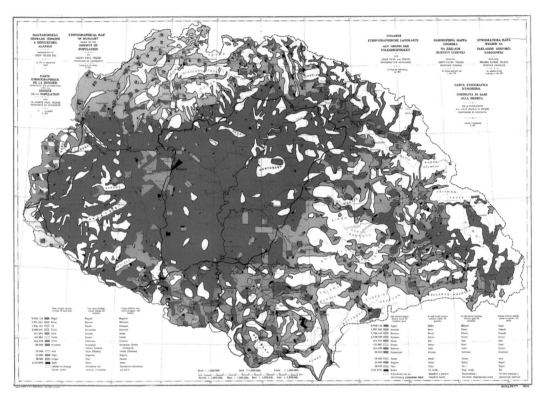

PLATE 5 Count Pál Teleki's famous "Carte Rouge": *Magyaroszág néprajzi térképe a népsűrüség alapján. Ethnographical Map of Hungary Based on Density of Population. Carte Ethnographique de la Hongrie consruite en accordance avec la densitè [sic] de la Population* (Budapest, 1919). It shows 9.9 million Hungarians in bright red, here on a scale of 1:1 million. Courtesy of the American Geographical Society Library, University of Wisconsin-Milwaukee Libraries, AGSL 642 C-1919 (one of three copies).

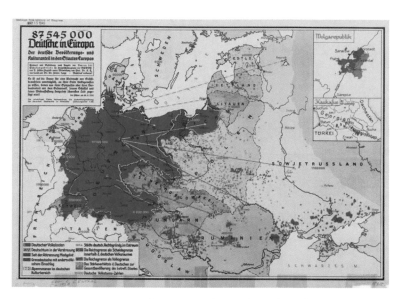

PLATE 6 *87,545,000 Germans in Europe: The German Population and Its Cultural Share in the States of Europe*, around 1938. Drawn in Berlin by Arnold Hillen Ziegfeld, who also visualized Penck's "Volks- und Kulturboden" in 1925. Noting open Eurasian frontiers to the east, the two inset maps depict German settlements in the Volga ("Volgarepublik") and the Caucasus. Courtesy of the American Geographical Society Library, University of Wisconsin-Milwaukee Libraries, 600-C Europe (Central) C-1938.

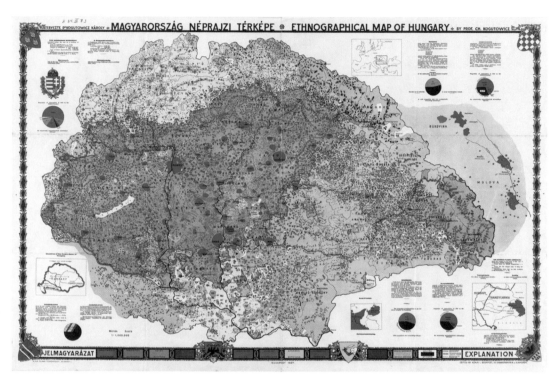

PLATE 7 Dr. Károly (Charles) Kogutowicz, "Ethnographical Map of Hungary" (*Magyarország néprajzi térképe*) (Budapest: Kókai Lajos, IV. Kamermayer, 1927). Poster map by Count Teleki's Budapest colleague after the "Carte Rouge," on a scale of 1:1,000,000. Károly the geographer (1886-1948) was the son of Manó the cartographer (1851-1908); the family actually moved to Hungary from Polish Galicia. The map reflects Habsburg census data in 1910 by nationality and refers to Hungary as the "Mother-Country." The insert map at the bottom-left shows an outline of the St. Stephen crownlands. Courtesy of the Library of the Herder Institute for East Central European Research, Map Collections, Marburg, Germany, K 54 III B 3.

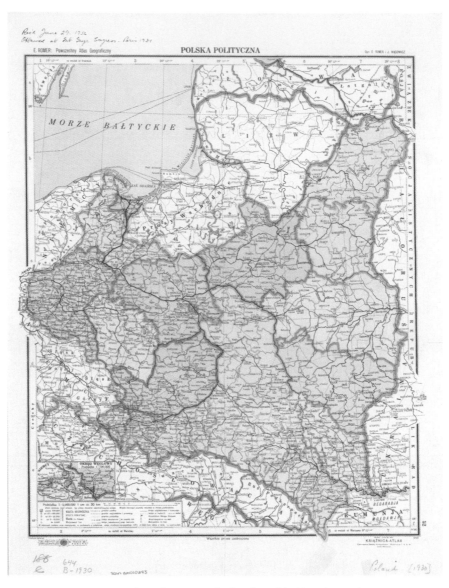

PLATE 8 Eugeniusz Romer and Józef Wąsowicz. "Political Map" (Polska Polityczna), in E. Romer, *Powszechny Atlas Geograficzny* (Lwów-Warszawa: Książnica-Atlas, 1930). Scale of 1:3,000,000. One of many Polish political maps passed by Romer to Bowman and the AGS, first in Paris in 1919, then (here) at the IGU in Paris 1931, again in Warsaw in 1934. Note Bowman's handwriting at the top, "Rec'd June 29, 1932, Obtained at Int. Geogr. Congress, Paris 1931," and the atlas logo for Romer's Cartographic Institute at the bottom. The map refers to past battlefield sites and joins all of Poland's voievodships (*województwa*) into a mosaic whole. Courtesy of the American Geographical Society Library, University of Wisconsin-Milwaukee Libraries, Courtesy of AGSL 644-B-1930.

CONCLUSION

Isaiah Bowman owned at least two suits in his life. The map man had many roles in the making of East Central Europe and the world. Yet he was the last person in the world to admit his flaws as an academic, a father, or a geographer. He followed the lead of President Wilson, who hoped in Paris for a lasting scientific peace that never happened. Upon his retirement from Johns Hopkins in December 1948, Bowman was still the Bowman of 1919, a person of rank and intense ambition. He was the same person as the president of the IGU in 1934, his pet League of Nations, giving a speech before a captive audience. In the presence of Romer but not Penck in Warsaw, he declared, "With all our diversity of interest and endeavor in the wide field of geography, we have a common dependence upon the map. It is the symbol of our profession. At one time or another, every geographer seeks to make a contribution to the map of the world—to survey still unmeasured portions of land and sea, to compile surveys into useful base maps, to display and interpret distributional phenomena, to deepen the understanding of the spatial elements of our physical world and its life relationships by invoking the highest standards of graphic art."[1]

FIGURE 8.1. Photo of Isaiah Bowman in his office, probably in 1935, the first of his thirteen years (until 1948) as the president of Johns Hopkins University. Courtesy of Special Collections, The Milton S. Eisenhower Library, The Johns Hopkins University, Baltimore, Maryland. Isaiah Bowman Papers (IB-P Ms. 58), Series I, Box 3, Item 7.

Prophet Isaiah liked to remind people of the heroic map, of brave new worlds of geographers. He belonged to another place and time (figure 8.1).

Bowman's biography, like those of our other men, is nothing if not problematic as a source. It is incomplete. He retold his life as a struggle. He manicured but did not destroy his archive. Our map men were emotional men, and they had many things to hide: blueprints for an American-led "new world" grounded in biological racism, binaries of sexual difference, colonial projects, and Europe's pre-1914 civilizing institutions. Bowman spun a tale of himself, characteristic of the bourgeois age in which he lived, part of an unceasing effort to breed his sons for noble deeds and impress others in academia and politics. The Ontarian outsider made himself into the centerpiece of a German-American story, minus the German, but his memoirs never saw the light of day. He built a myth of exceptionalism, just as Wilson in Paris advertised his own. Bowman's skills made him an ideal manager in the enterprise of maps, tools of education, the work of a con artist and university president. A map man of qualities, he could have starred in Musil's fictional Kakania or Kafka's Spindlermühle. Among Professor Penck's pupils in Ostmitteleuropa, he and other geographers lived out their fantasies of America and the world through their actions. A punchline is in order—Isaiah Bowman *was* a map.

What can we really learn from studying Professor Penck's pupils, the *homo geographicus* of the 1870s to the 1950s? Let's conduct a thought experiment. You gain access, for a moment, to a digital archive of every map ever made, and of knowledge of transit networks, every channel for emails, PMs, data, and metadata by which maps are passed around. Then imagine that you gain intelligence on those doing the trafficking. Such an archive, dystopian or not, may soon exist. In fact, it may already exist, in support of publicly accessible knowledge, in violation of trust, privacy rights, and confidentiality—all in the present. Maps give us insight into these moving worlds of fantasies by human exchanges. Maps, like letters, were scripts for behavior, neither rational nor plainly of nations. Geographers left behind their Rorschach blots, the psychological clues that detail the character formation of fragile aspirational professionals who believed in science, yet dreamt up and participated in illiberal projects. Through the lens of epistolary geography, we can now summarize the book's main story of map men and East Central Europe in four takeaway points:

1. *Maps are made powerful by mobile and transnational, not merely geopolitical, men.* The *homo geographicus* was a new professional. These men took part in roving adventure projects of a mobile intelligentsia, before they became stateside experts in diplomacy on both sides of the Atlantic. In Europe's grand explorer tradition, these men aspired to make geography a study of far-flung global frontiers and landscapes, a privileged kind of mega-discipline. They forged their lives from local ideologies of sexual and ethnic difference, as well as the imagined paternal authority of a male-dominated, scientific, German-speaking East Central Europe. "Geopolitics," the word coined in 1899 by the Swedish politician Rudolf Kjellén that came to have such lasting impact in England, Germany, and Russia, alone is inadequate to describe their enterprise out of Ostmitteleuropa. From there, hyphenated Anglophile German scientists dreamt of imperial frontiers and colonial spaces as a way to draw up new priorities of travel and multigenerational grounds for scientific advancement.

2. *Maps are not modern.* If geopolitics does not fit well as a heuristic tool, neither does a European culturalist understanding of modernity as planning, the advancement of technology, or time-space compression. Maps involved

more than grids or technique. "Modern" map men often followed assumptions of development, but they were not one-dimensional representations of order from chaos, or groups into nationalities and nation-states. Just because nineteenth-century geographers made maps in greater volume after Europe's revolutions of 1848–49 or the U.S. Civil War of 1861–65, this did not make them modern or give them a fixed identity. Our Transylvanian prime minister, our Saxon and Galician academics, and others of a provincial European sort fell in love with nature, distrusted cities, and sought out frontier spaces of global contact as a wellspring in their lives. Modern maps appeared in rational guises or as educational tools and props. Maps, however, had an antique aura. They were surreal, sentimental, and subjective in the tensions they manifested during the twentieth century and beyond.

3. *Maps alert us to confraternal bonds that run deep, very deep.* Map men made places seem more permanent and lives seem professional, but not necessarily in a modern mass political or group-identitarian, postmodern single-issue sort of way. They adhered to polite norms of civility, forged by common expeditionary travel. For the most part, their careers started out collegially. During World War I, friends watched and kept their enemies close. When they sent maps abroad as tools and gifts, modes of personal exchange after 1918, they used those maps to rework a confraternal network into new channels of informal diplomacy. With families in long tow, they opted after Paris in 1919 into or out of a new Wilsonian liberal international order of states. Geographers' attachments were mannered by civilizing missions in education and by biological conventions (none could be called feminist), but identities are not easily reduced to labels. When our aristocratic-bourgeois men took risks to revise postimperial, postdynastic lands of Ostmitteleuropa, they recalibrated professional worlds into what remained of their personal bonds. As if starring in a buddy film across America, they bonded and looked for global paths for their nations. Maps became the artifacts of these bonds.

4. *Beyond literal interpretation, maps are texts, moods, and graphic sagas in which messy multigenerational plans and affairs of life and death are laid out.* Lives and deaths of the map men always intertwined. Many of them and their families did not survive war, occupation, or the effects of tragic loss. They attached maps to articles, and academic articles, no matter how boring or trivial, were, in fact, letters also intended to build contacts and sort through diplomatic channels. Some ordered their private collections destroyed. No one was untouched by politics. One (Count Pál Teleki) committed suicide in 1941; Stalin's purges took away another (Stepan Rudnyts'kyi) in 1937. Moods were disguised by professionals who "spoke map," all struggling in a high era of territorial nationalism

to preserve a veneer of objectivity and respectability. Maps divulge buried emotional worlds of love, privilege and authority, revenge and resentment, retooled into professional power grabs and matters of life and death.

We can discern such moods from a useful premise, that the disciplines of geography and cartography are not canonical, inherently democratic, or outside history itself. Biographical writing works best when its subjects are unbronzed, outside the nation.[2] Professional academics, never short on ego, considered themselves scientists, but notably, our geographers did not identify by geopolitics or principally within human or cultural geography traditions.[3] Letters from life stories draft larger plans, much like maps. Maps as exchanged letters get us closer to the intimacy of spatial thinking and networks of individuals who craved acknowledgment, leaving behind trace evidence of lives.[4] Their lives arc along the risky, bumpy road of professionalization that inheres to imperial biography, for all these men had to *look good*, defending the integrity of nineteenth-century science on a high ground of objective values and state service.[5] Gendered practices of men at work persisted after World War I, when further opportunities were lost and previous ties of intellectual and political cooperation severed.[6] Nevertheless, our subjects entertained high ambitions for geography as science, in the study of the natural world through geomorphology, hydrology, climatology, glaciology, alpinology, geology, and other subdisciplines—all within the natural world they had emotionally encoded.[7]

———————

My transnational love story warrants some final thoughts regarding the past and ongoing presence of map men. In the digital age, cartophilia is everywhere. Maps seem like sublime tools of progress. We remain wary of them, but we also think they can do anything, if only designed more persuasively. Maps are on Facebook and other social media, in the future of GIS studies, in the gaming industry, on phone apps. These too are fads, or illusions as fads. Smart investigators must do more than click the bait, blindly share images, and read maps literally. Nor can we say, with diplomatic partnerships being so delicate, that all maps are rational constructs, crafted for nations in a forever plural, prepared for the conspiratorially manipulated, and just leave it there. Categories and plans fail. Lines are not fixed. Nations are not eternal. Maps precede territory and war. Maps of East Central Europe are dangerously shipped around on media channels, with purpose and unintended consequences. Countries get erased when no one is watching.

Humor is one effective antidote. Humor can be counterhegemonic, to use a fancy word. Such talented graphic designers as the Bulgarian anarchist Yanko

Tsvetkov, the launcher in 2012 of the immensely popular *Atlas of Prejudice* series, resists labels by nationality and describes himself as "a misfit" and "a humanoid living on a planet called Earth."[8] The avant-garde Czech sculptor David Černý, commissioned to produce a tribute in 2009 in Brussels to mark the Czech Republic's accession to the EU presidency after Nicholas Sarkozy, caused a major stir with his twenty-seven-country "Entropa" exhibit when he refused to play by politically correct rules. His sculpted map happening depicted Sweden as a (can't) do-it-yourself box of Ikea furniture, Romania as a Dracula-themed amusement mark, Italy as a collectively fetishized football field with players masturbating on it, and Polish priests raising the gay pride flag, in the image of the photo of American soldiers at Iwo Jima. Further calls for alternative maps and an "affective turn" in history and geography have grown.[9] Philippe Rekacewicz, the outstanding cartographer of *Le Monde Diplomatique* and Visionscarto.net, the innovative media-and-maps platform, urges us to view them not as clashes of civilizations or exact science, but radically as part of an ongoing experimental workshop across borders.[10] Denis Wood, part anarchist, part iconoclast, suggests a fresh look at design aesthetics in cartography.[11] These savvy postmodern map men, insofar as they have to be labeled, are Professor Penck's distant pupils, great-great-grandchildren of Ostmitteleuropa's legacy of nineteenth-century geography and cartography. They are adroit at breaking down (or "mashing") maps into parts, as a scientist might do, then piecing them together again.

It might be better to say that all maps are epistemically groundless, nihilistic, or surreal, as in the 1929 "Surrealist Map of the World," which flipped the Mercator projection and centered globally the Pacific Ocean and Bering Strait, between Russia and Alaska. Getting out of Eurocentric frames is essential, for neither multilingualism nor travel fully civilized the modern geographer into a pure scientist (always a misnomer). Although professionals missionized for cartographic literacy and traveled as explorers, this did not make them any less provincial out of Saxony, East or West Galicia, Transylvania, Ontario, or Michigan. One might imagine that Count Teleki, a failed politician and diplomat, knew this well. His Europe was provincial. He did not leave a legacy without taint. He seems to have destroyed most of his correspondence before he killed himself in Budapest in 1941. Penck experienced similar problems. His writings after 1945 were guarded posthumously in a box during de-Nazification first by his family, then by his chosen acolytes. After the Holocaust and in an era when Karl Haushofer's name was associated with Hitler and Hess, geopolitics fell quickly out of favor in West Germany. Penck's *Heimat* in Leipzig fell on the other side of the Iron Curtain. "Scientists" such as Emil Meynen were members of the Nazi party (joined in 1937), and as such implicated in the Third Reich's crimes. Bowman, too, was protective of confidential letters up to his death in 1950. In fact, Bowman preferred that a scholar of about thirty-five years old and international repute should examine his archive—in other words, a civilizing hero, a Penck like himself.

It is worth recalling how wars turned people into disappointed lovers. The making of ethnographic maps was another totemic symbol in Ostmitteleuropa's history of nationalism. Reactive twentieth-century print media (and now twenty first-century digital media) fell into tribalistic echo chambers of this nature. With the flood of maps in Paris of 1919, it became a challenge to remember the tensions inherent to Europe's pan-colonial cooperation. As Cold War complexes for area studies evolved out of Erd-, Volks-, and Landeskunde traditions and started in the late 1940s in the United States, civilizational clashes became more predictable. American Sovietologists, mostly men for far too many decades, gathered maps and census data in their capacity of functional experts, just as late nineteenth-century explorers served imperial powers, exploited land, and colored in the planet's face.[12] Political science and economics, not geography, emerged dominant. Emigrés from the 1940s to the 1960s sought to explain the recent past. That did not preclude settling scores with fascism and communism, or theorizing mass violence that often appears unique to modernity. Attentive to geopolitics, they developed concepts of totalitarianism or authoritarianism, as evidenced by histories not only of tragic events, but also of masked personality disorders and inner, supposedly darker lives.[13]

Exile from Ostmitteleuropa meant loss, and loss deepened a symbolic sense of place. "Fathers" lost and reproduced "sons." Many hesitated to alienate patrons whom they needed for support and accolades. Others were accused of treason. Breaks with history were astounding in the postwar epoch when historians, fixated on the nation-state (both real and imagined), often forgot how technical intelligentsias worked and how mobile multilingual experts met their demise. Stepan Rudnyts'kyi and the leading geographer of Belarus, the soil scientist Arkadz Smolich (1891–1938), faced arrest, imprisonment, exile, and death. Recalled as heroic men by nations against communism in Europe's East, their transnational lives were marginalized.[14] All were enticed to join a new managerial class, to become handmaidens of empire, part of a fast-changing, educated bourgeoisie.

So Anglophile German geographers, not quite cosmopolitan, were map men of a Non-Republic of Letters. These bearers of the norms of a mannered confraternity, consultants held suspect, were dedicated to the natural sciences. Ideologues salaried their talents and purchased their silence. Not geopoliticians either, the geographers made maps into tools of literacy and propaganda, and it was hard to separate the two. In search of comfort, mobile colonial men thought in terms of character and the advancement of knowledge. Skilled cartographers learned techniques from each other and pressured states to develop geography. They admired their father-mentors, forgotten scientists such as William Morris Davis, Antoni Rehman, and Lajos Lóczy. They loved to be outdoors with men of their kind. They obsessed over unity. They took the fate of millions into their hands. They were the information mashers of their age.

Unfortunate associations of maps with modernity, identity, or nationalism do not give us a full picture of map men. Geographers move around. Our scientists outlived the Ostmitteleuropa they once made, and their technology survived. They cast their kin and compatriots as torchbearers of Western culture, a dreamt-up world of higher, moral politics. They believed in environmental activism and gender difference simultaneously. Map men were unlikeable heroic explorers, intolerant antiheroic careerists, and privileged transcultural racists. Our men fantasized about frontiers, feared closure, hitched lives to declining states, and fed data to metropoles. Leftover secrets are a decoder's challenge. As Timothy Brook puts it, the venture to explain any map is "about the people whose stories intersected with it."[15] Maps offer more than pretty pictures or techno-utopian tools or apps. If maps reduced citizens to pixels, geography to geopolitics, language to nationality, and race to space, that was the mark of an era's fragile, contingent plans. Maps were a love story, a hate story, a means of conversation and generational contact. Today, maps give us tales of broken men. They allow us into distant worlds, if only for a glimpse.

ABBREVIATIONS

A-AGS   Archives of the American Geographical Society, materials in transit from New York City to the AGSL in Milwaukee (beginning in 2011)

A-MFT   Archives of the Hungarian Geographical Society (Magyar Földrajzi Társaság Levéltára), Budapest, Hungary

AGSL   American Geographical Society Library, Milwaukee, Wisconsin

AP-P   Albrecht Penck Papers, Nachlass Albrecht Penck, K. 871–878

ER-P   Collected Papers of Eugeniusz Romer, Archives of Jagiellonian University Library (Biblioteka Jagiellońska), Department of Manuscripts (Oddział Rękopisów), Kraków, Poland

IB-P   Isaiah Bowman Papers, Ms. 58, Special Collections of the Sheridan Libraries, Johns Hopkins University, Baltimore, Maryland

OSZK-T   National Széchényi Library (Országos Széchényi Kön-
                yvtár), Térképtára (Map Collections), Budapest,
                Hungary
PT-P   Pál Teleki Papers, Hungarian National Archives (Magyar
                Nemzeti Levéltár), changed in 2011 from Hungarian
                State Archives (Magyar Országos Levéltár), Buda-
                pest, Hungary

NOTES

INTRODUCTION

1. Isaiah Bowman to James Lee Love, 5 July 1939, Isaiah Bowman Papers (hereafter IB-P), Ms. 58, Series I, Box 1.1, Folder 5.

2. Neil Smith, *American Empire: Roosevelt's Geographer and the Prelude to Globalization* (Berkeley: University of California Press, 2003), 38; Geoffrey J. Martin, *The Life and Thought of Isaiah Bowman* (Hamden: Archon, 1980), 12.

3. On persistent provincialism, see Robert Nemes's excellent *Another Hungary: The Nineteenth-century Provinces in Eight Lives* (Stanford: Stanford University Press, 2016); also Catherine Evtuhov, *Portrait of a Russian Province: Economy, Society, and Civilization in Nineteenth-century Nizhnii Novgorod* (Pittsburgh: University of Pittsburgh Press, 2011); and Celia Applegate, *A Nation of Provincials: The German Idea of Heimat* (Berkeley: University of California Press, 1990); on bourgeois "tensions," Ann Laura Stoler and Frederick Cooper, eds., *Tensions of Empire: Colonial Cultures in a Bourgeois World* (Berkeley: University of California Press, 1997).

4. Richard J. Evans, *The Pursuit of Power: Europe, 1815–1914* (New York: Viking, 2016), 380.

5. Anderson's influential "Census, Map, Museum" chapter is in *Imagined Communities: Reflections on the Origins and Spread of Nationalism*, rev. ed. (London: Verso, 1991). The late historian's timeframe for studying maps as an invention of tradition is limited to the period after the 1790s.

6. For socioinstitutional studies, see Anne Goldgar, *Impolite Learning: Conduct and Community in the Republic of Letters, 1680–1750* (New Haven: Yale University Press, 1995); Goldgar and Robert I. Frost, *Institutional Culture in Early Modern Society* (Boston: Brill, 2004); Norbert Elias, *The Germans: Power Struggles and the Development of Habitus in the Nineteenth and Twentieth Centuries* (New York: Columbia University Press, 1996); Gary B. Cohen, *Education and Middle-class Society in Imperial Austria, 1848–1918* (West Lafayette: Purdue University Press, 1996); Fritz Ringer, *Fields of Knowledge: French Academic Culture in Comparative Perspective, 1890–1920* (Oxford: Oxford University Press, 1992); Ringer, *The Decline of the German Mandarins: The German Academic Community, 1890–1933* (Cambridge: Harvard University Press, 1969); Konrad H. Jarausch and Geoffrey Cocks, eds., *German Professions, 1800–1950* (Oxford: Oxford University Press, 1990); Jarausch, *Students, Society, and Politics in Imperial Germany: The Rise of Academic Illiberalism* (Princeton: Princeton University Press, 1982); and Jarausch, *The Unfree Professions: German Lawyers, Teachers, and Engineers, 1900–1950* (New York: Oxford University Press, 1990).

7. Mitchell G. Ash and Jan Surman, eds., *The Nationalization of Scientific Knowledge in the Habsburg Empire, 1848–1918* (New York: Palgrave Macmillan, 2012); Martin Kohlrausch, Katrin Steffen, and Stefan Wiederkehr, eds., *Expert Cultures in Central Eastern Europe: The Internationalization of Knowledge and the Transformation of Nation States since World War I* (Osnabrück: Fibre, 2010).

8. Adam Hochschild, *King Leopold's Ghost: A Story of Greed, Terror, and Heroism in Africa* (Boston: Houghton Mifflin, 1998); also the innovative, if uneven, Klaus Theweleit, *Male Fantasies*, 2 vols. (Minneapolis: University of Minnesota Press, 1987–89).

9. As in Stephen Kotkin, *Stalin: Paradoxes of Power, 1878–1928*, vol. 1 (New York: Penguin, 2014).

10. See the important issue of "Queering German History," guest ed. Jennifer Evans, *German History* 34:3 (September 2016): 371–528; for the modern period, Evans and Matt Cook, eds., *Queer Cities, Queer Cultures: Europe since 1945* (New York: Bloomsbury, 2014), and Clayton J. Whisnant, *Queering Identities and Politics in Germany: A History, 1880–1945* (New York: Harrington Park Press, 2016). For an American study, John Howard, *Men Like That: A Southern Queer History* (Chicago: University of Chicago Press, 1999).

11. Sidonie Smith and Julia Watson, eds., *Reading Autobiography: A Guide for Interpreting Life Narratives*, 2nd ed. (Minneapolis: University of Minnesota Press, 2010); Phil Hubbard and Rob Kitchin, eds., *Key Thinkers on Space and Place*, 2nd ed. (London: Sage, 2010); Linda McDowell, *Gender, Identity, and Place: Understanding Feminist Geographies* (Cambridge: Polity Press, 1999); Lise Nelson and Joni Seager, eds., *A Companion to Feminist Geography* (Maiden, Mass.: Blackwell, 2005); Doreen B. Massey, *For Space* (London: Sage, 2005); and Massey, *Space, Place, and Gender* (Minneapolis: University of Minnesota Press, 1994). On the canon and its tensions, David N. Livingstone and Charles W. J. Withers, eds., *Geographies of Nineteenth-century Science* (Chicago: University of Chicago Press, 2011); Felix Driver, *Geography Militant: Cultures of Exploration and Empire* (Oxford: Blackwell, 2001); and Livingstone, *The Geographical Tradition: Episodes in the History of a Contested Enterprise* (Oxford: Blackwell, 1993).

12. Stephen Kern, *The Culture of Time and Space 1880–1918*, rev. ed. (Cambridge: Harvard University Press, 2000); John Urry, *Mobilities* (London: Polity, 2007); Peter Adey et al., eds., *The Routledge Handbook of Mobilities* (London: Routledge, 2013); John Randolph and Eugene M. Avrutin, eds., *Russia in Motion: Cultures of Human Mobility since 1850* (DeKalb: University of Illinois Press, 2012).

13. Mark Mazower, *Governing the World: The History of an Idea* (New York: Penguin, 2012); Mazower, *No Enchanted Palace: The End of Empire and the Ideological Origins of the United Nations*

(Princeton: Princeton University Press, 2009); Susan Pedersen, *The Guardians: The League of Nations and the Crisis of Empire* (Oxford: Oxford University Press, 2015); Helen Tilley, *Africa as a Living Laboratory: Empire, Development, and the Problem of Scientific Knowledge, 1870–1950* (Chicago: University of Chicago Press, 2011).

14. Larry Wolff, *The Idea of Galicia: History and Fantasy in Habsburg Political Culture* (Stanford: Stanford University Press, 2010); Pieter M. Judson, *Inventing Germanness: Class, Ethnicity, and Colonial Fantasy at the Margins of the Habsburg Monarchy* (Minneapolis: University of Minnesota Press, 1993); Vejas Gabriel Liulevicius, *The German Myth of the East, 1800 to the Present* (Oxford: Oxford University Press, 2009); Liulevicius, *War Land on the Eastern Front: Culture, National Identity, and German Occupation in World War I* (Cambridge: Cambridge University Press, 2000); Gregor Thum, "Megalomania and Angst: The Nineteenth-century Mythicization of Germany's Eastern Borderlands," in Omer Bartov and Eric D. Weitz, eds., *Shatterzone of Empires: Coexistence and Violence in the German, Habsburg, Russian, and Ottoman Borderlands* (Bloomington: Indiana University Press, 2013), 42–60; Thum, ed., *Traumland Osten: Deutsche Bilder vom östlichen Europa im 20. Jahrhundert* (Göttingen: Vandenhoeck & Ruprecht, 2006); Fritz Stern, *Dreams and Delusions: The Drama of German History* (New York: Knopf, 1987).

15. On how this turn related to East Central Europe's modern Jewish populations, Michael L. Miller and Scott Ury, eds., *Cosmopolitanism, Nationalism, and the Jews of East Central Europe* (London: Routledge, 2015); for older political contexts, Joseph Rothschild, *East Central Europe between the Two World Wars*, rev. ed. (Seattle: University of Washington Press, 1998); Ezra Mendelsohn, *The Jews of East Central Europe between the Wars* (Bloomington: Indiana University Press, 1987).

16. David Thomas Murphy, *The Heroic Earth: Geopolitical Thought in Weimar Germany 1918–1933* (Kent: Kent State University Press, 1997); Mark Bassin, Sergey Glebov, and Marlene Laruelle, eds., *Between Europe and Asia: The Origins, Theories, and Legacies of Russian Eurasianism* (Pittsburgh: University of Pittsburgh Press, 2015); Michael Heffernan, *The Meaning of Europe: Geography and Geopolitics* (London: Arnold, 1998); Klaus Dodds and David Atkinson, eds., *Geopolitical Traditions: A Century of Geopolitical Thought* (London: Routledge, 2000); Gearóid Ó Tuathail [Gerard Toal], *Critical Geopolitics: The Politics of Writing Global Space* (Minneapolis: University of Minnesota Press, 1996). On revision and revisionism, Marina Cattaruzza, Stefan Dyroff, and Dieter Langewiesche, eds., *Territorial Revisionism and the Allies of Germany in the Second World War: Goals, Expectations, Practices* (New York: Berghahn Books, 2013); Holly A. Case, *Between States: The Transylvanian Question and the European Idea during World War II* (Stanford: Stanford University Press, 2009); Miklós Zeidler, *A revíziós gondolat* (Pozsony: Kalligram, 2009).

17. Beth Holmgren, ed., *The Russian Memoir: History and Literature* (Evanston: Northwestern University Press, 2003); Jochen Hellbeck, *Revolution on My Mind: Writing a Diary under Stalin* (Cambridge: Harvard University Press, 2006); Sheila Fitzpatrick and Yuri Slezkine, eds., *In the Shadow of Revolution: Life Stories of Russian Women from 1917 to the Second World War* (Princeton: Princeton University Press, 2000); Maria Bucur, Rayna Gavrilova, Wendy Goldman, Maureen Healy, and Mark Pittaway, "Forum on Everyday Life: Six Historians in Search of *Alltagsgeschichte*," *Aspasia: The International Yearbook of Central, Eastern, and Southeastern European Women's and Gender History* 3 (2009): 189–212.

18. See Paul Hanebrink, *In Defense of Christian Hungary: Religion, Nationalism, and Antisemitism, 1890–1944* (Ithaca: Cornell University Press, 2006), 165–69. Hanebrink classifies Teleki as a "conservative" in the mold of Count István Bethlen, which is well supported in the interwar

period, but he does not deal as much with Teleki's private life or colonial entanglements of race, space, and geography.

19. Judith Butler, *Dispossession: The Performative in the Political* (London: Polity, 2013); Butler, *Gender Trouble: Feminism and the Subversion of Identity* (London: Routledge, 1989); Erving Goffman, *The Presentation of Self in Everyday Life* (New York: Anchor Books, 1959).

20. As noted in subsequent chapters, the classic study of Weimar and Third Reich cartography is Guntram Henrik Herb, *Under the Map of Germany: Nationalism and Propaganda, 1918–1945* (London: Routledge, 1997). Herb's sharp analysis of maps in *Under the Map of Germany* was also the first work in English to examine the Penck-Romer conflict. He mentions Romer but does not analyze Polish-language sources, and thus overlooks that the Penck-Romer conflict was examined earlier, in 1980, by the Polish geographer Aniela Chałubińska, who was a student of Romer's. Chałubińska, "Kontakty Eugeniusza Romera i Albrechta Pencka," *Studia i materiały z dziejów nauki polskiej*, Seria C, vol. 24 (1980): 15–33. Bowman's two main biographers, the geographers Geoffrey J. Martin and the late Neil Smith, also did not use any East European languages for research. Martin named Romer three times in his 1980 biography and labeled him "an ardent Polish nationalist," but spells his name inconsistently as "Eugeniusz" and "Eugenius." Smith, in his award-winning biography of Bowman and empire, devoted an entire subsection to the Polish Question in 1919 without any note of Romer. In fact, he mentions Romer's name just once in the entire book, Germanizing him into "Eugen Romer." See Martin, *The Life and Thought of Isaiah Bowman* (Hamden: Archon, 1980), 33, 93, and 199; and Smith, *American Empire: Roosevelt's Geographer and the Prelude to Globalization* (Berkeley: University of California Press, 2003), 280.

21. Peter Haslinger, *Nation und Territorium im tschechischen politischen Diskurs 1880–1938* (München: R. Oldenbourg, 2010); Haslinger and Vadim Oswalt, eds., *Kampf der Karten: Propaganda- und Geschichtskarten als politische Instrumente und Identitätstexte* (Marburg: Herder-Institut, 2012).

22. Pieter M. Judson, *Guardians of the Nation: Activists on the Language Frontiers of Imperial Austria* (Cambridge: Harvard University Press, 2006); Judson, *Inventing Germanness*. Also Tara Zahra, "Imagined Non-communities: National Indifference as a Category of Analysis," *Slavic Review* 69 (Spring 2010): 93–119; James Bjork, *Neither German nor Pole: Catholicism and National Indifference in a Central European Borderland* (Ann Arbor: University of Michigan Press, 2008).

23. Deborah R. Coen, *Vienna in the Age of Uncertainty: Science, Liberalism, and Private Life* (Chicago: University of Chicago Press, 2011).

24. Guido Hausmann, "Maps of the Borderlands: Russia and Ukraine," in Róisín Healy and Enrico Dal Lago, eds., *The Shadow of Colonialism on Europe's Modern Past* (Houndmills: Palgrave Macmillan, 2014), 194–210; Hausmann, "Die Kultur die Niederlage: Der erste Weltkrieg in der ukrainischen Erinnerung," *Osteuropa* 64, no. 2–4 (2014): 127–40; Hausmann, "Das Territorium der Ukraine: Stepan Rudnyc'kyjs Beitrag zur Geschichte räumlich-territorialen Denkens über die Ukraine," in *Die Ukraine: Prozesse der Nationsbildung*, ed. Andreas Kappeler (Köln: Böhlau, 2011), 145–57.

25. Willard Sunderland, *The Baron's Cloak: A History of the Russian Empire in War and Revolution* (Ithaca: Cornell University Press, 2014), 1–11.

26. Kristin Kopp, *Germany's Wild East: Constructing Poland as Colonial Space* (Ann Arbor: University of Michigan Press, 2012); Kopp, "Gray Zones: On the Inclusion of 'Poland' in the Study of German Colonialism," in Michael Perraudin and Jürgen Zimmerer, eds., *German Colonialism and National Identity* (New York: Routledge, 2011), 33–42; and Kopp, "Cartographic

Claims: Colonial Mappings of Poland in German Territorial Revisionism," in Gail Finney, ed., *Visual Culture in Twentieth-century Germany: Text as Spectacle* (Bloomington: Indiana University Press, 2006), 199–213.

27. See Amir Weiner, ed., *Landscaping the Human Garden: Twentieth-century Population Management in a Comparative Framework* (Stanford: Stanford University Press, 2003); and Peter Holquist, "'Information Is the Alpha and Omega of Our Work': Bolshevik Surveillance in Its Pan-European Context," *Journal of Modern History* 69:3 (September 1997): 415–50.

28. Matthew H. Edney, "People, Places and Ideas in the History of Cartography," *Imago Mundi* 66 (2014): 83–106.

29. Christine D. Worobec, ed., *The Human Tradition in Imperial Russia* (Lanham, Md.: Rowman & Little.eld, 2009); Marci Shore, *The Taste of Ashes: The Afterlife of Totalitarianism in Eastern Europe* (New York: Crown, 2013); Shore, *Caviar and Ashes: A Warsaw Generation's Life and Death in Marxism, 1918–1968* (New Haven: Yale University Press, 2006); Willard Sunderland and Stephen Norris, eds., *Russia's People of Empire: Life Stories from Eurasia, 1500 to the Present* (Bloomington: Indiana University Press, 2012); Donald G. Ostrowski and Marshall Poe, eds., *Portraits of Old Russia: Imagined Lives of Ordinary People, 1300–1725* (Armonk, N.Y.: M. E. Sharpe, 2011). On biography and life story in Russian history, Fitzpatrick and Slezkine, *In the Shadow of Revolution: Life Stories of Russian Women from 1917 to the Second World War.*

30. I build here on scholarship of a minimalist style among human geographers: Gunnar Olsson, *Abysmal: A Critique of Cartographic Reason* (Chicago: University of Chicago Press, 2007); John Pickles, *A History of Spaces: Cartographic Reason, Mapping, and the Geo-coded World* (London: Routledge, 2004); and Yi-Fu Tuan, *Space and Place: The Perspective of Experience* (Minneapolis: University of Minnesota Press, 2001). The Thai historian Thongchai Winichakul offered an insightful way of merging emotion and nation by introducing the interpretive neologism he called the "geo-body." Geo-bodies, he proposed, layered in half-rational cognition and perception, are "not merely space or territory . . . [but] a component of the life of a nation . . . a source of pride, loyalty, love, passion, bias, hatred, reason, unreason." Winichakul, *Siam Mapped: A History of the Geo-body of a Nation* (Honolulu: University of Hawai'i Press, 1988), 17.

31. Pamela Moss, ed., *Placing Autobiography in Geography* (Syracuse: Syracuse University Press, 2000); Moss and Karen Falconer Al-Hindi, eds., *Feminisms in Geography: Rethinking Space, Place, and Knowledges* (Lanham, Md.: Rowman & Littlefield, 2008); Lynn A. Staeheli, "Place," in John Agnew, et al., eds., *A Companion to Political Geography,* 2nd ed. (London: Blackwell, 2008), 158–70; and Staeheli et al., eds., *Mapping Women, Making Politics: Feminist Perspectives on Political Geography* (Routledge: London, 2004).

32. Jürgen Osterhammel, *Die Verwandlung der Welt: Eine Geschichte des 19. Jahrhunderts* (Munich: Beck, 2009); trans. Patrick Camiller as *The Transformation of the World: A Global History of the Nineteenth Century* (Princeton: Princeton University Press, 2014).

33. Bauman develops the concept in *Liquid Modernity* (Cambridge: Polity, 2000); *Liquid Love: On the Frailty of Human Bonds* (Cambridge: Polity, 2003); *Liquid Times: Living in an Age of Uncertainty* (London: Polity, 2006); and *Culture in a Liquid Modern World* (Cambridge: Polity, 2011). On performative selfhood, Goffman, *The Presentation of Self in Everyday Life.* For a poststructuralist approach, Butler, *Dispossession*; and Butler, *Gender Trouble.*

34. As in Kate Brown's collection, *Dispatches from Dystopia: Histories of Places Not Yet Forgotten* (Chicago: University of Chicago Press, 2015); also Jane T. Costlow, *Heart-Pine Russia: Walking and Writing the Nineteenth-century Forest* (Ithaca: Cornell University Press, 2013); Serguei Alex.

Oushakine, *The Patriotism of Despair: Nation, War, and Loss in Russia* (Ithaca: Cornell University Press, 2009); Omer Bartov, *Erased: Vanishing Traces of Galicia in Present-day Ukraine* (Princeton: Princeton University Press, 2007); and Gregor Thum, *Uprooted: How Breslau Became Wrocław during the Century of Expulsions* (Princeton: Princeton University Press, 2011).

35. Sven Beckert, "AHR Conversation: On Transnational History," *American Historical Review* 111:5 (2006): 1445; for applications, Heinz-Gerhard Haupt and Jürgen Kocka, eds., *Comparative and Transnational History: Central European Approaches and New Perspectives* (New York: Berghahn Books, 2009); Charles Maier, "Transformations of Territoriality, 1600–2000," in Gunilla Budde, Sebastian Conrad, and Oliver Janz, eds., *Transnationale Geschichte: Themen, Tendenzen und Theorien* (Vandenhoeck & Ruprecht, 2010), 32–55.

36. Oleh Kupchyns'kyi, ed., *Lystuvannia Stepana Rudnyts'koho* (Lviv: Shevchenko Scientific Society, 2006); Mykola Mushnyk, ed., *Lysty Stepana Rudnyts'koho do Sofii ta Stanislava Dnistrians'kykh: 1926–1932* (Edmonton: Canadian Institute of Ukrainian Studies, 1991).

37. Rogers Brubaker develops this insight in *Ethnicity without Groups* (Cambridge: Harvard University Press, 2006).

38. Maciej Górny, *Wielka Wojna profesorów, 1912–1923* (Warszawa: Instytut PAN, 2014); and Górny, "War on Paper? Physical Anthropology in the Service of States and Nations: 'Krieg der Geister,' Eastern Europe, and Physical Anthropology," in Jochen Böhler, Włodzimierz Borodziej, and Joachim von Puttkamer, *Legacies of Violence: Eastern Europe's First World War* (München: Oldenbourg, 2014), 131–67.

CHAPTER ONE

1. A recent exception is Andrea Wulf's gripping *The Invention of Nature: Alexander von Humboldt's New World* (New York: Knopf, 2015). For critical analysis of the German tradition, Michael Perraudin and Jürgen Zimmerer, eds., *German Colonialism and National Identity* (London: Routledge, 2015).

2. Primary details are in the *Memorial Volume of the Transcontinental Excursion of 1912 of the American Geographical Society of New York* (New York: American Geographical Society, 1915); and William Morris Davis, *The Development of the Transcontinental Excursion of 1912* (New York: American Geographical Society, 1915).

3. Jenő Cholnoky, *Utazásom Amerikában Teleki Pál gróffal* (Budapest: Vajda–Wichmann Kiadás, 1943), 3–6, 273–301. Cholnoky included in this wartime tribute to his old friend their personal photos of American West. Cholnoky's favorite images were of San Francisco, the Mormon temple in Salt Lake City, and the city of Provo, Utah.

4. Geoffrey J. Martin, *American Geography and Geographers: Toward Geographic Science* (Oxford: Oxford University Press, 2015), 493–94.

5. Pieter M. Judson, "Frontiers, Islands, Forests, Stones: Mapping the Geography of a German Identity in the Habsburg Monarchy, 1848–1900," in Patricia Yeager, ed., *The Geography of Identity* (Ann Arbor: University of Michigan Press, 2005), 382–406; Krista O'Donnell, Renate Bridenthal, and Nancy Reagin, eds., *The Heimat Abroad: The Boundaries of Germanness* (Ann Arbor: University of Michigan Press, 2005); Applegate, *A Nation of Provincials*.

6. Albrecht Penck, "Erinnerungen von Albrecht Penck," K. 871, S. 3, 71, Nachlass Albrecht Penck (hereafter Albrecht Penck Papers, or AP-P).

7. Ibid., K. 871, S. 3, 70. In Saxony, part of the Lutheran branch of the Penck family came from Odenwald, where Protestant identity was legally prescribed.

8. On German *bürgerlich* habitus and ethos, Elizabeth A. Drummond, "From 'verloren gehen' to 'verloren bleiben': Changing German Discourses on Nation and Nationalism in Poznania," in Charles W. Ingrao and Franz A. J. Szabo, eds., *The Germans and the East* (Lafayette, Ind.: Purdue University Press, 2009), 226–40; David Blackbourn and Richard J. Evans, eds., *The German Bourgeoisie: Essays on the Social History of the German Middle Class from the Late Eighteenth to the Early Twentieth Century* (London: Routledge, 1993).

9. Emil Meynen, "Albrecht Penck, 1858–1945," in T. W. Freeman, ed., *Geographers: Biobibliographical Studies*, vol. 7 (London: Mansell, 1983), 101–8.

10. Sarah K. Danielsson, *The Explorer's Roadmap to National-Socialism: Sven Hedin, Geography, and the Path to Genocide* (London: Ashgate, 2012).

11. Anne Marie Claire Godlewska, *Geography Unbound: French Geographic Science from Cassini to Humboldt* (Chicago: University of Chicago Press, 2000); and Godlewska and Neil Smith, eds., *Geography and Empire* (London: Blackwell, 1994).

12. Jeremy W. Crampton and Stuart Elden, eds., *Space, Knowledge, and Power: Foucault and Geography* (London: Ashgate, 2007); and Mitchell Dean, *Governmentality: Power and Rule in Modern Society* (London: Sage, 1999).

13. Mott T. Greene, "Geology," in Peter J. Bowler and John V. Pickstone, eds., *The Cambridge History of Science: The Modern Biological and Earth Sciences*, vol. 6 (Cambridge: Cambridge University Press, 2009), 178.

14. Heinz-Peter Brogiato, *"Wissen ist Macht—Geographisches Wissen ist Weltmacht": Die schulgeographischen Zeitschriften im deutschprachigen Raum (1880–1945) unter besonderer Berücksichtigung des Geographischen Anzeigers*, 2 vols. (Trier: Geographische Gesellschaft Trier, 1998).

15. See Richard J. Smith, *Mapping China and Managing the World: Culture, Cartography, and Cosmology in Late Imperial Times* (London: Routledge, 2013); Kären Wigen, Sugimoto Fumiko, and Cary Karacas, eds., *Cartographic Japan: A History in Maps* (Chicago: University of Chicago Press, 2016); Laura Hostetler, *Qing Colonial Enterprise: Ethnography and Cartography in Modern China* (Chicago: University of Chicago Press, 2005); and Jordana Dym and Karl Offen, eds., *Mapping Latin America: A Cartographic Reader* (Chicago: University of Chicago Press, 2011).

16. On May's legacy and popularity, Rivka Galchen, "Wild West Germany," *New Yorker*, 9 April 2012, 40–45; and Helmut Schmiedt, *Karl May oder Die Macht der Phantasie: Eine Biographie* (München: C. H. Beck, 2011). On German literary fantasies of America, Jeffrey L. Sammonds, *Ideology, Mimesis, Fantasy: Charles Sealsfield, Friedrich Gerstäcker, Karl May, and Other German Novelists of America* (Chapel Hill: University of North Carolina Press, 1998); Colin G. Galloway, Gerd Gemunden, and Suzanne Zantop, eds., *Germans and Indians: Fantasies, Encounters, Projections* (Lincoln: University of Nebraska Press, 2002). On Native American culture and scenography, Philip J. Deloria, *Playing Indian* (New Haven: Yale University Press, 1998); and for a more extensive take on Karl May and Native American history, H. Glenn Penny, *Kindred by Choice: Germans and American Indians since 1800* (Chapel Hill: University of North Carolina Press, 2013).

17. On the concept of "transcultural," I have in mind Mary Louise Pratt's poststructuralist critique of Alexander von Humboldt, *Imperial Eyes: Travel Writing and Transculturation* (London: Routledge, 1992), 111–43. For applications, Larry Wolff, *Inventing Eastern Europe: The Map of Civilization on the Mind of Enlightenment* (Stanford: Stanford University Press, 1994); and Wolff, "The Traveler's View of Central Europe: Gradual Transitions and Degrees of Difference in

European Borderlands," in Bartov and Weitz, eds., *Shatterzone of Empires*, 23–41. See also David Blackbourn, "'The Garden of Our Hearts': Landscape, Nature, and Local Identity in the German East," in Blackbourn and James Retallack, eds., *Localism, Landscape, and the Ambiguities of Place: German-speaking Central Europe, 1860–1930* (Toronto: University of Toronto Press, 2007), 149–64.

18. Eszter B. Gantner and Péter Varga, eds., *"Transfer—Interdiziplinär!": Akteure, Topographien und Praxis der Wissenstransfer* (Frankfurt: Peter Lang, 2013); Kapil Raj, *Relocating Modern Science: Circulation and the Construction of Knowledge in South Asia and Europe, 1650–1900* (New York: Palgrave Macmillan, 2007), 223–34.

19. Heffernan, *The Meaning of Europe, 63–64.*

20. For a history of the BAAS, Charles W. A. Withers, *Geography and Science in Britain, 1831–1939: A Study of the British Association for the Advancement of Science* (Manchester: Manchester University Press, 2010); and Withers, "Scale and the Geographies of Civic Science: Practice and Experience in the Meetings of the British Association for the Advancement of Science in Britain and in Ireland, c. 1845–1900," in Livingstone and Withers, *Geographies of Nineteenth-century Science, 99–122.*

21. Albrecht Penck and Eduard Brückner, *Die Alpen im Eiszeitalter*, 3 vols. (Leipzig, 1901–9).

22. Osterhammel, *Die Verwandlung der Welt*, translated as *The Transformation of the World.*

23. Józef Babicz, "Eugeniusz Romer, 1871–1954," in T. W. Freeman, Marguerita Oughton, and Philippe Pinchemel, eds., *Geographers: Biobibliographical Studies*, vol. 1 (Salem, N.H.: Mansell, 1977), 89–96.

24. On early modern to modern transformations, Eagle Glassheim, *Noble Nationalists: The Transformation of the Bohemian Aristocracy* (Cambridge: Harvard University Press, 2005); Jeremy King, *Budweisers into Czechs and Germans: A Local History of Bohemian Politics, 1848–1948* (Princeton: Princeton University Press, 2005); Rita Krueger, *Czech, German, and Noble: Status and National Identity in Habsburg Bohemia* (New York: Oxford University Press, 2009); Janusz Tazbir, *Kultura szlachecka w Polsce: rozkwit, upadek, relikty* (Poznań: Wydawnictwo Poznańskie, 1998); and Ivo Banac and Frank E. Sysyn, eds., *Concepts of Nationhood in Early Modern Eastern Europe* (Cambridge: Ukrainian Research Institute, Harvard University, 1987).

25. In 1543, King Sigismund I Vasa of Sweden and Polish-Lithuania bestowed titles to the father of Stefan Romer (his name is unknown). One branch settled in southern Poland. Mateusz Romer fought against the Cossacks in Ukraine during the Khmel'nyts'kyi Uprising of 1648–54; Adam Romer was a Renaissance humanist and professor at Jagiellonian University who wrote books on Cicero. His grandfather's cousin Aleksander was a historian and eighteenth-century philanthropist who founded a library in Zabełcze, close to Nowy Sącz. On the genealogy, Stanisław Marian Brzozowski, "Eugeniusz Romer," *Polski słownik biograficzny*, 31/4, no. 131 (Wrocław: Polska Akademia Umiejętności, 1989), 636–45.

26. For context, Iryna Vushko, *The Politics of Cultural Retreat: Imperial Bureaucracy in Austrian Galicia, 1772–1867* (New Haven: Yale University Press, 2015).

27. In 1913, Jan was placed in charge of the 30th Artillery Force in Lwów. He was promoted to general during the war, in 1917. By 1918–19, in the Polish-Ukrainian War, he commanded Poland's Thirteenth Regiment on the outskirts of Lwów.

28. Romer's choice unmade a complex Polish-Lithuanian and "liberal" Habsburg multicultural past into a Polish and modern European ethnonational one. See Yohanan Petrovsky-Shtern, *The Anti-imperial Choice: The Making of the Ukrainian Jew* (New Haven: Yale University

Press, 2009); Timothy Snyder, *The Red Prince: The Secret Lives of a Habsburg Archduke* (New York: Basic Books, 2008); Serhiy Bilenky, *Romantic Nationalism in Eastern Europe: Russian, Polish, and Ukrainian Political Imaginations* (Stanford: Stanford University Press, 2012); Serhii Plokhii, *Unmaking Imperial Russia: Mykhailo Hrushevsky and the Writing of Ukrainian History* (Toronto: University of Toronto Press, 2005).

29. Heffernan, *The Meaning of Europe*, 71–76; Naumann is covered comprehensively in the *Begriffsgeschichte* by Henry Cord Meyer, *Mitteleuropa in German Thought and Action, 1815–1945* (The Hague: Martinus Nijhoff, 1955). For uses of the term, Johann P. Arnason, Petr Hlaváček, Stefan Troebst, et al., eds., *Mitteleuropa? Zwischen Realität, Chimäre und Konzept* (Prague: Charles University Department of Philosophy, 2015).

30. Eugeniusz Romer, *Pamiętniki paryski, 1918–1919* (Wrocław: Zakład Narodowy im. Ossolińskich, 1989), 21–29.

31. See Antoni Jackowski, Stanisław Liszewski, and Andrzej Richling, eds., *Historia geografii polskiej* (Warszawa: PWN, 2008), 42–47; and Seegel, *Mapping Europe's Borderlands: Russian Cartography in the Age of Empire* (Chicago: University of Chicago Press, 2012), 177–85. In 1877 at Jagiellonian University, the Department of Geography was reestablished after a twenty-five-year suspension. It was founded initially in 1849 with the poet-naturalist Wincenty Pol as its first chair. Habsburg authorities in Vienna dissolved it in 1852, on fears of Polish separatism.

32. Eugeniusz Romer, "Notebooks 1891–1900," Collected Papers of Eugeniusz Romer (hereafter ER-P), Archive of Jagiellonian University Library (hereafter A-JUL), Department of Manuscripts (hereafter DoM), Kraków, 10211 II, T. 1-4; 10212 II, T. 1-4; 10216 II, T. 1-3.

33. In line with Poland's nationalizing policies after 1918, the school was renamed for Stefan Batory (Báthory), the sixteenth-century king.

34. Deborah R. Coen, "Imperial Climatographies from Tyrol to Turkestan," *Osiris* 26 (2011): 45–65; Coen, "Scaling Down: The 'Austrian' Climate between Empire and Republic," in James R. Fleming, Vladimir Jankovic, and Coen, eds., *Intimate Universality: Local and Global Themes in the History of Weather and Climate* (Sagamore Beach, Mass.: Science Publications, 2006), 115–40.

35. On the brother's family and career, Jan Edward Romer, *Pamiętniki* (Warszawa: Muzeum Historii Polski, 2011), 23–32.

36. Edmund Romer, *Geograf trzech epok: wspomnienia o ojcu* (Warszawa: Czytelnik, 1985), 28–29.

37. Dobiesław Jędrzejczyk, "Geopolitical Essence of Central Europe in Writings of Eugeniusz Romer," *Miscellanea Geographica* 11 (2004): 199–206.

38. Quoted in Stanisław Eile, *Literature and Nationalism: Literature and Nationalism in Partitioned Poland, 1795–1918* (London: MacMillan Press, 2000), 16.

39. Oleh Shablii, *Akademik Stepan Rudnyts'kyi: fundator Ukrains'koi heohrafii*, 2nd ed. (Lviv: Vydavnychyi Tsentr LNU im. Ivana Franka, 2007), 9.

40. On the intelligentsia's roles on both sides of the Habsburg-Russian border, Serhy Yekelchyk, *Ukraine: Birth of a Modern Nation* (Oxford: Oxford University Press, 2007); Timothy Snyder, *The Reconstruction of Nations: Poland, Ukraine, Lithuania, Belarus, 1569–1999* (New Haven: Yale University Press, 2003).

41. Seegel, *Mapping Europe's Borderlands*, 253–58; Ihor Stebelsky, *Placing Ukraine on the Map: Stepan Rudnyts'kyi's Nation-Building Geography* (self-published by the author, 2014), 20–25.

42. Shablii, *Akademik Stepan Rudnyts'kyi*, 18–19.

43. As in Ann Laura Stoler, *Along the Archival Grain: Epistemic Anxieties and Colonial Common*

Sense (Princeton: Princeton University Press, 2009); and *Carnal Knowledge and Imperial Power: Race and the Intimate in Colonial Rule* (Berkeley: University of California Press, 2002).

44. The seminal social history is Martha Bohachevsky-Chomiak, *Feminists Despite Themselves: Women in Ukrainian Community Life, 1884–1939* (Edmonton: Canadian Institute of Ukrainian Studies Press, 1988).

45. On spaces of continuity, Coen, *Vienna in the Age of Uncertainty*; and Stephen Lovell, *Summerfolk: A History of the Dacha, 1710–2000* (Ithaca: Cornell University Press, 2003).

46. Volodymyr Potul'nyts'kyi, "Galician Identity in Ukrainian Historical and Political Thought," in Chris Hann and Paul Robert Magocsi, eds., *Galicia: A Multicultured Land* (Toronto: University of Toronto Press, 2005), 82–102.

47. The Shevchenko Scientific Society published the work in three articles in 1905, 1907 and 1913. After World War I, Rudnyts'kyi compiled these into his two-part *Osnovy morfolohii i geolohii Pidkarparts'koi Rusy i Zakarpattia vzahali*. This came out in Uzhhorod in 1925 and 1927, after his move to Kharkov/Kharkiv.

48. The original Romer-Rudnyts'kyi exchange was in *Kosmos* 32 (1907): 219–20, 367–78, 462–63. On this scientific quarrel, Shablii, *Akademik Stepan Rudnyt'skyi*, 19–20. The venue was as important as the political confrontation itself, in the Polish-language journal *Kosmos*, titled after Humboldt's acclaimed study of geography in 1842. Rudnyts'kyi's previous work was in Ukrainian and German in the NTSh journal edited by Mykhailo Hrushevs'kyi, and Rudnyts'kyi served as one of the coeditors of its special section on research in mathematics, natural sciences, and medicine. Wars changed memory, a lot: in a curious unpublished fragment typed on the back of a 31 May 1920 draft of a set of instructions, Romer advised the Second Republic's Ministry of Public Works in Warsaw to explain to the government what cartographers should be doing. It is clear that Romer associated "German" Penck and Rudnyts'kyi together. Looking back through the lens of the World War I, he criticized Rudnyts'kyi's work in 1907 as unscientific and his habilitation at the University of Lemberg in 1908 as not merited. ER-P, 10193 IV, k. 11–13 and reverse side of k. 13.

49. Stepan Rudnyts'kyi, *Nacherk heohrafichnoi terminolohii* (Lviv: NTSh, 1908).

50. Larry Wolff, *The Idea of Galicia: History and Fantasy in Habsburg Political Culture* (Stanford: Stanford University Press, 2010).

51. Smith, *American Empire*, 32.

52. Bowman to Dr. John C. French, 5 January 1946, IB-P Ms. 58, Series I, Box 1.1, Folder 7.

53. On "self-made man" myths in Gilded Age America, a shrewd analysis is offered in Scott A. Sandage, *Born Losers: A History of Failure in America* (Cambridge: Harvard University Press, 2005).

54. "The Name and Family of Bowman," IB-P Ms. 58, Series I, Box 1.1, Folder 6.

55. Susan Schulten, *Mapping the Nation: History and Cartography in Nineteenth-century America* (Chicago: University of Chicago Press, 2012); Schulten, *The Geographical Imagination in America, 1880–1950* (Chicago: University of Chicago Press, 2001); and Martin Brückner, *The Geographic Revolution in Early America: Maps, Literacy, and National Identity* (Chapel Hill: University of North Carolina Press, 2006).

56. See Smith, *American Empire*, 36, where the author insightfully calls it "the biggest journey of [Bowman's] life."

57. Ibid., 38–39, 275–76.

58. As quoted in Martin, *The Life and Thought of Isaiah Bowman*, 12. The date is not specified, but it is likely that the two Harvard letters from Bowman to Jefferson were typed in fall 1904.

59. Ibid., 13.

60. Penck to Bowman, 11 December 1905, IB-P Ms. 58, Series II, Box 35.

61. On Bowman's expeditions and aims in the 1900s–1910s to South America, Smith, *American Empire*, 53–82.

62. Penck to Bowman, 24 July 1909, IB-P Ms. 58, Series II, Box 35.

63. Martin, *The Life and Thought of Isaiah Bowman*, 31–32.

64. For chronology and background, George Kish, "Paul Teleki, 1879–1941," in T. W. Freeman, ed., *Geographers: Biobibliographical Studies*, vol. 11 (London: Mansell, 1987): 139–43.

65. The family's mysterious genealogy and Teleki's early life is detailed in Ablonczy, "Egy szépreményű fiatalember: száarmazáas, család, ifjúkor," *Rubicon 15:2* (2004): 8–9; and *Pál Teleki*, 1–4. See also the file, "Teleki Andor hagyatéka: Teleki családra vonatkozó eredeti iratok, 1764–1966," National Archives of Hungary (Magyar Nemzeti Levéltár, hereafter MNL), Budapest, P 1881, 41, 1-3.

66. See Ablonczy, *Pál Teleki*, 6 and 164, for confirmation of Teleki's passion for Karl May.

67. László Bassa, "Teleki Pál és a térkép," *Tér és társadalom* 3–4 (1990): 175–83.

68. Ratzel's works are probably the most influential single current in the development of modern Hungarian human/cultural geography. See Zoltan Hajdú, "Friedrich Ratzel hatása a Magyar földrajztudományban," *Tér és társadalom*, no. 3 (1998): 96–99.

69. As quoted in Ablonczy, *Pál Teleki*, 13.

70. In the MFT, Cholnoky was general secretary from 1905 to 1911. Teleki directly succeeded him and served from 1911 to 1922. Cholnoky was elected after Lóczy and served as president from 1914 all the way until 1945.

71. Pál Teleki, *Atlasz a Japani szigetek cartográphiájának történetéhez* (Budapest: Kilián Figyes Utóda Magy. Kir. Egyetemi Könyvkereskedő, 1909). Published also in German as the *Atlas zur Geschichte der Kartographie der Japanischen Inseln* (Leipzig: K. W. Hiersemann, 1909).

72. On Europe's colonial mapping of Japan, Kären Wigen, *A Malleable Map: Geographies of Restoration in Central Japan, 1600–1912* (Berkeley: University of California Press, 2010).

73. J. M. Blaut, *The Colonizer's Model of the World: Geographical Diffusionism and Eurocentric History* (New York: Guilford Press, 1992); for skillful postcolonial readings of maps, Kopp, *Germany's Wild East*, especially chapter 4, 124–59.

74. Named for the explorer Otto von Nordenskjöld, the Nordenskjöld Collection of maps is stored today at the Helsinki University Library.

75. Not much was produced: Teleki and Károly Kogutowicz, son of the Hungarian-Polish cartographer Manó Kogutowicz, cooperated to publish maps for the project between 1912 and 1914 in *Del-Magyarország térképe* (Budapest: Magyar Földrajzi Intézet, 1912), but the two men disagreed about basic things like schedules and the mechanics of production. Teleki's aims for a better-developed patriotic school geography were derailed by the war's stalemates, and only three maps were finally made. The original pre-1914 Teleki-Kogutowicz maps (scale of 1:100,000) are in the map collection of the National Széchényi Library (Országos Széchényi Könyvtár), Budapest, 912/439, 2-13/1912, TM 4 147.

76. Teleki, *Rapport sur les voyages et les travaux géographiques exécutés par les explorateurs et les savants hongrois depuis l'année 1889* (Roma: Tip. dell'Unione Editrice, 1914).

77. On maps of cognition and counterhegemonic strategies, G. Malcolm Lewis, ed., *Cartographic Encounters: Perspectives on Native American Mapmaking and Map Use* (Chicago: University of Chicago Press, 1998); Martin Brückner, ed., *Early American Cartographies* (Chapel Hill:

University of North Carolina Press, 2012); and the superb review of map scholarship by Camilo Arturo Leslie, "Territoriality, Map-mindedness, and the Politics of Place," *Theory and Society* 45 (2016): 169–201.

CHAPTER TWO

1. Michael Heffernan, "Professor Penck's Bluff: Geography, Espionage, and Hysteria in World War I," *Scottish Geographical Journal* (2000) 116:4, 267–82, here 269–70, 280n27. Original sources are in Leipzig: Albrecht Penck, "Professoren als Spione. Abwehr von Prof. Dr. Albrecht Penck, Direktor des geographischen Instituts der Berliner Universität" (1917), AP-P, K. 877, S. 36/4.

2. Albrecht Penck, *Von England festgehalten: Meine Erlebnisse während des Krieges im britischen Reich* (Stuttgart: J. Engelhorn, 1915), 115–19. The Engelhorn family founded the publisher of Penck's volume in Stuttgart in 1860. It catered to a German middlebrow readership and printed other so-called *Kriegsbücher* in 1914–15, similar to Penck's. These were Gottfried Traub's *From the Forging of Weapons* (*Aus der Waffenschmiede*), Heinrich Ehotzen's *Faith of the Courageous* (*Der Glaube des Tapferen*), and Paul Rohrbach's *On the Way to a Global People* (*Zum Weltvolk hindurch*).

3. Penck, *Von England festgehalten*, 166.

4. On knowledge and conflict in German-Polish relations, Gernot Briesewitz, *Raum und Nation in der polnischen Westforschung, 1918–1948: Wissenschaftsdiskurse, Raumdeutungen und geopolitische Visionen im Kontext der deutsch-polnischen Beziehungsgeschichte* (Osnabrück: Fibre, 2014); and Alexandra Schweiger, *Polens Zukunft liegt im Osten: Polnische Ostkonzepte der späten Teilungszeit, 1890–1918* (Marburg: Herder-Institut, 2014).

5. Penck, *Von England festgehalten*, 8, 13–15, 23–24, 140–43, 200, 209.

6. Penck, "Sven Hedin über England und Deutschland," *Zeitschrift der Gesellschaft für Erdkunde zu Berlin* 50 (1915): 243–45. In the review, Penck praised Hedin's prowar book *A Nation in Arms* (*Ein Volk in Waffen*) and his calls for German unity.

7. Penck, *Was wir im Kriege gewonnen und was wir verloren haben* (Berlin: Carl Behmanns Verlag, 1915), 25–26.

8. Penck, "Der Krieg und das Studium der Geographie," *Zeitschrift der Gesellschaft für Erdkunde zu Berlin* 51 (1916): 158–76, 222–48.

9. Penck, "Die Ukraina," *Zeitschrift der Gesellschaft für Erdkunde zu Berlin* 51 (1916): 346–61, 458–77.

10. Penck, "Die Ukraina, das Land der Steppen und des Lösses" (1916). The quote here is from the original manuscript, AP-P, K. 876, S. 7.

11. Mroczko, *Eugeniusz Romer*, 72; Timothy Snyder, *The Red Prince: The Secret Lives of a Habsburg Archduke* (New York: Basic Books, 2008), 83–90.

12. Penck to Hans Hartwig von Beseler, 21 April 1916. The original is in the ER-P, 10197 IV, T. IV, No. 26, k. 331-337. It is evidenced here that Penck alerted the Warsaw governor-general to Romer's 1916 *Geographical-Statistical Atlas of Poland*, but he did not call for an arrest.

13. Aniela Chałubińska, "Kontakty Eugeniusza Romera i Albrechta Pencka," *Studia i materiały z dziejów nauki polskiej*, Seria C, vol. 24 (1980): 15–33. Chałubińska was the first to treat the Penck-Romer conflict, though in communist-era Poland she lacked access to Romer's political writings as well as both of the geographers' private papers.

14. Postcard from Penck to Romer, 6 September 1902, ER-P, 10323 III, T. 44, k. 227. The postcard featured a snapshot of the Alps on the front, published by a Zurich print company. Penck thanked Romer in a friendly way for sending him his published work on the Carpathians.

15. Romer to Penck, 22 May 1916, ER-P, 10378 III, k. 111–13.

16. Ibid.

17. Penck to Romer, 5 June 1916, ER-P, 10323 III, T. 44, k. 229.

18. On objectivity as a set of shared values and professional standards, Peter Novick, *That Noble Dream: The "Objectivity Question" and the American Historical Profession* (Cambridge: Cambridge University Press, 1988).

19. The term suggests the effectiveness of Germany's eastern soils to produce crops. From Penck's vantage, it also denotes an open frontier and the need to "tame" nature. The male experts' biological and European colonial gaze encoded the German developmentalist stereotype of the "Polish economy."

20. For recent theoretical debates, Thomas Nail, *Theory of the Border* (Oxford: Oxford University Press, 2016); Jordan Branch, *The Cartographic State: Maps, Territory, and the Origins of Sovereignty* (Cambridge: Cambridge University Press, 2014); Denis Wood and John Fels, *The Natures of Maps: Cartographic Constructions of the Natural World* (Chicago: University of Chicago Press, 2009); and Christian Jacob, *The Sovereign Map: Theoretical Approaches to Cartography throughout History* (Chicago: University of Chicago Press, 2006).

21. Romer to Penck, 31 December 1916, ER-P, 10378 III, k. 112.

22. Mroczko, *Eugeniusz Romer*, 62–63. Romer was referring to Ratzel's classic *Politische Geographie* (München and Berlin: Oldenbourg, 1903), 271. He also cited the prewar work of Alfred Hettner, who wrote *Das europäische Russland: Eine Studie zur Geographie des Menschen* (Leipzig and Berlin: Oldenbourg, 1905). Ratzel and Hettner were not identical in their visions: Hettner argued for a German empire elsewhere, because he thought it unlikely that any of the three partitioning powers of Poland would relinquish their lands. If Germany restored Poland as a buffer state against Russia, it would have to pay all the land-based costs.

23. On Polish uses and discourses of civilization, Jerzy Jedlicki, *A Suburb of Europe: Nineteenth-century Polish Approaches to Western Civilization* (Budapest: Central European University Press, 1999); and Brian A. Porter, *When Nationalism Began to Hate: Imagining Modern Politics in Nineteenth-century Poland* (Oxford: Oxford University Press, 2002).

24. The boundary expert drew from his expertise and empirical work in the (North/South) Americas, which he mapped *into* a fantasy of Eastern Europe. His borderland-themed works during the war were "Results of an Expedition to the Central Andes," *Bulletin of the American Geographical Society* 46 (1914): 161–83; "Non-existence of Peqary Channel," *Geographical Review* 1 (1916): 448–56; and "Frontier Region of Mexico: Notes to Accompany a Map of the Frontier," *Geographical Review* 3 (1917): 16–27.

25. No joke here: Bowman literally owned a black book, stored today with his personal papers (IB-P Ms. 58), Series I (Biographical Materials), at the Sheridan Libraries, Johns Hopkins University.

26. Referring to "Polish" Romer and his place in Lemberg/Lwów/Lviv, Geoffrey Martin gets a few small details incorrect about East Central Europe in a single sentence: "Bowman sought to bring Eugenius [*sic*] Romer, living temporarily in Vienna, from the recently destroyed University of Lwow [Lwów, to be exact] in war-devastated Lemburg [*sic*]. But that plan too failed." Martin, *The Life and Thought of Isaiah Bowman*, 33.

27. On Teleki and race science, Ablonczy, *Pál Teleki*, 33–46.

28. On race and eugenics here, Christian Promitzer, Sevasti Trubeta, and Marius Turda, eds., *Health, Hygiene, and Eugenics in Southeastern Europe to 1945* (Budapest: Central European University Press, 2010); Turda and Paul J. Weindling, eds., *"Blood and Homeland": Eugenics and Racial Nationalism in Central and Southeast Europe, 1900–1940* (Budapest: Central European University Press, 2007); Maria Bucur, *Eugenics and Modernization in Interwar Romania* (Pittsburgh: University of Pittsburgh Press, 2002); and Emese Lafferton, "The Magyar Moustache: The Faces of Hungarian State Formation, 1867–1918," *Studies in History and Philosophy of Biological and the Medical Sciences* 38 (2007): 706–32.

29. Ablonczy, "Útkeresés: A turáni társaságban," *Rubicon* 15:2 (2004): 12–14.

30. The original article was published in May 1916. It is reprinted in full in Pál Teleki, "Táj és faj," *Válogatott politikai írások és beszédek* (Budapest: Osiris, 2000), 14–26.

31. The MFT remained a quite aristocratic, male-dominated fraternity well into the interwar period of the 1920s and 1930s. According to internal lists of Hungarian and foreign corresponding members, the first woman in the society seems to have been Dr. Károlyné Papp, wife of the geologist Dr. Károly Papp. She was elected in 1929. Archive of Hungarian Geographical Society (hereafter A-MFT), Document 9/1932.

32. For Herder's catastrophism about Hungary, Lórant Czigány, *The Oxford History of Hungarian Literature* (Oxford: Clarendon Press, 1984), 103, 114–16.

33. Teleki, "A Turán földrajzi fogalom," *Turán* 3 (1918): 44–83.

34. Teleki contributed over one hundred pages to the *Geographical Bulletin*. In addition, he wrote for the *Földrajzi évkönyv* (*Geographic Yearbook*), also based in Budapest.

35. Pál Teleki, review of (Baron) Dr. Franz Nopcsa, "Zur Geschichte der Kartographie Nordalbaniens," *Mitteilungen der Geologischen Gesellschaft in Wien* (1916), 520–85, in *Földrajzi Közlemények* 44 (1916): 474–75.

36. The count called for attention to the county unit (*megye*) in Hungarian and Transylvanian history, focusing on Vidal de la Blache. From Blache, Teleki regarded France's regionalism as a full modern political unity. He focused, as a potential model for Hungary, on landscape, rural productivity, and the durability of local identity (i.e., peasants *already* as Frenchmen) outside the capital. Teleki heralded the scientific study of the geography of France's provinces as part of modernization and a point of interest for Magyar settlement into St. Stephen crownlands. Eugen Weber, the historian of modern French history and Europe's right (he was born in Bucharest in 1925) famously challenged the modernization thesis in *Peasants into Frenchmen: The Modernization of Rural France, 1870–1914* (Stanford: Stanford University Press, 1976). See Teleki, review of Paul Vidal de la Blache, *Les division régionales de la France* (Paris: F. Alcan, 1913), in *Földrajzi Közlemények* 44 (1916): 474–75.

37. Teleki, *A földrajzi gondolat története* (Budapest: Szerző kiadása, 1917).

38. Teleki, "Főtitkari jelentes (Keleti torekveseink): A Magyar Földrajzi Társaság 1918 Április 25-en tartott közgyűlésen elmondotta," *Földrajzi Kozlemények* 46 (1918): 249. This was an address to the MFT main assembly in Budapest from his post of general secretary. He sought to harmonize Hungary's economic priorities with its scientific institutions such the MTA, the MFT, and the Earth Sciences Institute. Proudly he announced that the MTA and MFT were planning to publish a massive volume highlighting Hungarians' contributions in geography, zoology, botany, ethnography, history, architecture, and archaeology, including those by the luminaries Lóczy and Cholnoky. Teleki thus placed geography highest on the pedestal of knowledge, for it could mold the best and strongest Hungarian citizens.

39. Clemens Kaps and Jan Surman, eds., "Post-colonial Perspectives on Habsburg Galicia," *Historyka: Studia metodologiczne* 42 (2012): 7–35; and Mitchell G. Ash and Surman, eds., *The Nationalization of Scientific Knowledge in the Habsburg Empire, 1848–1918* (New York: Palgrave Macmillan, 2012).

40. Rudnyts'kyi, *Ukraina und die Ukrainer* (Wien: Verlag des Allgemeinen Ukrainischen Nationalrates, 1914); *The Ukraine and the Ukrainians* (Jersey City: Ukrainian National Council, 1915);

41. The full title for the 1916 edition was *Ukraina: Land und Volk eine gemeinfassliche Landeskunde* (Wien: Bund zur Befreiung der Ukraina, 1916). This was later rendered into English, as *Ukraine: The Land and Its People: An Introduction to Its Geography* (New York: Rand McNally, 1918).

42. Guido Hausmann, "The Ukrainian Moment of World War I," in Gearóid Barry, Enrico Dal Lago, and Róisín Healy, eds., *Small Nations and Colonial Peripheries in World War I* (Leiden: Brill, 2016), 177–91; Hausmann, "Das Territorium der Ukraine: Stepan Rudnyc'kyjs Beitrag zur Geschichte räumlich-territorialen Denkens über die Ukraine," in Andreas Kappeler, ed., *Die Ukraine: Prozesse der Nationsbildung* (Köln: Böhlau, 2011), 145–157; and Stebelsky, *Placing Ukraine on the Map*, 20–25.

43. Sh. Levenko [pseud. Stepan Rudnyts'kyi], *Chomu my khochemo samostiinoi Ukrainy* (Wien, 1915).

44. For instance, Eric Lohr, Vera Tolz, Alexander Semyonov, and Mark von Hagen, *The Empire and Nationalism at War* (Bloomington, Ind.: Slavica, 2014); Alexander Statiev, *The Soviet Counterinsurgency in the Western Borderlands* (Cambridge: Cambridge University Press, 2010); and Lohr, *Nationalizing the Russian Empire: The Campaign against Enemy Aliens during World War I* (Cambridge: Harvard University Press, 2003).

45. Penck to Romer, 10 January 1917, ER-P, 10323 III, T. 44. k. 232; and 10378 III, T. 13, k. 113.

46. Anonymous, [no title], *Ukrainische Korrespondenz*, 16 February 1917, no. 6.

47. Romer, "Über die kriegspolitische Karte Polens: Aus Anlass der Unzufriedenheit, welche sie in den 'ukrainischen Sphären' hervorgerufen hat," 15 March 1917, *Polen* 116 [no page], at the AGSL, uncatalogued Pamphlet File. Romer signed and sent the off-print to New York City in July 1920, during the Polish-Soviet War and while Piłsudski's forces in Ukraine and Belarus were in retreat. That was one month before the "miraculuous" turnaround for Piłsudski and the Poles at the Battle of Warsaw.

48. Tomasz E. Romer, "Eugeniusz Romer we wspomnieniach wnuka," in Jerzy Ostrowski, Jacek Pasławski, and Lucyna Szaniawska, eds., *Eugeniusz Romer: Geograf i kartograf trzech epok* (Warsaw: Biblioteka Narodowa, 2004), 7–10.

49. Penck, "Polnisches," 27 March 1917, *Posener Tagblatt* 116: 22–29. Penck published the review eleven days earlier, as part of a pamphlet distributed in Berlin. Penck, *Zeit- und Streitfragen. Korrespondenz des Bundes und deutscher Gelehter Künstler* no. 10, 16 March 1917.

50. Penck, "Polnisches," 23.

51. Penck to Romer, 5 April 1917, ER-P, 10323 III, T. 44, k. 232.

52. Evidence of their encounter on 19 July 1934 appears in Romer's private notebooks. Penck still refused to attend the IGC in Warsaw after their meeting in Berlin. ER-P, 10259, T. 3, k. 52–53.

53. I am strongly indebted here to Michael Gordin, *The Pseudoscience Wars: Immanuel Velikovsky and the Birth of the Modern Fringe* (Chicago: University of Chicago Press, 2012).

54. Romer, "Albrecht Penck über den 'Atlas von Polen,'" *Polen*, 4 May 1917 [no page number].

55. Romer, "Albrecht Penck o atlasie Polski," *Kurier Lwówski* no. 219 (1917). A copy is in the ER-P, 10185 III. no. 6, k. 31.

56. The seminal study on *Ostforschung* is Michael Burleigh, *Germany Turns Eastwards: A Study of Ostforschung in the Third Reich* (Cambridge: Cambridge University Press, 1988), printed again in 2002. The bulk of the book was devoted to 1933 to 1945. On interwar *Ostforschung*, Winson Chu, *The German Minority in Interwar Poland* (Cambridge: Cambridge University Press, 2012), 40–49; Jan M. Piskorski, Jörg Hackmann, and Rudolf Jaworski, eds., *Deutsche Ostforschung und polnische Westforschung im Spannungsfeld von Wissenschaft und Politik: Disziplinen im Vergleich* (Poznań: Poznańskie Towarzystwo Przyjaciół Nauk, 2002). For geography as a discipline in the Nazi period, Mechtild Rössler, "Wissenschaft und Lebensraum": *Geographische Ostforschung im National-sozialismus: ein Beitrag zur Disziplingeschichte der Geographie* (Berlin: D. Reimer, 1990).

57. Joseph Partsch, "Die Festgaben zu Albrecht Pencks sechzigsten Geburtstage," *Zeitschrift der Gesellschaft für Erdkunde zu Berlin* no. 7/8 (1918): 326–35.

58. Aniela Chałubińska, "Kontakty Eugeniusza Romera i Albrechta Pencka," *Studia i materiały z dziejów nauki polskiej*, Seria C, vol. 24 (1980): 28. The 438-page tribute as a book is *Festband Albrecht Penck: Zur Vollendung des sechzigsten Lebensjahrs gewidmet von seinen Schülern und der Verlagsbuchhandlung* (Stuttgart: J. Engelhorn, 1918). Penck's own copy is in the AP-P, K. 877, S. 37.

59. Stepan Rudnyckyi, "Die podolische Platte in Galizien," in *Festband Albrecht Penck*, 198–211. This part of Rudnyts'kyi's doctoral work was published by the Shevchenko Scientific Society (NTSh) in Lemberg/Lviv, in Ukrainian and German before the war.

60. On Cvijić and the mapping of Southeastern Europe by race, Jeremy W. Crampton, "The Cartographic Calculation of Space: Race Mapping and the Balkans at the Paris Peace Conference of 1919," *Social and Cultural Geography* 7:5 (2006): 731–52.

61. From the preface to the *Festband Albrecht Penck*, vii–xi.

62. For a postpositivist analysis of fact-fetishized surveys of this sort, Tong Lam, *A Passion for Facts: Social Surveys and the Construction of the Chinese Nation-state, 1900–1949* (Berkeley: University of California Press, 2011).

63. Dr. E[rich] Wunderlich, ed., *Handbuch von Polen (Kongress-Polen): Beiträge zu einer allgemeinen Landeskunde, Veröffentlichungen der Landeskundlichen Kommission beim Kaiserl. Deutschen Generalgouvernement Warschau* (Berlin: D. Reimer, 1918). References to a formidable "E. v. Romer" appear on pages 9, 24, 71, 142, 144, 145, 191, 324, 359, 360, and 382 of the compilation.

64. Penck, "Rede zur Gedächtnisfeier des Stifters der Berliner Universität König Friedrich Wilhelms III in der Aula am 3. Aug. 1918 (Die erdkundlichen Wissenschaften an der Universität Berlin)" (Berlin, 1918), 3. Original is in AP-P, II A 216.

65. Penck, "Rede," 4.

66. For the context of the Spett map, Richard Blanke, *Polish-speaking Germans? Language and National Identity among the Masurians since 1871* (Köln: Böhlau, 2001), 123.

67. On the revisionist geographers, Astrid Mehmel, "Deutsche Revisionspolitik in der Geographie nach dem ersten Weltkrieg," *Geographische Rundschau* 47 (1995): 498–505.

68. Guntram Henrik Herb, *Under the Map of Germany: Nationalism and Propaganda, 1918–1945* (London: Routledge, 1997), 124–29. Penck published the piece in *Deutsche Allgemeine Zeitung* on 9 February 1919, which according to Herb was the first of his public quarrels with Poland and its geographers. Without sources in Polish, however, one can only infer when quarrels were "public" based on if and when they appeared in German print.

69. A longer sidebar here: the writings of Hannah Arendt and Martin Heidegger certainly come to mind, or Jürgen Habermas's *Technik und Wissenschaft als Ideologie* (Frankfurt: Suhrkamp, 1968). Outside of questions about *techne* (in ancient Greek) or critiques of instrumental reason, one could even take Herb's detailed study here on point. Its strength and weakness lie in its in-depth deconstructive discussions of maps as textual sources and map praxis, which rehearse a post-Versailles echo chamber of charges from the German side against Polish use of "propaganda," based on misuse of sources and especially census data. As modern political geography, to call every border "constructed" and all maps "propaganda" might be correct, but it also omits biography and multivectored transfer of knowledge (*Wissenstransfer*). None of this discourse was transparent; both "young angry men" and older decolonial *völkisch* men who became revisionists invested more in personal contacts than simple science. In classic studies by Fritz Stern and George Mosse, each historian identified the "moderate" ordinary channels through which German middle-class frustrations flowed: Fritz Stern, *The Politics of Cultural Despair: A Study in the Rise of the Germanic Ideology* (Berkeley: University of California Press, 1961); and George Mosse, *The Crisis of German Ideology: Intellectual Origins of the Third Reich* (New York: Grosset & Dunlap, 1964). The "German" method of Quellenkritik *itself*, so essential to nineteenth-century scholarship, was not about "objective" assessment of sources and data alone; it excluded the worlds of all sorts of knowledge seekers, especially women, from dialogue among "professionals" in a modern polity. Thus, rational worlds of science were a national front to guard. See Bonnie Smith on "men and facts," *The Gender of History: Men, Women, and Historical Practice* (Cambridge: Harvard University Press, 2000), 130–56.

70. Herb, *Under the Map of Germany*, 35–36.

71. On the evolving fears of migrants and borders, Annemarie Sammartino, *The Impossible Border: Germany and the East, 1914–1922* (Ithaca: Cornell University Press, 2010).

72. Herb, *Under the Map of Germany*, 132–33.

73. Albrecht Penck and Herbert Heyde, *Karte der Verbreitung von Deutschen und Polen längs der Warthe-Netze-Linie und der unteren Weichsel, sowie an der Westgrenze in Posen* (Berlin, 1919), A-AGSL, 64-C-1:100K.

74. Häberle, who praised Spett's work previously, now turned to Penck in order to refute it; he objected to the use of source material for Poland that was drawn from the 1897 All-Russian Census. He is examined in depth by Herb. See Dietrich Häberle, "Der Anteil der Deutschen und Polen an der Bevölkerung von West-Preussen und Posen (nach A. Penck): Mit einer Kartenskizze," *Geographische Zeitschrift* 25 (1919): 124–27; and Hans von Präsent, "Beiträge zur polnischen Landeskunde: Das Quellenmaterial zur Bevölkerungsstatistik Polens," *Zeitschrift der Gesellschaft für Erdkunde zu Berlin*, no. 3 (1917): 245–49; Review of Dietrich Schäfer's "Sprachenkarte der deutschen Ostmarken" and Jakob Spett's "Nationalitätenkarte der östlichen Provinzen des deutschen Reiches nach den Ergebnissen der amtlichen Volkszählung vom Jahre 1910," *Geographische Zeitschrift* 25 (1919): 128–29.

75. Herb, *Under the Map of Germany*, 29.

76. Literature on the modern German question is vast. On borders and frontiers, see Brian Vick, *Defining Germany: The 1848 Frankfurt Parliamentarians and National Identity* (Cambridge: Harvard University Press, 2002); Sammartino, *The Impossible Border*; Kopp, *Germany's Wild East*.

77. Albrecht Penck, "Protest der Gesellschaft für Erdkunde gegen die Ausstossung Deutschlands aus der Reihe der kolonisierenden Mächte," *Zeitschrift der Gesellschaft für Erdkunde zu Berlin*, no. 1–2 (1919): 24–29.

78. Herb, *Under the Map of Germany*, 27. Herb argues on the basis of the experts' political success/failure in preparations that the Germans had failed to realize the pro-Polish stance of maps of the corridor until other countries made note of it. He infers (logically) that because of the high esteem around the world for German geographic scholarship and journals before 1914, "German geographic publications would have been noticed" if Penck and company were able to get them out before the finalization of the Versailles settlement. Idealizing Poland, he faults German geographers for not recognizing the pro-Polish stance of maps like those of Romer and Spett, while noting that Penck did not succeed in getting the geographic work out on time before May 1919. While these things may well have been true in the moment in mid-1919, and in light of Allied commissions, appeals to fame and authority also reflect the canonical past, and especially for Europe's cult of explorers and coveted nineteenth-century experts.

79. Teleki, review of Dr. Th. H. Engelbrecht, *Landwirtschaftlicher Atlas des Russischen Reiches in Europa und Asien* (Berlin: D. Reimer, 1916), in *Földrajzi Közlemények* 46 (1918): 140–42. Engelbrecht's work on British colonial India was from *Die Feldfrüchte Indiens in ihrer geographischen Verbreitung* (1914).

80. Teleki, "Főtitkari jelentes (Keleti torekveseink): A Magyar Földrajzi Társaság 1918 Április 25-en tartott közgyűlésen elmondotta," *Földrajzi Kozlemények* 46 (1918): 251–52.

81. Some versions were acquired by the "American Geographical Society for the Inquiry and Peace Conference in 1918–1919," in AGSL, 64206 C-[1918?] Peace Conf. [no. 01] 45. In Hungary, original versions in multiple languages are in the Map Collection of the National Széchényi Library (Országos Széchényi Könyvtár), Kézirattára (Manuscripts Division), in Budapest. See Paul Teleki, *Carte ethnographique de la Hongrie basée sur la densité de la population* (Budapest: V. Hornyánszky, 1919), TI 148–TM 6 309; *An Ethnographical Map of Hungary Based on the Density of Population* (Budapest: Hungarian Geographical Institute, 1919), TI 150–TM 6 626; and *Magyarország néprajzi térképe a népsürüseg alapjan. Ethnographical Map of Hungary Based on Density of Population. Carte ethnographique de la Hongarie contruite en accordance avec la densité de la population* (Budapest: M[agyar] Földr[ajzi] Int[ézet], 1919), TM 6 308–TI 148.

82. An English version is in the Library of the American Geographical Society in Milwaukee. Count Paul Teleki, *Maps of Hungary, showing (1) Ethnography; (2) Communes, showing persons speaking Hungarian; (3) Communes, showing distribution of religion; (4) Communes, showing persons able to read and write*, 1:200,000 (Budapest: Hungarian Geographical Society, 1918), AGSL, 160-F.

83. Árpad Papp-Váry, *Magyarorszag története térképeken* (Budapest: Kossuth Kiadó-Cartographia, 2002), 252–53. The author adheres to the "heroic" ethnocentric model and compliments the "novelty and objectivity" of the "Carte Rouge." It was "fully credible from a scientific point of view," he argues, in Teleki's defense of Hungary's integrity and particularly against Romanian claims against it in the 1930s. For a more circumspect take on Hungary's claims in lieu of its minority Jewish populations and Hungarian-Romanian geopolitics, see Holly Case's lucid analysis in *Between States*, 39–48.

84. Papp-Váry, *Magyarorszag története térképeken*, 56–57.

85. A fine short bio of Kiepert is by Dirk Hänsgen, "Heinrich Kiepert: ein Handwerker unter den Geographie-Ordinarien der ersten Stunde," in Nitz, Schultz, and Schulz, eds., *1810–2010: 200 Jahre Geographie in Berlin*, 51–57.

86. On Masaryk's outlook and the journal's impact, Roman Szporluk, *The Political Thought of Thomas G. Masaryk* (Boulder, Colo.: East European Monographs, 1981), 131–46.

87. Liliana Riga and James Kennedy, "*Mitteleuropa* as Middle America?: 'The Inquiry' and the Mapping of Central Europe in 1919," *Ab Imperio*, no. 4 (2006): 284.

88. Quoted in Margaret Macmillan, *Paris 1919: Six Months that Changed the World* (New York: Random House, 2001), 42. Macmillan covers Eastern Europe in a vast sweep, but she does not utilize primary sources in the languages of the region. Lloyd George's telling comment foreshadowed Neville Chamberlain's at Munich in September 1938, when the minister boasted of keeping England out of "a quarrel in a far-away country [Czechoslovakia] between people of whom we know nothing."

89. Hugh and Christopher Seton-Watson, *The Making of a New Europe: R. W. Seton-Watson and the Last Years of Austria-Hungary* (Seattle: University of Washington Press, 1981). Robert Seton-Watson endorsed Polish independence and wrote the preface to August Zaleski's *Poland's Case for Independence Being a Series of Essays Illustrating the Continuance of Her National Life* (New York: Dodd, Mead & Co., 1916), which was disseminated by the Polish Information Committee in Lausanne during the war. On conflations in Eastern Europe of race and nationality, David I. Kertzer and Dominique Arel, "Census, Identity Formation, and Political Power," in Kertzer and Arel, eds., *Census and Identity: The Politics of Race, Ethnicity, and Language in National Censuses* (Cambridge: Cambridge University Press, 2002), 12.

90. György Litván, *A Twentieth-century Prophet: Oscar Jászi 1875–1957* (Budapest: Central European University Press, 2006).

91. On "heroic" contours of modern Czech nationalism, Andrea Orzoff, *Battle for the Castle: The Myth of Czechoslovakia in Europe, 1914–1948* (New York: Oxford University Press, 2009); Nancy M. Wingfield, *Flag Wars and Stone Saints: How the Bohemian Lands Became Czech* (Cambridge: Harvard University Press, 2007).

92. Ablonczy, *Pál Teleki*, 64.

93. Teleki, "A földrajz,—tudomany es tantárgy" *Földrajzi Kozlemények* 49 (1919): 20–41. Here was a plain German-to-Hungarian colonial syncretism in Europe: Teleki's engagement with German geographers and the volume "Die Geographie als Wissenschaft und Lehrfach," published in 1918–19 by the Berlin Zentralinstitute für Erziehung und Unterricht.

94. Details are in Mroczko, *Eugeniusz Romer*, 71; Władysław Pawłak, "Eugeniusz Romer jako geograf i kartograf," in Lucyna Szaniawska et al., eds., *Eugeniusz Romer: geograf i kartograf trzech epok* (Warszawa: Biblioteka Narodowa, 2004), 42–44; Chałubińska, "Kontakty," 22.

95. See the essays in Bartov and Weitz, *Shatterzone of Empires*; Michael Fahlbusch and Ingo Haar, *Völkische Wissenschaften und Politikberatung im 20. Jahrhundert: Expertise und "Neuordnung" Europas* (Paderborn: Schöningh, 2010); Haar and Fahlbusch, eds., *Handbuch der völkischen Wissenschaften: Personen–Institutionen–Forschungsprogramme–Stiftungen* (München: K. G. Saur, 2008); and Haar and Fahlbusch, eds., *German Scholars and Ethnic Cleansing, 1919–1945* (New York: Berghahn Books, 2005).

96. Ferenc Fodor and Ferenc Tibor Szávai, *Teleki Pál: Egy "bujdosó könyv": megjelent Teleki Pál halálának 60. évében* (Budapest: Mike és Társa Könyvkiadó, 2001), 273–80. The disciple Fodor adhered to the tragic script of the Teleki's "national death" (*nemzethalál*), rooted in German romantic nationalism and Herder's dire predictions. An alternative analysis is offered by Zoltán Hajdú, "Geográfus politikus avagy politikus geográfus? A tudomány és a politika kölcsönhatása Teleki Pál életművében," *Földrajzi Közlemények* 39:1–2 (1991): 1–9; and "A magyar földrajztudomány szerepvállalása a trianoni békeszerződésre való tudományos felkészülésben," *Debreceni Egyetem Természetföldrajzi és geoinformatikai tanszék* (2010): 125–32.

CHAPTER THREE

1. Albrecht Penck, "Begrüssung der heimgekehrten Ostafrikaner," *Zeitschrift der Gesellschaft für Erdkunde zu Berlin*, no. 1–2 (1919): 18.

2. Goffman, *The Presentation of the Self in Everyday Life*; Charles Taylor, *Sources of the Self: The Making of the Modern Identity* (Cambridge: Harvard University Press, 1989).

3. Isaiah Bowman, "Diary (original)" (hereafter Bowman's Paris Diary), 4 January–9 May 1919, IB-P Ms. 58, Series XIII, Box 2, Paris Peace Conference Files. The Bowman-Romer correspondence up to 1935 is stored at the old AGS Archive in New York City; I had the good fortune to review it in 2010 in transport to the AGSL in Milwaukee (the archive began moving there from New York City in 2011), thanks to Geoffrey J. Martin.

4. Bowman's Paris Diary, 28 January 1919, IB-P Ms. 58, Series XIII, Box 2.

5. Bowman's diary entry of 21 February 1919 even gives this sense of voyeurism. He wrote, "Storck took photographs of the room littered with maps and papers." One might even suggest something erotic to the homosocial encounters of these men. (Lord chose celibacy and became a Catholic priest in 1924.) In her critical section on men's scientific history and "sex in the archives," the historian Bonnie Smith incisively investigated the seminar form's erotics—which may apply here to "workshops" for drawing of maps as heightened spaces of arousal at the prospect of drawing new worlds and geo-bodies. Smith, *The Gender of History*, 103–29; and Smith, ed., *Women's History in Global Perspective*, 3 vols. (Urbana: University of Illinois Press, 2004–5).

6. On such "moving" sensory geography, Rebecca Solnit, *Wanderlust: A History of Walking* (New York: Penguin Books, 2001); Will Self, *Psychogeography: Disentangling the Modern Conundrum of Psyche and Place* (New York: Bloomsbury USA, 2007); and Merlin Coverley, *Psychogeography* (Harpenden: Oldcastle Books, 2010).

7. Mazower, *Governing the World*, 127. In the previous chapter, Mazower provides a clear overview of scientific internationalism—how the nineteenth- to early twentieth-century cult of the expert in Europe and America became fused with Wilson's distinctively American mix of missionary zeal and pragmatism.

8. Romer to Bowman, 19 February 1919, IB-P Ms. 58, Series XIII, Box 9, page 119, items 1–2.

9. Bowman's Paris Diary, 20 February 1919.

10. In hindsight, the "problem" or question by nationality is examined by Piotr S. Wandycz, "The Polish Question," in Manfred F. Boemeke, Gerald D. Feldman, and Elisabeth Glaser, eds., *The Treaty of Versailles: A Reassessment after 75 Years* (Cambridge: Cambridge University Press, 1998), 313–35; and Kay Lundgreen-Nielsen, *The Polish Problem at the Paris Peace Conference: A Study of the Policies of the Great Powers and the Poles, 1918–1919* (Odense: Odense University Press, 1979).

11. Bowman's Paris Diary, 21 February and 27 February 1919.

12. Geoffrey J. Martin, *The Life and Thought of Isaiah Bowman* (Hamden, Conn.: Archon, 1980), 93.

13. Bowman's Paris Diary, 27–28 February 1919.

14. For an overview, Alan Sharp, *The Versailles Settlement: Peacemaking After the First World War, 1919–1923*, 2nd ed. (London: Palgrave Macmillan, 2008).

15. Bowman's Paris Diary, 28 February 1919.

16. Bowman, IB-P Ms. 58, Series XIII, Box 9, page 133, items 1–2.

17. On Poland's diplomacy, Jerzy Borzęcki, *The Soviet-Polish Peace of 1921 and the Creation of Interwar Europe* (New Haven: Yale University Press, 2008); Anna M. Ciencała and Titus Komarnicki, *From Versailles to Locarno: Keys to Polish Foreign Policy, 1919–25* (Lawrence: University of Kansas Press, 1984); Piotr S. Wandycz, *France and Her Eastern Allies, 1919–1925: French-Czechoslovak-Polish relations from the Paris Peace Conference to Locarno* (Minneapolis: University of Minnesota Press, 1962).

18. Bowman's Paris Diary, 1 March 1919.

19. Ibid., 3 March 1919.

20. Ibid., 4 March 1919.

21. Bowman, IB-P Ms. 58, Series XIII, Box 10, page 158, item 1.

22. Bowman's Paris Diary, 15 March 1919.

23. Ibid., 8 April 1919.

24. Ibid., 5 April 1919.

25. For context and effect, Neal Pease, " 'This Troublesome Question': The United States and the 'Polish Pogroms' of 1918–9," in Mieczysław Biskupski, ed., *Ideology, Politics, and Diplomacy in East Central Europe* (Rochester: University of Rochester Press, 2003), 58–79; William W. Hagen, "The Moral Economy of Popular Violence: The Pogrom in Lwow, November 1918," in Robert Blobaum, ed., *Antisemitism and Its Opponents in Modern Poland* (Ithaca: Cornell University Press, 2005), 124–47.

26. Bowman's Paris Diary, 27 April 1919.

27. Czesław Miłosz, *The Captive Mind* (New York: Knopf, 1953); for time mapping inspired by Milan Kundera, see Shore, *The Taste of Ashes*; also the critique of Maria Todorova, "The Trap of Backwardness: Modernity, Temporality, and the Study of Eastern European Nationalism," *Slavic Review* 64:1 (Spring 2005): 140–64.

28. Penck's articles from 1918 to 1921 include "Polen: eine Anzeige," *Zeitschrift der Gesellschaft für Erdkunde zu Berlin* 53 (1918): 97–131; "Die deutsch-polnische Sprachgrenze," *Zeitschrift der Gesellschaft für Erdkunde zu Berlin* (1919): 108–9; "Deutsche und Polen in Westpreussen und Posen," *Aus dem Ostlande* 14 (1919): 65–70; and "Die Deutschen im polnischen Korridor," *Zeitschrift der Gesellschaft für Erdkunde zu Berlin* 56 (1921): 169–85.

29. Penck, "Deutsche und Polen in Westpreussen und Posen," *Deutsche Allgemeine Zeitung*, 9 February 1919, no. 27, 3.

30. Ibid., 4.

31. Penck's other key maps and map-related works are *Deutschen im polnischen Korridor: Karte der Verbreitung der deutsch und polnisch sprechenden Bevölkerung auf Grund der Volkszählen vom 1. Dezember 1910*, coauthored with Wilhelm Volz and published in Berlin; *Das Deutschtum und Polentum in Oberschlesien, nach den Ergebnissen der Abstimmung am 20. März 1921*, in a German-language compilation of data on "the Upper Silesia Question"; and *Die völkische Struktur Oberschlesiens*, pertaining to the plebiscite in Upper Silesia, also printed in Breslau in 1921.

32. Penck, "Die Deutschen im polnischen Korridor," 184; Herb, *Under the Map of Germany*, 36.

33. On the persistence of the "Volksgruppe paradigm" and German revisionist discourses of cohesion and regional particularism through the Weimar period, see the historiographical overview in Chu, *The German Minority in Interwar Poland*, 1–11, 21–61.

34. TEVÉL and the Hungarian Geographical Society distributed their publications widely and sent them to the U.S. throughout 1919 and 1920. Take the collaborative *La Hongrie: Cartes et notions géographiques, historiques, etnographiques, economiques et intellectuelles* (Budapest: Hungarian

Geographical Society, [1919]), AGSL, Pamphlet Box 129. Translated text was prepared in French, English, German, and Czech. It included a "unified" and very accessible chronological table of all of Hungarian history.

35. Ablonczy, *Pál Teleki*, 51.

36. Mazower rightly points out that the "trusteeship" logic of the Great Powers after Vienna in 1815 was in support of Europe's imperial-colonial dominance, and integral to the blueprints for the League of Nations in 1919. See his emphasis on geographic expertise and nineteenth-century continuity in *Governing the World* and *No Enchanted Palace*.

37. Teleki's 27 March 1919 memo to the Hungarian Foreign Office is cited in Ablonczy, *Pál Teleki*, 53.

38. Miklós Zeidler, *A revíziós gondolat* (Pozsony: Kalligram, 2009); Cattaruzza, Dyroff, and Langewiesche, *Territorial Revisionism and the Allies of Germany in the Second World War*.

39. Paul [Pál] Teleki, *Short Notes on the Economical and Political Geography of Hungary* (Budapest: Hornyánszky, 1919), 1–22. At the AGSL, Pamphlet Box 131. This gift to the AGS "from Dr. Bowman" was received on 30 July 1919.

40. Ibid., 1–2.

41. Ibid., 6.

42. Ibid., 10–12.

43. Ibid., 14.

44. Worried about ethnic decline, Pivány ferried along Teleki's message that Hungary was the "most compact in Europe, as a glance at the map will show," that "racially the Hungarian or Magyar race predominates," and that the population is "more than three times as strong as the next race in numbers, the Rumanians." He concurred that the "predominance [of the Magyar race] . . . has been due not to mere numbers, but mainly to the fact that it has founded, built up and maintained the Hungarian State for a thousand years, and put a distinctly Magyar stamp on the civilization of the whole country." Eugene Pivány, *Some Facts about the Proposed Dismemberment of Hungary, with a Map, Statistical Table and Two Appendices* (Cleveland: Hungarian American Federation, 1919), AGSL vertical files [no call number]. Pivány was a Hungarian-American author of popular nationalist books in English, including *Hungarians in the American Civil War*, published in Cleveland in 1913.

45. On minority rights and anti-Semitism in 1919, Carole Fink, *Defending the Rights of Others: The Great Powers, The Jews, and International Minority Protection, 1878–1938* (Cambridge: Cambridge University Press, 2006); and Murray Baumgarten, Peter Kenez, and Bruce Thompson, eds., *Varieties of Antisemitism: History, Ideology, Discourse* (Newark: University of Delaware Press, 2009).

46. Pál Teleki, "A zsidokerdes," *Teleki Pál programbeszéde, melyet Szeged első választó-kerületének Keresztény Nemzeti Egyesülés Pártja alakuló és képviselő nagygyűlésén mondott el, Szegeden, a Tisza-szallo nagytermeben 1919, ev december havanak 14 napjan* (Szeged: TEVÉL, 1919), 3–5.

47. Baudrillard's *Simulacra and Simulation* (Ann Arbor: University of Michigan Press, 1994) was first published in France in 1981.

48. Andrew S. Curran, *The Anatomy of Blackness: Science and Slavery in an Age of Enlightenment* (Baltimore: Johns Hopkins University Press, 2013); Dorinda Outram, *The Enlightenment*, 3rd ed. (Cambridge: Cambridge University Press, 2013), 99–113; and Larry Wolff and Marco Cipolloni, eds., *The Anthropology of the Enlightenment* (Stanford: Stanford University Press, 2007).

49. Macmillan, *Paris 1919*, 257–70. Additional accounts are in Francis Deák, *Hungary at the Paris Peace Conference: The Diplomatic History of the Treaty of Trianon* (New York: Columbia

University Press, 1942); and Arno J. Mayer, *Politics and Diplomacy of Peacemaking: Containment and Counterrevolution at Versailles, 1918–1919* (New York: Knopf, 1967).

50. For context, Ignác Romsics, *The Dismantling of Historic Hungary: The Peace Treaty of Trianon, 1920* (Boulder, Colo.: East European Monographs, 1990); on Teleki's take, Ablonczy, *Pál Teleki*, 69.

51. Works by Hungarians during the 1920s were often printed abroad in English and characterized by these tropes of "dismemberment" and "justice for Hungary," as in Ladislaus Buday, *Dismembered Hungary* (London: Grant Richards, 1923); Charles Tisseyre, *An Error in Diplomacy: Dismembered Hungary* (Paris: Mercure, 1924); and Count Albert Apponyi, *Justice for Hungary* (London: Longmans, Green & Co., 1928).

52. Teleki wrote the preface in July 1920. As when Romer at Bowman's urging in 1921 revised his 1916 *Geographical-Statistical Atlas of Poland* to spin a heroic tale of knowledge, Teleki detailed how he took on the task of drawing maps and assembling information in October 1918. See *The Hungarian Peace Negotiations: Report on the Activity of the Hungarian Peace Delegation at Neuilly* (Budapest: Magyar Külügyminisztérium, 1920), in OSZK-T, TA 3159, Budapest. Maps and information produced by the Hungarian Foreign Ministry were reprinted by the firm of Viktor Hornyánsky in Budapest in 1920, 1921, and 1922, then disseminated to libraries all around the world.

53. Ablonczy, *Pál Teleki*, 112. Essentially, the school was set up for what in German would be called *Erziehung* (education or upbringing, as building of male character), to cultivate a neo-colonial, Magyar Christian middle-class elite against what Teleki and others perceived to be "oppressive" and preponderant Jewish power in the capital.

54. Distributed widely, the work appeared in several languages. Aladár Edvi Illés, ed., *The Economies of Hungary in Maps: To the Commission of Count Paul Teleki, Chief of the Office for the Preparation of Peace Negotiations* (Budapest: Hornyánsky, 1921). The first edition of early 1920 is held at OSZK-T, TA 1645-1648, Budapest. There were three editions in 1920 and three more in 1921. Other original maps from the compilation are at the AGSL, At. 642 E-1921.

55. Teleki, *Magyarország néprajzi térképe a népsűrűség alapján. Ethnographical Map of Hungary Based on the Density of Population* (Budapest: Hungarian Geographical Institute, 1920); and Teleki, *Ethnographical Map of Hungary, Based on the Density of Population* (The Hague: W. P. van Stockum & Son, 1920). AGSL 160-C, Milwaukee.

56. Pál Teleki, *La Hongrie du Sud: Questions de l'Europe Orientale* (Paris: H. le Soudier, 1920; Budapest: Ferd. Preifer, 1920); Paul Teleki, *Southern Hungary* (Budapest: Hornyánszky, 1922).

57. The count's geographic notions of unity were picked up by many of his contemporaries and successors. See Teleki, *Magyarország gazdaság földrajzi térképe dr. Teleki Pál gróf és dr. Cholnoky Jenő közreműködésével hivatalos adatok alapján szerkesztette dr. Fodor Ferenc. Carte de géographie économique de la Hongrie . . . Economic-geographical map of Hungary* (Budapest, Magyar Földrajzi Intézet, 1920). See Fodor's Teleki-inspired history of Hungarian geography, *A Magyar földrajztudomány története* (Budapest: Magyar Tudományos Akademia Földrajztudományi Kutatóintézet, 2006), written largely for the drawer in communist Hungary. Grievances have long trajectories: Fodor in 1920 wrote a pamphlet declaring that the Czech state was a "geographical impossibility."

58. The original is in Hungarian, "A trianoni békeszerződésről," in Antal Papp, ed., *Grof Teleki Pál országgyűlési beszédei* (Budapest, 1942), 54–73; reprinted in Teleki, *Válogatott politikai írások és beszédek* (Budapest: Osiris, 2000), 65–88.

59. Another sidebar here: the significant body of work by Jan T. Gross on the history of

Jedwabne and twentieth-century Polish-Jewish relations taps very well into this emotion. In Polish political cosmology, fear is often affixed to national narratives of victimhood, or struggles to recover a golden or "Edenic" age of imagined multiethnic harmony. Challenges are offered by Gross, *Fear: Anti-Semitism in Poland after Auschwitz: An Essay in Historical Interpretation* (Princeton: Princeton University Press, 2006); Gross, István Deák, and Tony Judt, eds., *The Politics of Retribution in Europe* (Princeton: Princeton University Press, 2000); and Gross, *Revolution from Abroad: The Soviet Conquest of Poland's Western Ukraine and Western Belorussia*, new ed. (Princeton: Princeton University Press, 2002).

60. Eugeniusz Romer, *Pamiętnik paryski: 1918–1919* (Wrocław: Ossolineum, 1989), 383.

61. Romer to Bowman, 17 December 1919, Archive of the American Geographical Society (hereafter A-AGS), New York City and Milwaukee.

62. On Bowman's research on indigenous peoples, Smith, *American Empire*, 53–82.

63. Romer to Bowman, 30–31 December 1919, A-AGS.

64. Romer to Bowman, 2 March 1920, A-AGS.

65. Bowman to Romer, 20 April 1920, A-AGS.

66. Ibid.

67. Bowman to Romer, 30 April 1920, A-AGS.

68. Ibid.

69. Bowman to Romer, 19 June 1920, A-AGS.

70. Bowman to Romer, 23 July 1920, A-AGS.

71. Bowman to Romer, 23 December 1920, A-AGS.

72. On the cultural geography, Antoni Kroh, *Tatry i Podhale* (Wrocław: Wydawnictwo Dolnośląskie, 2002); Patrice M. Dabrowski, "Borderland Encounters in the Carpathian Mountains and Their Impact on Identity Formation," in Bartov and Weitz, *Shatterzone of Empires*, 193–208; and " 'Discovering' the Galician Borderlands: The Case of the Eastern Carpathians," *Slavic Review* 64:2 (Summer 2005): 380–402.

73. The original source is W. M. Davis, "The Geographical Cycle," *Geographical Journal* 14 (1899): 481–504; see also R. J. Chorley, Robert P. Beckinsale, and Anthony J. Dunn, *The History of the Study of Landforms, or the Development of Geomorphology*, vol. 2 (London: Methuen, 1973), 498.

74. Geoffrey J. Martin, "A Fragment on the Penck-Davis Conflict," *Geography and Map Division: Special Libraries Association Bulletin* 98 (1974): 11.

75. From his more or less unreconstructed Soviet Marxist standpoint, the late Neil Smith pointedly blamed this exchange on the "arrogance" of Davis, Bowman, and U.S. progenitors of global capitalism more generally in the 1920s. Smith, *American Empire*, 277–78.

76. On Davis's "universal" cycle of erosion, and for letters between Davis and the two Pencks, see Chorley, Beckinsale, and Dunn, eds., *The History of the Study of Landforms*, vol. 2, 516, 537–54; W. M. Davis, "The Cycle of Erosion and the Summit Level of the Alps," *Journal of Geology* 31 (1923): 1–41; Davis, "The Penck Festband: A Review," *Geographical Review* 10 (1920): 249–61; and Davis, "Passarge's Principles of Landscape Description," *Geographical Review* 8 (1919): 266–73. On the legacy of Walther Penck, see Isaiah Bowman, "The Analysis of Land Forms: Walther Penck on the Topographic Cycle," *Geographical Review* 16:1 (January 1926): 122–32; and "Symposium: Walther Penck's Contribution to Geomorphology," *Annals of the Association of American Geographers* 30:4 (December 1940): 219–80. Walther Penck's *Die morphologie Analyse: Ein Kapitel der physikalischen Geologie* (Stuttgart: Geographische Abhandlungen, 1924) was finally

translated into English as *Morphological Analysis of Land Forms: A Contribution to Physical Geology* (London: Macmillan, 1953), several decades too late.

77. Martin, "A Fragment," 12. Martin concludes that "Bowman remained faithful to Davis's teaching till the end of his days," and speculates that their other exchanges between 1905 and 1917 probably have been lost.

78. Davis to Bowman, 2 February 1920, A-AGS.

79. Martin, *The Life and Thought of Isaiah Bowman*, 189–90. Martin sees it as a diplomatic rift that could have become much worse had not a faithfully objective Bowman intervened; unlike Smith, he treats geographic science and national prejudice not as two bourgeois fictions but as two things that are possible to disentangle.

80. Szumański worked loyally with and for Romer, who found a position for him in 1923 at Jan Kazimierz University so that the two could cooperate more closely.

81. On Romer's Poland in the early 1920s, Marian Mroczko, *Eugeniusz Romer: biografia polityczna* (Słupsk: Wydawnictwo Naukowe Akademii Pomorskiej, 2008), 122–69.

82. Bowman was sure to credentialize himself in the title page of the first edition: *The New World: Problems in Political Geography. By Isaiah Bowman, Ph.D., Director of the American Geographical Society. Illustrated with 215 Maps and with 65 Engravings from Photographs* (Yonkers-on-Hudson, N.Y.: World Book Company, 1921).

83. See Gerry Kearns's tour de force biography of Mackinder, *Geopolitics and Empire: The Legacy of Halford J. Mackinder* (Oxford: Oxford University Press, 2009).

84. Neil Smith, "Bowman's New World and the Council on Foreign Relations," *Geographical Review* 76:4 (Oct. 1986): 438–60.

85. Smith, *American Empire*, 280–81.

86. Bowman to Romer, 31 December 1921, A-AGS.

87. Romer to Bowman, 1 October 1921, A-AGS.

88. Ibid.

89. Romer to Bowman, 28 December 1921, postcard from Zakopane, A-AGS.

90. Romer to Bowman (from Zakopane), 24 July 1922; Bowman to Romer, 20 March 1923; Romer to Bowman, 13 October 1925, A-AGS.

91. Aniela Chałubińska (1901–98), Romer's former student and a pioneering scholar in Poland on pedagogical methods for teaching modern geography, was one of these notable women.

92. Romer to Bowman, 22 April 1922, A-AGS.

93. This national Enlightenment-era canon set up by Romer persists in Polish historical geography to this day. See Jackowski, Liszewski, and Richling, *Historia geografii polskiej*, 32–42.

94. Formalized in 1924, the Książnica-Atlas was based in Lwów and also opened an office in Warsaw. Abiding by Romer's directives and zeal for didactic geography, it was intended to serve as a publishing film for scientific-educational maps for the Society of High School Teachers (Towarzystwo Nauczycieli Szkół Średnich i Wyższych), which had been formed from two bodies in 1919, the Society of High School Teachers in Galicia and the Polish Teachers Association of the former Kingdom of Poland.

95. Władysław Pawlak, "Eugeniusz Romer jako geograf i kartograf," in *Eugeniusz Romer: geograf i kartograf trzech epok* (Warszawa: Biblioteka Narodowa, 2004), 11–63.

96. The Polish geographers' output was overwhelming. Romer's colleagues Szumański, Pawłowski, and Józef Wąsowicz got Polish maps out very quickly to Bowman and to international audiences. Romer and Szumański coauthored physical maps such as the *Mapa Polski* (1921),

reprinted often in Lwów and Vienna. The pace of production quickened as Romer himself authored nine wall maps and two relief maps in 1922—the *Black Sea Regions* (*Kraje Czarnomorskie*) and *Baltic Sea Regions* (*Kraje Bałtyckie*) were published in the *Atlas of Contemporary Poland* (*Atlas Polski współczesnej*) of 1923. With Szumański and Pawłowski, Romer started a series of travel and tourism maps for all of Poland's provinces. Between 1923 and 1925, Książnica-Atlas published five complex atlases and three successive parts of the *General Geographical Atlas* (*Powszechny atlas geograficzny*). By the end of the 1920s, a comprehensive school atlas was printed for use by every student in all the secondary schools of the interwar Polish republic.

97. It is forgotten that Romer as a scientist actually had good relations with Soviet geographers in the era of the New Economic Policy from 1921 to 1928. He also took part in the Congress of Slavonic Geographers and Ethnographers in Kraków in 1924, a concrete result of which was Polish-Soviet geographic rapprochement.

98. Romer to Bowman, 24 August 1922, A-AGS.

99. Bowman to Romer, 20 March 1923, A-AGS.

100. Romer to Bowman, 14 April 1923, A-AGS.

101. Bowman to Romer, 7 May 1923, A-AGS.

102. Bowman to H. N. MacCracken, 12 May 1923; Ferris J. Meigs to Bowman, 19 May 1923, A-AGS.

103. H. N. MacCracken to Bowman, 19 May 1923, A-AGS.

104. Bowman to Romer, 11 June 1923, A-AGS.

105. One germane point here: the Darülfünun, by 1900 officially called the Darülfünun-ı Şahane, was a European-style "house of sciences" set up in the Ottoman Empire during the Tanzimat (1839–76) era. It became the basis for Istanbul University in 1933.

106. Albrecht Penck, "Penck, Walther, 1888–1923," AP-P, K. 865-70; and "Walther Penck 1888–1923," preface to Walther Penck, *Die morphologische Analyse* (Stuttgart: J. Engelhorn, 1924), vii–xvii, published after his son's death.

107. Martin, *The Life and Thought of Isaiah Bowman*, 188-89; Smith, *American Empire*, 277–80; also Martin, "A Fragment."

108. Smith, *American Empire*, 509–10n23.

109. Isaiah Bowman, "The Analysis of Land Forms: Walther Penck on the Topographic Cycle," *Geographical Review* 16:1 (January 1926): 122–32.

110. Bowman to Penck, 26 July 1923, A-AGS.

111. Bowman to Davis, 26 July 1923, A-AGS

112. Davis to Bowman, 27 July 1923, A-AGS.

113. Bowman to Davis, 31 July 1923, A-AGS.

114. Penck to Bowman, 22 August 1923, IB-P Ms. 58, Series II, Box 35. The translated version curiously adds scare quotes for "moment" in the text.

115. Bowman to Penck, 25 September 1923, IB-P Ms. 58, Series II, Box 35.

116. Bowman to Penck, 18 October 1923, A-AGS.

117. Cited in Martin, *The Life and Thought of Isaiah Bowman*, 19. The original is Penck to Bowman, 30 October 1923, A-AGS, as a reply to the lengthy missive of 25 September 1923.

118. Penck to Bowman, 30 October 1923; Bowman to Penck, 28 November 1923, AGS Archive; as noted in Smith, *American Empire*, 282, 509–10n23.

119. Bowman to Penck, 28 November 1923, A-AGS.

120. Franz Kafka, *The Castle*, trans. Mark Harman (New York: Schocken Books, 1998), 1.

121. Scott Spector, *Prague Territories: National Conflict and Cultural Innovation in Franz Kafka's Fin de Siècle* (Berkeley: University of California Press, 2002).

CHAPTER FOUR

1. Michael Fahlbusch and Ingo Haar have admirably interpreted the co-optation of geographers by their complicity in the darker aspects of Europe up to 1945: German colonialism, imperialism, racism, militarism, fascism, anti-Semitism, and genocide and ethnic cleansing. This begs the question when their "honor" was ever achieved—in the nineteenth century? See the compilation in Haar and Fahlbusch, eds., *German Scholars and Ethnic Cleansing, 1919–1945* (New York: Berghahn Books, 2005).

2. The German and twentieth-century European tendency to read back from 1945, 1941, and 1933 is also present in Danielsson, *The Explorer's Roadmap to National-Socialism: Sven Hedin, Geography, and the Path to Genocide* (Burlington, Vt.: Ashgate, 2012).

3. Diary of Albrecht Penck, 24 April 1927, AP-P, K. 871, S. 3. Penck again traveled to the United States in April 1927, this time with Albert Haushofer (1903–45), Karl's brother and a writer and diplomat who later opposed the NSDAP. Before embarking, Penck took with him a prewar map (no. 85) from Steiler's Hand-Atlas, featuring a survey of frontiers in the United States and Mexico.

4. On geopolitics, Ó Tuathail, *Critical Geopolitics*; for an excellent study of tensions of *Beruf* and *Wissenschaft* in Serbian academic discourses including geography and cartography, see Nenad Stefanov, *Wissenschaft als nationaler Beruf: Die Serbische Akademie der Wissenschaften 1944–1992: Tradierung und Modifizierung nationaler Ideologie* (Wiesbaden: Harrasowitz, 2011).

5. On Penck's engagements and political geography, Hans-Dietrich Schultz, " 'Ein wachsendes Volk braucht Raum': Albrecht Penck als politischer Geograph," in Nitz, Schultz, and Schulz, eds., *1810–2010: 200 Jahre Geographie in Berlin*, 99–153.

6. On Karl Haushofer's activities in the 1920s and the enduring myth of his influence on Hitler, see David Thomas Murphy, "Hitler's Geostrategist? The Myth of Karl Haushofer and the Institut für Geopolitik," *The Historian* 76:1 (Spring 2014): 1–25.

7. For *Volkskunde* as science, Roland Cvetkovski and Alexis Hofmeister, eds., *An Empire of Others: Creating Ethnographic Knowledge in Imperial Russia and the USSR* (Budapest: Central European University Press, 2014); see also Francine Hirsch, *Empire of Nations: Ethnographic Knowledge and the Making of the Soviet Union* (Ithaca: Cornell University Press, 2005); Juliette Cadiot, *Le laboratoire imperial: Russie-URSS, 1870–1940* (Paris: CNRS Editions, 2007); and Emily D. Johnson, *How St. Petersburg Learned to Study Itself: The Russian Idea of Kraevedenie* (University Park: Pennsylvania State University Press, 2009).

8. On professional identity and academic politics, Vera Tolz, *Russian Academicians and the Revolution: Combining Professionalism and Politics*, 2nd ed. (New York: Palgrave Macmillan, 2014); and Michael David-Fox and Gyorgy Peteri, eds., *Academia in Upheaval: Origins, Transfers, and Transformations of the Communist Academic Regime in Russia and East Central Europe* (Westport, Conn.: Bergin & Garvey, 2000).

9. For the long continuum of Ukraine's multiple revolutionary events, Paul Robert Magocsi, *A History of Ukraine: The Land and Its Peoples*, 2nd ed. (Toronto: University of Toronto Press, 2011), 490–556.

10. Stepan Rudnyts'kyi, *Ohliad natsional'noi terytorii Ukrainy* (Berlin: Ukrains'ke Slovo, 1923).

11. See here Seegel, "Remapping the Geo-Body: Transnational Dimensions of Stepan Rudnyts'kyi and His Contemporaries," in Serhii Plokhy, ed., *The Future of the Past: New Perspectives on Ukrainian History* (Cambridge: Harvard Ukrainian Research Institute, 2016), 205–29; and also Georgiy Kasianov and Philipp Ther, eds., *A Laboratory of Transnational History: Ukraine and Recent Ukrainian Historiography* (Budapest: Central European University Press, 2008).

12. Heorhii [Georgiy] Kasianov, *Ukrains'ka intelihentsiia 1920-kh-30-kh rokiv: sotsial'nyi portret ta istorychna dolia* (Kyiv: Hlobus; VIK, 1992); and Kasianov and V. M. Danylenko, *Stalinizm i ukrains'ka intelihentsiia: 20-30-i roky* (Kyiv: Naukova dumka, 1991); Ihor Stebelsky, "Putting Ukraine on the Map: The Contribution of Stepan Rudnytsky to Ukrainian Nation-building," *Nationalities Papers* 39:4 (July 2011): 587–613; Shablii, *Akademik Stepan Rudnyts'kyi* (Lviv: Ivan Franko National University Publishing Center, 2007); Shablii, *Ukraina prostovora v kontseptsiynomu Stepan Rudnyts'koho: monografiia* (Kyiv: Ukrains'ka Vydavnycha Spilka, 2003); and Pavlo Shtoiko, *Stepan Rudnyts'kyi, 1887–1937: Zhyttepysno-bibliohrafichnyi narys* (Lviv: NTSh, 1997).

13. Stanislav Dnistrians'kyi, *Ukraina and the Peace Conference* (*L'Ukraine et la Conférence de la paix*), was published first in 1919 and reprinted by Nabu Press in 2010. Rudnyts'kyi's letters are in Mushnyk, ed., *Lysty Stepana Rudnyts'koho*.

14. On the movement's diversity, Mark Bassin, Sergey Glebov, and Marlene Laruelle, eds., *Between Europe and Asia: The Origins, Theories, and Legacies of Russian Eurasianism* (Pittsburgh: University of Pittsburgh Press, 2015).

15. In his seminal work, the historian Per Rudling rightly points out that Rudnyts'kyi's arguments from the 1920s in *On the Basis of Ukrainian Nationalism* were against mixed marriages with Ukraine's neighboring peoples (i.e., Poles and Jews). These were appropriated by both the OUN and UPA, and used tactically by collaborating fascists, at least as early as 1943, to argue for Ukrainian national "health" against multiethnic diversity and so-called biological race-mixing into a mythic nation-state territorial homeland. See Per Anders Rudling, *The OUN, the UPA, and the Holocaust: A Study in the Manufacturing of Historical Myths* (Pittsburgh: University of Pittsburgh, 2011), 5.

16. Terry Martin, *The Affirmative Action Empire: Nations and Nationalism in the Soviet Union, 1923–1939* (Ithaca: Cornell University Press, 2001).

17. On Stalinist terror and the purges in Kharkov/Kharkiv, Ivan Drach, ed., *Ostannia adresa: do 60-richchia solovets'koi trahedii*, 3 vols. (Kyiv: Sfera, 1997–99), and the updated edition, Iurii Shapoval and Serhii Bohunov, eds., *Ostannia adresa: rozstrily solovets'kykh v'iazniv z Ukrainy u 1937–1938 rokakh*, 2nd ed. (Kyiv: Sfera, 2003).

18. For key debates in the 1930s, Hiroaki Kuromiya, *The Voices of the Dead: Stalin's Great Terror in the 1930s* (New Haven: Yale University Press, 2007); Timothy Snyder, *Bloodlands: Europe between Hitler and Stalin* (New York: Basic Books, 2010); Orlando Figes, *The Whisperers: Private Life in Stalinist Russia* (New York: Metropolitan Books, 2007); J. Arch Getty and Roberta Manning, eds., *Stalinist Terror: New Perspectives* (Cambridge: Cambridge University Press, 1993); and Sheila Fitzpatrick, *Everyday Stalinism: Ordinary Lives in Extraordinary Times: Soviet Russia in the 1930s* (Oxford: Oxford University Press, 1999).

19. Ablonczy, *Pál Teleki*, 267.

20. On issues of nationality, modern censuses funneled lives into ethnic categories. Teleki followed Penck and Romer in 1921 by appealing to national science, as expressed by modern Hungarian census data. Pál Teleki, "A nemzetiségi kérdés—A geográfus szemszögéből," in Teleki, *Európáról es Magyarországról* (Budapest: Athenaeum, 1934), 30–50.

21. Teleki included the "Cum pars oceanus" quote as an epigraph to his *The Geographical Bases of Economic Life*, his university lectures in Budapest as collected by his students in 1920–21. His injunction to students to learn from the map is discussed in László Bassa, "Teleki Pál és a térkép," *Tér és Társadalom* 3–4 (1990): 175–83.

22. Schulten, *Mapping the Nation: History and Cartography in Nineteenth-century America* (Chicago: University of Chicago Press, 2012); Matthew G. Hannah, *Governmentality and the Mastery of Territory in Nineteenth- Century America* (Cambridge: Cambridge University Press, 2000).

23. Róbert Keményfi, "Grenzen—Karten—Ethnien: Kartenartige Konstituierungsmittel im Dienst des ungarischen nationalen Raums," in Jörn Happel and Christophe von Werdt, eds., *Osteuropa kartiert: Mapping Eastern Europe* (Münster: LIT Verlag, 2010), 201–14.

24. Teleki, "Statisztika és térkép a gazdasági földrajzban," *Földrajzi Közlemények* 50 (1922): 90.

25. Teleki, *Amerika gazdasági földrajza: különös tekintettel az Észak-Amerikai Egyesült Államokra* (Budapest: Centrum Kiadóvállalat részvénytársaság, 1922).

26. Teleki, *The Evolution of Hungary and Its Place in European History* (New York: Macmillan Company, 1923).

27. Teleki, *The Evolution of Hungary*, 140–42; Ablonczy, *Pál Teleki*, 99. On the construct and use of the "żydokomuna" stereotype in Poland, Jan T. Gross, *Fear: Antisemitism in Poland after Auschwitz* (New York: Random House, 2007). For a broader context, Murray Baumgarten, Peter Kenez, and Carole Fink, eds., *Varieties of Antisemitism: History, Ideology, Discourse* (Newark, Del.: University of Delaware Press, 2009).

28. See especially Miklós Zeidler, *Ideas on Territorial Revision in Hungary, 1920–45*, trans. Thomas and Helen DeKornfeld (Boulder, Colo.: Social Science Monographs, 2008).

29. An innovative take on Fodor's modern Hungarian selfhood is Steven A. Jobbitt, "Memory and Modernity in Fodor's Geographical Work on Hungary," in Steven Tötösy de Zepetnek and Louise Olga Vasvári, eds., *Comparative Hungarian Cultural Studies* (West Lafayette, Ind.: Purdue University Press, 2011), 59–71; and Jobbitt, "A Geographer's Tale: Nation, Modernity, and the Negotiation of Self in 'Trianon' Hungary, 1900–1960" (unpubl. Ph.D. diss., University of Toronto, 2008).

30. Teleki continued as the secretary general of the MFT into the summer of 1923, with Jenő Cholnoky serving as president. Other leading Hungarian figures were Gyula Prinz in ethnography, Gustav Thirring in statistics, Károly Kogutowicz (author of the 1927 poster map based on "Carte Rouge") in cartography, and György Vargha, Zsigmond Bátky, Aurél Littke, Gyula Halász, and Ferenc Fodor in subfields of geography. Teleki's own map men labored on; even while ill, he urgently prepared a textbook for high school geography teachers, sponsored by the didactic section of the MFT.

31. For the family's concerns with his health and the disease, Ablonczy, *Pál Teleki*, 112.

32. The key articles by Penck were "Mittenwald: Ein Grenzmarkt in den deutschen Alpen," in *Geographie der deutschen Alpen: Festschrift für Prof. Dr. Sieger zum 60. Geburtstage* (Wien: Seidel, 1924), 103–32; and "Das Hauptproblem der physischen Anthropogeographie," *Zeitschrift für Geopolitik* 2 (1925): 330–48.

33. Martin, *The Life and Thought of Isaiah Bowman*, 13.

34. Penck to Bowman, 21 February 1925; Bowman to Penck, 25 February 1925, A-AGS.

35. Kate Brown, *A Biography of No Place: From Ethnic Borderland to Soviet Heartland* (Cambridge: Harvard University Press, 2005), 2–3.

36. Bowman to Romer, 26 January 1925, A-AGS.

37. Romer to Bowman, 30 November 1925, A-AGS.

38. Bowman to Romer, 23 December 1925, A-AGS.

39. Davis to Bowman, 23 November 1925; Davis to Bowman, 4 December 1925; Bowman to Davis, 8 December 1925, A-AGS.

40. Martin, *The Life and Thought of Isaiah Bowman*, 22. Locating the tensions only as a scientific issue, Martin compliments Bowman's objectivity and actions as an unqualified diplomatic success. He suggests they were also a moment of clarity, on "the essential difference of physiographic posture between the Pencks and W. M. Davis. These two publications did much to clarify the physiographic issue, and place what had from time to time threatened to become personal and contentious onto an academic plane."

41. Penck to Bowman, *Geographical Review* 16:1 (Jan. 1926): 122–32; and *Geographical Review* 16:2 (April 1926): 350–52.

42. Bowman to D. W. Johnson, 19 November 1925; D. W. Johnson to Bowman, 25 November 1925, A-AGS. A summary is in Martin, "A Fragment on the Penck-Davis Conflict," 13.

43. For imperial context here, Sarah Shields, "Mosul, the Ottoman Legacy and the League of Nations," *International Journal of Contemporary Iraqi Studies* 3:2 (2009): 217–30. On Teleki's role in Mosul, Ablonczy, *Pál Teleki*, 115–21; and István Klinghammer and Gábor Gercsák, "The Hungarian Geographer Pál Teleki, Member of the Mosul Commission," *Cartographica Helvetica* 19 (1999): 17–25. See also D. K. Fieldhouse, ed., *Kurds, Arabs, and Britons: The Memoir of Wallace Lyon in Iraq, 1918–1944* (London: I. B. Tauris, 2002), on Lyon's work as Iraq's head administrator/inspector in managing Teleki and the three members of the Mosul commission.

44. Charles Tripp, *A History of Iraq* (Cambridge: Cambridge University Press, 2007), 58–60.

45. MFT librarian István Dubovitz to H. C. Rizer, chief clerk, Department of the Interior, USGS, Washington, D.C., 29 March 1924, A-MFT. The hard-pressed interwar MFT could not even afford to buy maps. On a regular basis, Dubovitz solicited recent maps, catalogs, and other literature from the USGS "to complete the lacunae" in Hungary's map collections. He wrote, "As we see from the price-list sent to us, that the purchase of these publications surpass the pecuniary means of our Society, as in the present hard times our Society has in general to struggle with great material difficulties." See also Dubovitz to Rizer, 15 April 1924; Rizer to Dubovitz, 12 May 1924; Rizer to Dubovitz, 29 October 1924, A-MFT, File N 146-328/1924.

46. Colonel Martin in 1924 [undated] sent Teleki photostat copies of his "Division of Turkey according to secret agreements 1915–1917," from H. W. V. Temperley's *History of the Peace Conference of Paris*, vol. 6, 1924, facing page 6. He wrote, "You asked for it and you shall have a larger copy of my own 'Secret Treaties' map within a few days, together with additional notes. With warm regards, Very sincerely yours, Lawrence Martin."

47. Lawrence Martin to Pál Teleki, 28 May 1925, Pál Teleki Papers (hereafter PT-P) (1924–1941), MNL (formerly MOL), Budapest, Cs. A. dosszie 1924-1925, K. 37-4, 26, and K. 37-9. The Martin-Teleki connection shows their relationship after the U.S. lectures at Williams College in 1921, and some motives behind Teleki's creation of the Institute of Political Sciences in Budapest in 1926. Starting on 1 September 1925, Martin himself took a new position as the curator of the division of maps at the U.S. Library of Congress.

48. Carl O. Sauer, chairman of the geography department, University of California, Berkeley, to Pál Teleki, University of Budapest, 11 November 1924, PT-P (1924–1941), MNL (Budapest), Cs. A. dosszie 1924-1925, K. 37-9, 29.

49. Bowman to Teleki, 17 September 1925, PT-P (1924–1941), MNL (Budapest), Cs. A. dosszie 1924–1925, K. 37-9, 58.

50. Eric Drummond to Teleki, League of Nations, Geneva Office, 24 December 1924. Drummond was replying to Teleki's (now lost) letter of 21 December 1924: "I am very sorry that the Commission should be finding a difficulty in the matter of an interpreter, but I have no doubt that by one means or another a satisfactory person or persons will be found. . . . Other things being equal, and in view of the extremely delicate task of the Commission and its position towards the two Governments concerned, I am doubtful as to the desirability of taking anybody who served during the War in the German forces in Turkish territory, but, of course, the matter is one in which the decision must rest entirely with the Commission itself." PT-P (1924–1941), MNL (Budapest), Cs. A. dosszie 1924-1925, K. 37-1, 13-14.

51. I refer here to Susan Pedersen's exhaustively researched *The Guardians: The League of Nations and the Crisis of Empire* (Oxford: Oxford University Press, 2015).

52. Róisín Healy and Enrico Dal Lago, eds., "Investigating Colonialism within Europe," in Healy and Dal Lago, *The Shadow of Colonialism on Europe's Modern Past* (Houndmills: Palgrave Macmillan, 2014), 3–22.

53. On the continuity/discontinuity issue in an *Ostforschung* context, Fahlbusch, "Volks- und Kulturbodenforschung in der Weimarer Republik: Der 'Grenzfall' Böhmen und Mähren," in Ute Wardenga and Ingrid Hönsch, eds., *Kontinuität und Diskontinuität der deutschen Geographie in Umbruchphasen: Studien zur Geschichte der Geographie* (Münster: Institut für Geographie der West-fälischen Wilhelms-Universität, 1995), 100–112.

54. Professor Dr. Albrecht Penck, "Deutscher Volks- und Kulturboden," in Dr. K. C. von Loesch, *Volk unter Völkern: Bücher des Deutschtums*, Band 1, für den Deutschen Schutzbund (Breslau: Ferdinand Hirt, 1925), 62–73. Original is in AP-P, K. 877, S. 12.

55. Ingo Haar, "German Ostforschung and Anti-semitism," in Haar and Michael Fahlbusch, eds., *German Scholars and Ethnic Cleansing*, 21, 21n6.

56. Ibid., 30–31.

57. Ibid., 37–38.

58. Jan Piskorski and Jörg Hackmann, "Polish myśl zachodnia and German Ostforschung: An Attempt at Comparison," in Haar and Fahlbusch, *German Scholars and Ethnic Cleansing*, 260–71.

59. Haar, "German Ostforschung and Anti-Semitism," 3.

60. Penck, "Deutscher Volks- und Kulturboden," 69.

61. Herb, *Under the Map of Germany*, 56.

62. Haar, "German Ostforschung and Anti-Semitism," 39.

63. Penck, "Deutscher Volks- und Kulturboden," 66.

64. Walter Benjamin, "Theses on the Philosophy of History," in *Illuminations: Essays and Reflections* (New York: Schocken Books, 1969), 253–64.

65. The idea is explored magnificently in Ben Kafka, *The Demon of Writing: Powers and Failures of Paperwork* (New York: Zone Books, 2012).

66. Herb, *Under the Map of Germany*, 63–65.

67. Haar, "Leipziger Stiftung für deutsche Volks- und Kulturbodenforschung," in Haar and Fahlbusch, eds., *Handbuch der völkischen Wissenschaften: Personen—Institutionen— Forschungsprogramme—Stiftungen* (München: K. G. Saur, 2008), 374-82. While Haar also reads back from the twentieth century and is appropriately critical of the German enterprise, here he

does not make empirical use of Czech, Polish, Ukrainian, Russian, or any other Slavic-language sources.

68. In defense of a Greater Germany, Loesch wrote frequently in the right-wing *Deutsche Rundschau* and *Zeitschrift für Geopolitik*. In addition to *Volk unter Völkern*, to which Penck contributed his "Volks- und Kulturboden" essay in 1925, Loesch was the author of *Staat und Volkstum* (1926), *Grenzdeutschland seit Versailles* (1930), *Das deutsche Volk: Sein Boden und seine Verteidigung* (1937), *Deutsches Grenzland* (1935–39), *Der polnische Volkscharakter: Urteile und Selbtszeugnisse aus vier Jahrhunderten* (1940), and *Die Völker und Rassen Südosteuropas* (1943). For Karl Haushofer's anti-Polish proposals on border revision, see *Grenzen in ihrer geographischen und politischen Bedeutung* (1927).

69. On the process of politicization, Peter J. Holquist, "To Count, to Extract, and to Exterminate: Population Statistics and Population Politics in Late Imperial and Soviet Russia," in Ronald Suny and Terry Martin, eds., *A State of Nations: Empire and Nation-Making in the Age of Lenin and Stalin* (Oxford: Oxford University Press, 2001), 11–44; also the essays in Amir Weiner, ed., *Landscaping the Human Garden: Twentieth-century Population Management in a Comparative Framework* (Stanford: Stanford University Press, 2003).

70. The issue of rationality/modernity looms here again. Enlightenment-derived maps as tools would be used instrumentally in the Third Reich for conquest of *Lebensraum* and frontiers to the east, but neither censuses nor maps are reducible to nationalism, mass politics, or bureaucratic rationality. Debates over borderlands/bloodlands seem particularly stuck on this issue into the 1930s and 1940s. For some of the quarrels, see Omer Bartov, "Locating the Holocaust," *Journal of Genocide Research* 13:1 (2011): 121–29; Mark Mazower, *Hitler's Empire: How the Nazis Ruled Europe* (New York: Penguin, 2009); Snyder, *Bloodlands*, as reviewed by Bartov, *Slavic Review* (July 2012): 424–28. On the association of maps/censuses with rationality and control, Götz Aly and Karl-Heinz Roth, eds., *The Nazi Census: Identification and Control in the Third Reich* (Philadelphia: Temple University Press, 2004).

71. Haar, "Leipziger Stiftung," 378.

72. A front-to-back challenge is offered by Thum, "Megalomania and Angst."

73. Penck, "Deutschland als geographische Gestalt," in Leopold Kaiserlich, ed., *Deutschland: Die natürlichen Grundlagen seiner Kultur* (Leipzig: Quelle & Meyer, 1928), 1–9.

74. Curiously, after Teleki's death, the Államtudományi Intézet became part of the Ministry for Religious Affairs and Education, led by Bálint Hóman. Under the Third Reich, it was reorganized again at the end of 1941 and renamed the Count Pál Teleki Research Institute.

75. Aladár Edvi Illés and Albert Halász, eds., *Magyarország a háboru elött és után gazdaságstatisztikai térképekben* [*Hungary before and after the War in Economic-Statistical Maps*] (Budapest: Institute of Political Sciences, 1926), found in the AGSL, 90-F.

76. Ablonczy, *Pál Teleki*, 106; Ignác Romsics, *Magyarország története a XX. Században* (Budapest: Osiris, 1999), and *Trianon és a magyar politikai gondolkozás 1920–1953* (Budapest: Osiris, 1998).

77. Teleki, "Die weltpolitische und weltwirtschaftliche Lage Ungarns in Vergangenheit und Gegenwart," *Zeitschrift für Geopolitik* 3, no. 6 (1926): 381–409, here 381. Translated into Magyar as "Magyarország világpolitikai és világgazdasági helyzete a múltban és a jelenben: A berlini egyetem Magyar Intézetében 1926. Január 26-án tartott előadás," this influential lecture on geography and national science was also published by the *Földrajzi Évkönyv* (*Geographical Yearbook*) in 1927, and reprinted in Teleki, *Európáról és Magyarországról* (Budapest: Athenaeum, 1934), 7–29.

78. Teleki, "Die weltpolitische und weltwirtschaftliche Lage Ungarns," 398.

79. Ibid., 398–99.

80. Teleki, review of Károly Kaán, *A Magyar Alföld: gazdasagpolitikai tanulmany* (Budapest: Magyar Tudomanyos Akadémia, 1927), *Földrajzi Kozlemények* 55, nos. 1–3 (1927): 254–57.

81. Teleki, *Általános gazdasági földrajz, különös tekintettel az óvilági* (Budapest: Magyar Tudomanyos Akadémia, 1927), 552.

82. Ibid., 553.

83. The Hungarian economic nationalist argument for European integration is offered in Pál Teleki, Ferenc Fodor, and Lászlo Gerő, *Közgazdasagi Enciklopédia* (*Encyclopedia of Economics*), with twelve post-Trianon maps excerpted and translated as *Some Economic Maps* (Budapest: Athenaeum, 1930).

84. President Jenő Cholnoky and Dr. Ludwig Diels (first chairman) to Prof. Dr. Albrecht Penck (second chairman), and [unnamed] (general secretary) of the Berlin Geographical Society, 23 October 1927, A-MFT, File 31/1928.

85. Teleki, "Nemzeti szellem—Nemzeti kultura," *Válgatott politikai írások és beszédek* (Budapest: Osiris, 2000), 206.

86. Paolo conte Teleki, *La Geografia dell' Ungheria* (Roma: Istituto per L'Europe Orientale, 1929).

87. Teleki, *Dékáni megnyitó* (Budapest: Élet Ny, 1929), 6–7.

88. Teleki, review of Haushofer, Obst, Lautensach, and Maull, *Bausteine zu Geopolitik* (Berlin: Vowinckel, 1929), in *Földrajzi Kozlemények* 57 (1929): 46–47.

89. Teleki, review of R[ichard] Hennig, *Geopolitik, die Lehre vom Staat als Lebenswesen*, 8th ed. (Leipzig: Teubner, 1931), in *Földrajzi Kozlemények* 60 (1932): 170–72.

90. Teleki, review of Isaiah Bowman, *The New World: Problems in Political Geography*, 4th ed. (New York: World Book Company, 1929), in *Földrajzi Közlemények* 57 (1929): 48–51.

91. Ibid., 51.

92. Romer to Bowman, 17 February 1927, A-AGS.

93. Bowman to Romer, 25 February 1927, A-AGS.

94. Romer to Bowman, 18 June 1927, A-AGS.

95. They were (1) cartography, geophysics, geodesy, meteorology, climatology, and hydrography; (2) geomorphology and geology; (3) phytogeography; (4) zoogeography; (5) human geography and economic geography; (6) anthropology, demography, ethnography, and sociology; (7) regional geography; and (8) historical geography, the history of geography, and didactics of geography. Ludomir Sawicki and Walery Goetel, "Programme Provisoire du II. Congrès des Géographes et ethnographes slaves en Pologne 1927" (Cracovie: Comité d'Organisation du II. Congrès de Géographes et Ethnographes Slaves en Pologne, 1927), A-MFT, File 7/1927.

96. On their personal antipathy, Herb, *Under the Map of Germany*, 69–75.

97. Andor Klay, "Hungarian Counterfeit Francs: A Case of Post-World War I Political Sabotage," *Slavic Review* 33:1 (March 1974): 107–13. For the "textbook" context, Peter F. Sugar, Péter Hanák, and Tibor Frank, *A History of Hungary* (Bloomington: Indiana University Press, 1994), 322–33, and Paul Lendvai, *The Hungarians: A Thousand Years of Victory in Defeat*, trans. Ann Major (Princeton: Princeton University Press, 2003), 396–98. On Klay's complex transnational life (he spoke four languages fluently, German, Italian, Hungarian and English), see the preface to his collection of Hungarian writings in Andor Sziklay, *Magyar lábnyomok: körkep Washingtonból* ([U.S.A.]: Liberty Media, 1988).

CHAPTER FIVE

1. Isaiah Bowman, "Address of the President of the International Geographical Congress," *Science* 81, no. 2164 (26 April 1935): 389.

2. Ibid., 390.

3. For the all-important letters in German from Penck to Rudnyts'kyi between 1918 and 1934, Pavlo Shtoiko, ed., *Lystuvannia Stepana Rudnyts'koho* (Lviv: NTSh, 2006), 279–84.

4. On the mobile intelligentsia there, Roman S. Holiat, *Short History of the Ukrainian Free University* (New York: Shevchenko Scientific Society, 1964).

5. Mushnyk, *Lysty Stepana Rudnyts'koho*, 67.

6. Ibid., 24–25; for the geographer's vulnerability to the intrigues of the late 1920s and non-commitment to the Bolshevik party, Shablii, *Akademik Stepan Rudnyts'kyi*, 141–42.

7. Mushnyk, *Lysty Stepana Rudnyts'koho*, 71, 73, 78.

8. Ibid., 71.

9. Ibid., 80.

10. On the charges, Shablii, *Akademik Stepan Rudnyts'kyi*, 130–64.

11. See Snyder, *Bloodlands*; Kuromiya, *The Voices of the Dead*; Wendy Z. Goldman, *Inventing the Enemy: Denunciation and Terror in Stalin's Russia* (Cambridge: Cambridge University Press, 2011); Serhy Yekelchyk, *Stalin's Empire of Memory: Russian-Ukrainian Relations in the Soviet Historical Imagination*, 2nd ed. (Toronto: University of Toronto Press, 2004); on the issue of Stalin's Russocentrism, see the debate between David Brandenberger, Andreas Umland, and David Marples in *Nationalities Papers* 38:5 (2010): 723–60.

12. On Velychko's significance in Ukraine, see Seegel, *Mapping Europe's Borderlands*, 201–3, 208; and Shablii, ed., *Doktor heohrafii Hryhorii Velychko* (Lviv: Ivan Franko University, 2012).

13. Bowman to Romer, 11 May 1933, A-AGS.

14. Romer to Bowman, 3 October 1933, A-AGS.

15. Bowman to Romer, 27 October 1933, A-AGS.

16. Bowman to Romer, 5 December 1933, A-AGS.

17. Romer to Bowman, 13 January 1934, A-AGS.

18. Largely leaving out geographers, analyses of pre-1933 political culture tend toward grander speculative discussions of modernity, with pan-European or German-centric debates over the *Sonderweg* or change/continuity leading to Hitler's rise and the Final Solution. For a sampling, Eric Weitz, *Weimar Germany: Promise and Tragedy* (Princeton: Princeton University Press, 2007); Panikos Panayi, ed., *Weimar and Nazi Germany: Continuities and Discontinuities*, new ed. (New York: Longman Press, 2014); Anthony McElligott, ed., *Weimar Germany* (Oxford University Press, 2009); Detlev Peukert, *The Weimar Republic: The Crisis of Classical Modernity* (New York: Hill and Wang, 1992); and going back a little farther, Geoff Eley and James Retallack, eds., *Wilhelminism and Its Legacies: German Modernities, Imperialism, and the Meanings of Reform, 1890–1930* (New York: Berghahn Books, 2004).

19. See the essays in Larry Eugene Jones and James Retallack, eds., *Between Reform, Reaction, and Resistance: Studies in the History of German Conservatism from 1789 to 1945* (Providence: Berg, 1993).

20. Herb, *Under the Map of Germany*, 75, 131–32. Herb again leaves out Polish sources, but he notes insightfully that a Nazi press campaign was initiated to discredit Volz, and that this campaign finally came to an end in 1933. The Third Reich drew from Penck's "Volks- und Kulturboden" concept of 1925, if partially.

21. On this valuable point, Fahlbusch, "The Role and Impact of German Ethnopolitical Experts in the SS Reich Security Main Office," in Haar and Fahlbusch, ed., *German Scholars and Ethnic Cleansing, 1919–1945*, 30. Though the central enquiry is a German one circling around the Final Solution and 1941, Fahlbusch offers a fine network analysis of the "professionalization of experts" and their continuities. He covers "scientific knowledge" from the German tradition of *Landeskunde* (regional geography), into Weimar and the Third Reich's population politics.

22. Fahlbusch, *"Wo der deutsche . . . ist, ist Deutschland!" Die Stiftung für deutsche Volks- und Kulturbodenforschung in Leipzig von 1920–1933* (Bochum: Universitätsverlag Dr. N. Brockmeyer, 1994), 144.

23. Haar and Fahlbusch, *Handbuch der völkischen Wissenschaften: Personen–Institutionen–Forschungsprogramme–Stiftungen* (München: K. G. Saur, 2008), 380.

24. Haar and Fahlbusch, *German Scholars and Ethnic Cleansing*, 9–15.

25. Herb, *Under the Map of Germany*, 41–45. Geisler was under the mistaken impression that Romer's maps were made in Kraków, not Vienna, during World War I. Penck, for all his conspiratorial thinking and accusations, actually had the location correct. The last known public display of Spett's map was in 1935, at an oceanography exhibition in Kraków. In 1944, the Nazis destroyed the collection of maps in Poland's National Library (Biblioteka Narodowa) in Warsaw.

26. They include Hugo Hassinger, Hans Rothfelds, Erich Maschke, Bruno Dammann, Franz Rathenau, Werner Essen, Wilhelm Stuckart, Georg Leibbrandt, Otto Scheel, Hans Schwalm, Theodor Oberländer, Erich Koch, Werner Conze, Walter Kuhn, Gunther Ipsen, Emil Meynen, Erich Maschke, and Hans Schwalm. Names are in Fahlbusch, *Wissenschaft im Dienste der nationalsozialistischen Politik? Die 'Volksdeutschen Forschungsgemeinschaften' von 1931–1945* (Baden-Baden: Nomos, 1999), 57.

27. As, for instance, in Konrad Jarausch, *After Hitler: Recivilizing the Germans, 1945–1995* (Oxford: Oxford University Press, 2008).

28. Teleki, "Lóczy Lajos, az ember és a professzor," *Földrajzi Kozlémenyek*, 58:7/8 (1930): 101–5.

29. For instance, Teleki, *Ungarns Wirtschaftslage: Die Vielseitigkeitihrer Schwierigkeiten* (München: A. Dreisler, 1930); Teleki, "Aktuelle Fragen internationaler Politik und die politische Geographie," *Zeitschrift für Geopolitik* 7:1 (1930): 179–81; Teleki, "Politische Geographie," *Geopolitik* (1930): 45–57.

30. Ferenc Fodor, *Teleki Pál: egy "bujdosó könyv"* (Budapest: Mike és Tarsa Antikvarium, 2001), 277–78.

31. Paolo Teleki, *Ungheria ed Europa*, trans. Paolo Ruzicska (Budapest: Federazione italo-ungherese, 1931). It was reprinted again in 1940.

32. Among his collected speeches and writings, see especially "A 'Donauraum' Problémája: A szellemi együttműködés német szövetségeiben 1933 decemberében tartott előadások"; "Európa Problémája;" "Időszerű nemzetközi kérdések a politiaki földrajz megvilágításában"; and "Az európai szellem jelene és jövője" (L'avenir de l'esprit européen), in Teleki, *Válogatott politikai írások és beszédek*, ed. Balázs Ablonczy (Budapest: Osiris, 2000), 153–66, 273–86, 273–86, 301–11.

33. Teleki and Zoltán de Magyary, "Introduction," in Gyula Hantos, *Administrative Boundaries and the Rationalisation of the Public Administration* (Budapest: Athenaeum, 1932), 3–9.

34. Ablonczy, *Pál Teleki*, 101. The biographer notes Teleki's inclination toward geographic determinism, that the "plan demonstrated Teleki's increasingly pronounced regional-deterministic

ideas. The geographic factors, both economic and physical, clearly predominated over the historical-administrative perspectives."

35. Edmund Romer, *Geograf trzech epok*, 283.

36. Martin, *The Life and Thought of Isaiah Bowman*, 199. Martin more or less retells the story from Bowman's point of view, insisting on the integrity of Anglo-American geographic science and America's Wilsonian friendship with Poland. Yet by making Bowman into Wilson's heroic proxy, Martin misses out on Romer's own agency as a man of Ostmitteleuropa—the logic behind his neocolonial and (trans)national support for Bowman in the first place.

37. [No author], "The Fourteenth International Geographical Congress, Warsaw, 1934," *Geographical Review* 25:1 (January 1935), 142–48. There is no author listed, but Bowman almost certainly wrote this. He includes a compilation of statistics and description of the venues and events, and a list of attendees at the IGC.

38. Smith, *American Empire*, 280. Smith resorts again to a blinkered geopolitical explanation, claiming that the German geographers were united in their objections to the corridor: "To be in Poland, for German geographers, was to confront directly the ignominy of lost Lebensraum."

39. Romer to Bowman, 8 April 1934, A-AGS.

40. "The Fourteenth International Geographical Congress," 142.

41. Ibid., 147.

42. Ibid.

43. Bowman to Romer, 13 September 1934, A-AGS.

44. Romer to Bowman, 2 November 1934, A-AGS.

45. Bowman to Romer, 15 November 1934, A-AGS.

46. Penck to Bowman, 10 May 1935, IB-P Ms. 58, Series II, Box 35.

47. Bowman to Penck, 22 May 1935, IB-P Ms. 58, Series II, Box 35.

48. Johns Hopkins University Board of Trustees, "Minutes of 22 February 1935," and letter of 28 February 1935 from Daniel Willard, president of the Board of Trustees to Dr. J. S. Ames, president of Johns Hopkins University, JHU Board of Trustees Minutes and Supporting Papers (January 1935–July 1935), the Ferdinand Hamburger, Jr. Archives, R.G. No. 01, Box No. 11.

49. Bowman to Penck, 22 March 1939, IB-P Ms. 58, Series II, Box 35. This was a direct reference back to Penck's international proposal in Bern in 1891.

50. Mark Jefferson to Bowman, 23 February 1935; Bowman to Jefferson, 7 March 1935, IB-P Ms. 58, Series II, Box 23.

51. Bowman to Teleki, 15 October 1935, IB-P Ms. 58, Series II, Box 43.

52. Teleki to Bowman, 29 October 1935, IB-P Ms. 58, Series II, Box 43.

53. Bowman to John Pelényi, 15 November 1935, IB-P Ms. 58, Series II, Box 43 (found in the Bowman-Teleki Correspondence File).

54. Isaiah Bowman, "The Business of Diplomacy," speech at Presbyterian Social Union, 28 October 1935, IB-P Ms. 58, Series III, Box 3.1, Folder 18, 3–6.

55. The early father-son letters are in the file marked "Robert G[oldthwait] Bowman (1919–1942)," IB-P Ms. 58, Series II, Box 5, beginning with the postcards Isaiah sent to his "Bunny-Boy" from Paris in 1919.

56. Bowman to Ernest Horn, 16 November 1935, IB-P Ms. 58, Series II, Box 21.

57. Bowman to Ernest Horn, 27 November 1935, IB-P Ms. 58, Series II, Box 21.

58. Bowman to W. L. G. Joerg, 27 November 1935, IB-P Ms. 58, Ms. 58, Box 23.

59. Bowman to W. L. G. Joerg, 16 February 1942, IB-P Ms. 58, Series XIII, Box 2 (Paris Peace Conference Files). Joerg and Bowman were similarly oriented concerning U.S. policy. Later in 1942, while curating the Division of Maps and Charts at the National Archives in D.C., Joerg asked Bowman for his version of events on the Paris Peace Conference, at the time when the National Archives was systematically compiling a handbook of all federal world war agencies.

60. Isaiah Bowman to Robert G. Bowman, 16 December 1935, IB-P Ms. 58, Series II, Box 5.

61. Isaiah Bowman, "Next Steps in American Universities," IB-P Ms. 58, Series III (Speeches), Box 3.1, Folder 21. Bowman wrote his first draft for the speech on 23 November 1935.

62. Richard Hartshorne to Bowman, 8 January 1936; Bowman to Hartshorne, 1936 [undated], IB-P Ms. 58, Series II, Box 19.

63. Richard Hartshorne, "The Polish Corridor," *Journal of Geography* 36:5 (1937): 161–76.

64. Francis Deák to Bowman, 17 February 1936; Bowman to Deák, 20 February 1936, IB-P Ms. 58, Series II, Box 13.

65. Isaiah Bowman, "A Design for Scholarship: Commemoration Day Address on the Occasion of the Sixtieth Anniversary of the Founding of the Johns Hopkins University," 22 February 1936, IB-P Ms. 58, Series III, Box 1, Folder 35, 4-5. This was developed by Bowman into a short book, *A Design for Scholarship* (Baltimore: Johns Hopkins University Press, 1936).

66. Baker divulged his position in a letter to Bowman on 28 September 1936, IB-P Ms. 58, Series XIII, Box 2. He expressed concern about "the whole problem of scientific professors accepting retainers from industries and undertaking to combine their research and teaching with specialized investigations for the private profit of industry." Baker further warned Bowman about "the difficulties which cannot be avoided when that double loyalty is assumed," for instance at Hopkins, and noted that "the relationship is fundamentally unsound."

67. Newton D. Baker to Bowman, 23 February 1936, IB-P Ms. 58, Series II, Box 3.

68. Bowman to Gladys Wrigley, 5 March 1936, IB-P Ms. 58, Series II, Box 47.

69. Bowman to Newton D. Baker, 23 March 1936, IB-P Ms. 58, Series II, Box 3.

70. Newton D. Baker to Bowman, 27 March 1936, IB-P Ms. 58, Series XIII, Box 2.

71. Newton D. Baker to Bowman, 26 September 1936, IB-P Ms. 58, Series XIII, Box 2.

72. Isaiah Bowman, "Political Geography," given at Cornell University [no date, fall 1936], IB-P Ms. 58, Series III, Box 3.1, Folder 18.

73. Isaiah Bowman, 13 May 1937, "Is There a Logic in International Situations?" IB-P Ms. 58, Series III, Box 1, Folder 31, 8. IB-P Ms. 58, Series XIII, Box 3.2, address given 13 May 1937 before the Institute of Public Affairs, University of Georgia.

74. Ibid., 2.

75. Isaiah Bowman to Robert G. Bowman, 23 October 1936, IB-P Ms. 58, Series II, Box 5.

76. Bowman to Ernest Horn, 25 June 1937, as found in the file of "Robert G[oldthwait] Bowman (1919–1942)," Series II, Box 5.

77. Isaiah Bowman to Robert G. Bowman, 25 June 1937, IB-P, Ms. 58, Series II, Box 5.

78. Isaiah Bowman to Robert G. Bowman, 27 July 1937, IB-P Ms. 58, Series II, Box 5. The original letter from Bob on 12 July 1937 is either missing or destroyed.

79. Isaiah Bowman to Robert G. Bowman, 19 October 1937, IB-P Ms. 58 Series II, Box 5.

80. Isaiah Bowman to Robert G. Bowman, 4 December 1937, IB-P Ms. 58, Series II, Box 5.

81. Isaiah Bowman to Robert G. Bowman, 10 November 1937, IB-P Ms. 58, Series II, Box 5.

82. Isaiah Bowman to Robert G. Bowman, 1 February 1938, IB-P Ms. 58, Series II, Box 5.

83. Isaiah Bowman to Robert G. Bowman, 30 March 1938, IB-P Ms. 58, Series II, Box 5.

84. Isaiah Bowman to Robert G. Bowman, 24 May 1939, IB-P Ms. 58, Series II, Box 5.

85. Dibrova was a graduate student of Rudnyts'kyi's who finished in 1932. He became the author of a seminal work in postwar Soviet Ukrainian geography, *Heohrafiia Ukrains'koi RSR*, first published in Kiev/Kyiv in 1954.

86. As Shablii points out, the main source for Rudnyts'kyi's unpublished late work seems to have been Semyon Pidhainyi (1907–65). He published "Akademik Stepan Rudnyts'kyi," *Ukrains'kyi visnyk* 152, no. 26, in Berlin on 11 November 1944. See Pidhainyi, *Ukrains'ka intelihentsia na Solovkakh: Spomnyky 1933–34* (Novyi Ul'm: Prometei, 1947), 60–62.

87. On the NKVD and Stalin's purges, Vasyl Marochko and Götz Hillig, eds., *Represovani pedahohy Ukrainy: zhertvy politychnoho teroru (1929–1941)* (Kyiv: Naukovyi svit, 2003); and Ivan Drach, *Ostannia adresa: do 60-richchia solovets'koi trahedii*, 3 vols. (Kyiv: Sfera, 1997–99).

CHAPTER SIX

1. Mroczko, *Eugeniusz Romer*, 205–13.

2. On the layers of place, Bożena Shallcross, *Framing the Polish Home: Postwar Cultural Constructions of Heart, Nation, and Self* (Athens: Ohio University Press, 2002); Sofia Dyak, *Home: A Century of Change* (Pittsburgh: University of Pittsburgh Press, 2013); Michel de Certeau, *The Practice of Everyday Life*, reprint ed. (Berkeley: University of California Press, 2011); Gaston Bachelard, *The Poetics of Space* (Boston: Beacon Press, 1994).

3. Romer to Bowman, 4 April 1935, A-AGS. Romer planned to receive even more bibliographic materials. He wanted to produce what he called a "Repertorium Cartographicum," in alphabetical order, of the names of all the countries from around the world. It would be published in January 1936 and "approximately contain the data concerning Germany."

4. Bowman to Romer, 11 February 1936, A-AGS.

5. Mroczko, *Eugeniusz Romer*, 188–99.

6. Eugeniusz Romer, *Rady i przestrogi, 1918–1938* (Lwów: Zarzewia, 1938), 66–67.

7. Ibid., 190–208. In Romer's "On the Biosociology of the Polish Republic" (1937), an argument for Polish assimilation, he contended that Poland's eastern *kresy* remained intact as a territorial unity, despite centrifugal tendencies by so-called nationalities (i.e., Ukrainians).

8. Ibid., 219.

9. Romer, *Pamiętniki: problemy sumienia i wiary*, 268–69.

10. Bowman to Franklin Delano Roosevelt, 18 December 1930; Roosevelt to Bowman; 8 January 1931, IB-P Ms. 58, Series II, Box 38.

11. Franklin Delano Roosevelt to Bowman, 2 November 1938, IB-P Ms. 58, Series II, Box 38.

12. Franklin Delano Roosevelt to Bowman, 25 November 1938, IB-P Ms. 58, Series II, Box 38.

13. See Tara Zahra, *The Great Departure: Mass Migration from Eastern Europe and the Making of the Free World* (New York: W. W. Norton, 2016); and Matthew Frye Jacobson, *Barbarian Virtues: The United States Encounters Foreign Peoples at Home and Abroad* (New York: Hill and Wang, 2000).

14. Bowman to Franklin Delano Roosevelt, 10 December 1938, IB-P Ms. 58, Series II, Box 38.

15. Romer to Bowman, 2 December 1938, IB-P Ms. 58, Series II, Box 38.

16. Bowman to Romer, 17 December 1938, IB-P Ms. 58, Series II, Box 38; Bowman to Hamilton Fish Armstrong, 17 December 1938; Armstrong to Bowman [undated, after 17 December 1938], (also found in the Bowman-Romer Correspondence).

17. Bowman to James Lee Love, 5 July 1939, IB-P Ms. 58, Series I, Box 1.1, Folder 5.

18. Isaiah Bowman to Robert G. Bowman, 31 July 1939, IB-P Ms. 58, Series II, Box 5.

19. Isaiah Bowman, "Inquiry, August 28, 1939," IB-P Ms. 58, Series XIII, Box 1.

20. Regarding his authority in the U.S. Inquiry, Bowman even corresponded with Upton Sinclair, who was working in 1939 on a novel about the Paris Peace Conference. The geographer bragged of his managerial talents among the U.S. experts: "I don't like to use so mouth-filling a phrase as the following but it happens to be fact that in addition to being the chief territorial specialist of the American delegation I was executive officer of the Section of Economic, Political and Territorial Intelligence. The division was composed of about one hundred men, divided into eighteen sections; and each section was headed by a scholar." Bowman to Upton Sinclair, 2 October 1939, Series XIII.2 (Paris Peace Conference Files).

21. Edmund Romer, *Geograf trzech epok*, 298–99.

22. Anna Cienciala, "The Foreign Policy of Józef Piłsudski and Józef Beck: Misconceptions and Interpretations," *Polish Review* 56:1–2 (2011): 111–53.

23. On Romer and the Ukrainians, see the entry of 30 June 1939 in his Paris recollections. Romer, *Pamiętnik paryski*, 495.

24. Mroczko, *Eugeniusz Romer*, 203.

25. Eugeniusz Romer, *Pamiętniki: problemy sumienia i wiary* (Kraków: Znak, 1988), 306.

26. "Erinnerungen von Albrecht Penck," Berlin entry of 15 October 1943 and Hindenburg-Stadt entry of 10 December 1943, AP-P, K. 871, S. 3, 68–70.

27. Penck, *Besinnliche Rheinreise*, 2nd ed. (Bielefeld: Velhagen & Klasing, 1937).

28. Penck to Hedin, [no date] November 1937, "Letters of Albrecht Penck to Sven Hedin, 1921–1944 (1948)," AP-P, K. 872, K. 3.

29. Hedin to Penck, 19 February 1938, AP-P, K. 872, K. 3.

30. "Erinnerungen von Albrecht Penck," AP-P, K. 871, S. 3, 58.

31. Penck to Hedin, 17 October 1938, AP-P, Leibniz-IfL Leipzig, K. 872, K. 3.

32. "Erinnerungen von Albrecht Penck," AP-P, K. 871, S. 3, 58.

33. Ibid., 67.

34. Penck to Hedin, 13 December 1939, AP-P, K. 872, K. 3.

35. See especially James Retallack, ed., *Saxony in German History: Culture, Society, and Politics, 1830–1933* (Ann Arbor: University of Michigan Press, 2000); Applegate, *A Nation of Provincials*.

36. "Erinnerungen von Albrecht Penck," AP-P, K. 871, S. 3, 66. Penck ranked Meinecke among the "real democrats" but did not see fit to mention Meinecke's liberalism or his anti-Semitism or anti-Slav sentiments. The octogenarian's point here rather stresses the long intellectual and service-oriented continuity of German academic experts from the Second Reich to the Weimar Republic and Third Reich.

37. Penck to Hedin, 25 April 1940, AP-P, K. 872, K. 3.

38. "Erinnerungen von Albrecht Penck," AP-P, K. 871, S. 3, 58.

39. For the Penck family, see the useful twelve-page "Personen-Register zu den 'Lebenserinnerungen von Albrecht Penck,'" compiled by his former student Karl Albert Habbé, attached at the very end of Penck's memoirs. "Erinnerungen von Albrecht Penck," AP-P, K. 871, S. 3.

40. Sven Hedin, *America im Kampf der Kontinente* (Leipzig: F. A. Brockhaus, 1942), reprinted in 1943.

41. Penck to Hedin, 24 April 1943, AP-P, L-IfL Leipzig, AfG, K. 872, K. 3.

42. On Horthy's dilemmas and diplomacy, Thomas Sakmyster, *Hungary's Admiral on Horseback: Miklós Horthy, 1918–1944* (Boulder: East European Monographs, 1984), and Sakmyster, *Hungary, the Great Powers, and the Danubian Crisis, 1936–1939* (Athens: University of Georgia Press, 1980).

43. Marina Cattaruzza and Dieter Langewiesche, "Contextualizing Territorial Revisionism in East Central Europe: Goals, Expectations, and Practices," in Cattaruzza, Dyroff, and Langewiesche, eds., *Territorial Revisionism and the Allies of Germany in the Second World War*, 1–16.

44. Romantic notions of Polish-Hungarian friendship were revived as "Central Europe" and in civilizational discourse by writers including Milan Kundera in 1984. Kundera, borrowing from the "three Europes" mental map by the Hungarian medievalist Jenő Szűcs (1928–98), argued for the modern political/cultural distinctiveness of "captive" nations of the Visegrád countries—Poland, Hungary, and Czechoslovakia (after 1993, the Czech and Slovak Republics). George Schöpflin and Nancy Wood, eds., *In Search of Central Europe* (Cambridge: Polity Press, 1989).

45. Colonel Lawrence Martin annotated Teleki's maps in the catalog carefully. The three men corresponded about it at the start of World War II in 1939–40. I am grateful to John Hessler, senior cartographic librarian of the geography and map division at the U.S. Library of Congress, for his help in piecing together this puzzle.

46. John Pelényi to Isaiah Bowman, 27 January 1940, IB-P Ms. 58, Series II, Box 43 (in the Bowman-Teleki Correspondence File).

47. Bowman to John Pelényi, 5 February 1940, IB-P Ms. 58, Series II, Box 43.

48. John Pelényi to Bowman, 8 February 1940, IB-P Ms. 58, Series II, Box 43.

49. Teleki to Bowman, 12 February 1940, IB-P Ms. 58, Series II, Box 43.

50. Ibid.

51. Teleki's inventory of materials was sent by Bowman to Mr. Mallory at the AGS, with a copy of the Teleki-to-Bowman letter of 12 February 1940 and the Bowman-to-Teleki letter of 13 March 1940, IB-P Ms. 58, Series II, Box 43 (in the Bowman–Teleki Correspondence File).

52. Bowman to John K. Wright, 13 April 1940, IB-P Ms. 58, Series II, Box 43.

53. Bowman to Teleki, 13 March 1940, IB-P Ms. 58, Series II, Box 43.

54. Ibid.

55. Bowman to Teleki, 9 April 1940, IB-P Ms. 58, Series II, Box 43.

56. Teleki to Bowman, 12 April 1940, IB-P Ms. 58, Series II, Box 43.

57. Bowman's secretary [unnamed] to Colonel Lawrence Martin, 16 April 1940, IB-P Ms. 58, Series II, Box 43.

58. Teleki to Bowman, 20 April 1940, IB-P Ms. 58, Series II, Box 43.

59. John Pelényi to Bowman, 25 April 1940, IB-P Ms. 58, Series II, Box 43.

60. Bowman to Hugh R. Wilson, 3 June 1940, IB-P Ms. 58, Series II, Box 43.

61. See Ignác Romsics, *Mítoszok, legendák, tévhitek a 20. századi magyar történelemrol* (Budapest: Osiris, 2002).

62. Teleki's full, cryptic suicide note in Hungarian on 3 April 1941: "Szószegők lettünk gyávaságból—a mohácsi beszéden alapuló örökbéke szerződéssel szemben. A nemzet érzi, és mi odadobtuk becsületét. A gazemberek oldalára álltunk—mert a mondvacsinált atrocitásokból egy szó sem igaz! Sem a magyarok ellen, de még a németek ellen sem! Hullarablók leszünk! a legpocsékabb nemzet. Nem tartottalak vissza. Bűnös vagyok."

63. For a summary of Bowman's role in resettlement issues, Smith, *American Empire*, 293–316.

64. Isaiah Bowman to Robert G. Bowman, 23 September 1939, IB-P Ms. 58, Series II, Box 5.

65. Isaiah Bowman to Robert G. Bowman, 25 September 1939, IB-P Ms. 58, Series II, Box 5.

66. Isaiah Bowman to Robert G. Bowman, 16 October 1939, IB-P Ms. 58, Series II, Box 5.

67. Isaiah Bowman to Robert G. Bowman, 9 November 1939, IB-P Ms. 58, Series II, Box 5.

68. Isaiah Bowman to Robert G. Bowman and Walter Bowman, 31 July 1940, "Annual Letter in Duplicate," IB-P Ms. 58, Series II, Box 5.

69. Robert G. Bowman to Isaiah Bowman, 30 September 1940, IB-P Ms. 58, Series II, Box 5.

70. George L. Warren, c/o Isaiah Bowman, 16 October 1940, IB-P Ms. 58, Series II, Box 5 (in the Isaiah Bowman–Robert G. Bowman Correspondence File).

71. A. C. Trowbridge to Bowman, 9 November 1940, IB-P Ms. 58, Series II, Box 5.

72. Isaiah Bowman to A. C. Trowbridge, 11 November 1940, IB-P Ms. 58, Series II, Box 5.

73. A. C. Trowbridge to Isaiah Bowman, 19 November 1940, IB-P Ms. 58, Series II, Box 5.

74. Isaiah Bowman to Robert G. Bowman, 23 November 1940, Ms. 58, Series II, Box 5.

75. Isaiah Bowman to Ernest Horn, 17 December 1940, IB-P Ms. 58, Series II, Box 5.

76. Isaiah Bowman to Robert G. Bowman, 29 May 1941, IB-P Ms. 58, Series II, Box 5.

77. Robert G. Bowman, "Prospects of Land Settlement in Western Australia," *Geographical Review* 32:4 (1942): 598–621.

78. Witold Romer to Bowman, 4 August 1940, IB-P Ms. 58, Series II, Box 38.

79. Bowman to W[itold] Romer, 23 August 1940, IB-P Ms. 58, Series II, Box 38.

80. Imprisoned in Budapest after the Hungarian Revolution in 1956, Kosáry served as past president of the Hungarian Academy of Sciences from 1990 to 1996.

81. Bowman to D. G. Kosáry, 15 September 1941; Isaiah Bowman to Countess Paul Teleki [as addressed], 18 September 1941, IB-P Ms. 58, Series II, Box 43.

82. Isaiah Bowman to Robert G. Bowman, 28 January 1942, IB-P Ms. 58, Series II, Box 5.

83. Robert G. Bowman to Isaiah Bowman, 10 March 1942; and Isaiah Bowman to Robert G. Bowman, 16 March 1942, IB-P Ms. 58, Series II, Box 5.

84. Ernest Horn to Isaiah Bowman, 15 April 1942; Robert G. Bowman to Isaiah Bowman, 17 April 1942; Robert G. Bowman to Isaiah Bowman, 22 April 1942, IB-P Ms. 58, Series II, Box 5.

85. Isaiah Bowman to Robert G. Bowman, 10 August 1942, IB-P Ms. 58, Series II, Box 5.

86. Robert G. Bowman to Isaiah Bowman, 7 January 1943, IB-P Ms. 58, Series II, Box 6.

87. Robert G. Bowman to Isaiah Bowman, 12 January 1943, IB-P Ms. 58, Series II, Box 6.

88. Isaiah Bowman to Robert G. Bowman, 24 February 1943, IB-P Ms. 58, Series II, Box 6.

89. Robert G. Bowman to Isaiah Bowman, 16 March 1943, IB-P Ms. 58, Series II, Box 6.

90. Isaiah Bowman to Robert G. Bowman, 2 May 1943; Isaiah Bowman to Major General Reckford, 20 May 1943, IB-P Ms. 58, Series II, Box 6.

91. Robert G. Bowman to Isaiah Bowman, 7 November 1943, IB-P Ms. 58, Series II, Box 6.

92. Richard Hartshorne to Isaiah Bowman, 15 October 1943, IB-P Ms. 58, Series II, Box 19.

93. Isaiah Bowman to Robert G. Bowman, 18 October 1943, IB-P Ms. 58, Series II, Box 6.

94. Robert G. Bowman to Isaiah Bowman, 7 November 1943, IB-P Ms. 58, Series II, Box 6.

95. Robert G. Bowman to Isaiah Bowman [undated, between 7 November and 12 December 1943], IB-P Ms. 58, Series II, Box 6.

96. Eugeniusz Romer, *Pamiętniki*, 357–58.

97. Mroczko, *Eugeniusz Romer*, 208–9.

98. On the city's history and people under occupation, Tarik Cyril Amar, *The Paradox of Ukrainian Lviv: A Borderland City between Stalinists, Nazis, and Nationalists* (Ithaca: Cornell University Press, 2015); Christoph Mick, *Lemberg, Lwów, L'viv: Violence and Ethnicity in a Contested City* (West Lafayette, Ind.: Purdue University Press, 2015); William Risch, *The Ukrainian West: Culture and the Fate of Empire in Soviet Lviv* (Cambridge: Harvard University Press, 2011).

99. Romer, "Spomniki IX 1943–IX 1944," ER-P, sygn. 1027 1/III, 6-7.

100. The Polish historian Marian Mroczko follows Romer's own gendered language literally, referring to Romer's path after Jadwiga's death as a return to the "bosom" (*lono*) of the Catholic Church. See Mroczko, *Eugeniusz Romer*, 212. As far as the conversion is concerned, not until after the end of Poland's independence in 1939 did the lifelong agnostic actually attend a mass. For Romer's account, *Pamiętniki*, 321.

101. Edmund Romer, *Geograf trzech epok*, 306. In the Polish original: "Nie mogli przecież zapomnieć o jego roli w Wersalu, o likwidacji niemieckiej kartografii w Polsce, wszak żył jeszcze mistrz i denunciant Ojca z 1916 roku, Albrecht Penck."

102. Romer, "Spomniki IX 1943–IX 1944," ER-P, sygn. 1027 1/III, 1.

103. Eugeniusz Romer, *Pamiętniki*, 470–1; Edmund Romer, *Geograf trzech epok*, 307.

104. Eugeniusz Romer, *Pamiętniki*, 272–73; Snyder, *The Reconstruction of Nations*, 148–49.

105. Eugeniusz Romer, *Pamiętniki*, 273–74; and "Spomniki IX, 1943-IX 1944," ER-P, sygn. 1027 1/III, 5-7.

106. Allen Paul, *Katyn: Stalin's Massacre and the Triumph of Truth* (DeKalb: Northern Illinois University Press, 2010); Anna M. Cienciala, Natalia S. Lebedeva, and Wojciech Materski, eds., *Katyn: A Crime without Punishment* (New Haven: Yale University Press, 2008).

107. Romer, "Spomniki IX 1943–IX 1944," ER-P, sygn. 1027 1/III, 7.

108. Ibid., 4. Romer endearingly called Wąsowicz his "shepherd" (*baca*) and "one of my bravest scientific colleagues and most faithful friends" who served as his principal link to the outside world.

109. Ibid., 4, 15.

110. Ibid., 5–6.

111. For this assessment of the Soviet and Nazi occupations, see the manuscript copy of Romer's conversion memoir [*Pamiętniki: Problemy sumienia i wiary*], ER-P, sygn. 10273/III, p. 331.

112. Romer, "Spomniki IX 1943–IX 1944," ER-P, A-JUL (Kraków), sygn. 1027 1/III, 5.

113. Ibid., 15–24.

114. Ibid., 16. All but forgotten today, Foerster too lived a transnational life. In fact, he was an ethicist and peace activist who also wrote a book called *Mein Kampf*—in 1920, as *My Struggle against a Militaristic and Nationalistic Germany*. A pacifist opponent of revision, he fled into exile in Switzerland in 1922 and moved to France in 1926. The Nazis considered him an enemy of the state and burned his books in public rituals in 1933. He fled again in 1940. When the Nazis occupied France, he was wanted by the Gestapo. Foerster tried to seek asylum in neutral Switzerland, but the Swiss government considered him a German and refused to accept him on account of his dubious citizenship. He fled to Portugal and eventually the United States, where he lived in New York until 1963. In his final years he moved back to Switzerland, where he died close to Zurich in 1966.

115. Romer, "Spomniki IX 1943–IX 1944," ER-P, sygn. 1027 1/III, 11.

116. Ibid., 2.

117. Ibid., 27.

CHAPTER SEVEN

1. "Erinnerungen von Albrecht Penck," entry of 10 December 1943, AP-P, K. 871, S. 3, 76-77.

2. Penck to Ilse [Tschermak] von Seysenegg [Penck], drafted letter of 9 December 1943, AP-P, K. 871, S. 3, 78.

3. "Erinnerungen von Albrecht Penck," AP-P, K. 871, S. 3, 77-78.

4. Penck to Hedin, 6 January 1944, AP-P, K. 872, K. 3.

5. Penck to Hedin, 18 April 1944, AP-P, K. 872, K. 3.

6. Penck to Hedin, 11 July 1944, AP-P, K. 872, K. 3.

7. Penck to Professor Dr. [Eduard] Spranger, 10 August 1944, AP-P, K. 872, S. 1/20/20a.

8. Ilse Tschermak von Seysenegg [Penck] to Hedin, 11 March 1945, AP-P, K. 872, K. 3 (copy). The original letter is in the Sven Hedin Arkhiv, vol. 485, of the Swedish State Archive (Riksarchivet) in Stockholm.

9. Isaiah Bowman to Robert G. Bowman, 26 April 1944, IB-P Ms. 58, Series II, Box 6.

10. Isaiah Bowman to Robert G. Bowman, 18 August 1944; Isaiah Bowman to Harry E. Neburn [the State University of Iowa], 18 August 1944, IB-P Ms. 58, Series II, Box 6.

11. Robert G. Bowman to Isaiah Bowman, 27 January 1945; 21 February 1945; and 23 April 1945, IB-P Ms. 58, Series II, Box 6.

12. Isaiah Bowman to Robert G. Bowman, 29 January 1945, IB-P Ms. 58, Series II, Box 6.

13. Edward Stettinius to Isaiah Bowman, 31 March 1945, IB-P Ms. 58, Series II, Box 6 (in the Correspondence file between Robert G. and Isaiah Bowman).

14. To say that Bowman's prejudices were not unusual for his historical time is, of course, not to excuse them. On Bowman's wartime and early Cold War obsessions with the getting the "best men" for American geography, and as an aspect of his anticommunism, anti-Semitism, and homophobia, see Smith, *American Empire*, 258, 442–43.

15. Robert G. Bowman to Isaiah Bowman, 23 April 1945, IB-P, Series II, Box 6.

16. Isaiah Bowman addressed to "Sons" [Walter and Robert G. Bowman], 5 July 1945, IB-P Ms. 58, Series II, Box 6.

17. Isaiah Bowman to Lieutenant Walter P. Bowman, Hq. 5th Air Force, 4 August 1945, IB-P Ms. 58, Series II, Box 6.

18. Robert G. Bowman addressed to "Family," 8 September 1945, IB-P Ms. 58, Series II, Box 6.

19. Mroczko, *Eugeniusz Romer*, 224. Romer researched the names of 5,000 known Poles from a book of biograms *Czy wiesz kto to jest?* [Do You Know Who This Is?], edited by S. Łzy in Warsaw in 1939.

20. Romer, "Spomniki IX 1943–IX 1944," ER-P, sygn. 1027 1/III, 28.

21. Ibid., 11.

22. Ibid., 39–41. For a detailed history of the issue, John Connelly, *From Enemy to Brother: The Revolution in Catholic Teaching on the Jews, 1933–1965* (Cambridge: Harvard University Press, 2012).

23. Romer, "Spomniki IX 1943–IX 1944," 29–30.

24. Edmund Romer, *Geograf trzech epok*, 312.

25. Romer to Prof. H[enryk] Arctowski, 14 June 1946, ER-P, sygn. 10366/III, 41.

26. On territorial legitimation in the PRL's early years, Marcin Zaremba, *Im nationalen Gewande Strategien kommunistischer Herrschaftslegitimation in Polen 1944–1980* (Osnabrück: Fibre, 2011).

27. Snyder, *The Reconstruction of Nations*, 180–81.

28. Bohdan Winid, "Profesor Eugeniusz Romer—jakiego znałem," in Dobiesław Jędzrejczyk and Waldemar Wilk, eds., *Eugeniusz Romer jako geograf społeczno-gospodarczy* (Warszawa, 1999), 16–18.

29. Mroczko, *Eugeniusz Romer*, 235. Mroczko contends that Romer was demoralized and therefore not engaged in postwar politics, but the exiled geographer was involved in redrawing Poland's "lost territories" in 1945–46. Romer's retreat is true only in the ideological sense of avoiding publicly—because he had to—the preset frames of Soviet Marxist followed by late Stalinist geography. On Romer's renewed Poland-centric mindset in 1946, see his letter to Prof. H[enryk] Arctowski, 14 June 1946, ER-P, sygn. 10366/III, 41-42.

30. David Harris, U.S. State Department, Washington, D.C., to Isaiah Bowman, 16 October 1943, IB-P Ms. 58, Series II, Box 43 (found in the Isaiah Bowman–Count Géza Teleki correspondence file).

31. Bowman to (Count) Géza Teleki, minister of education, Budapest, 17 July 1945, IB-P Ms. 58, Series II, Box 43.

32. (Count) Géza Teleki to Isaiah Bowman, 1 December 1945, IB-P Ms. 58, Series II, Box 43. Mistakenly, it seems, Teleki addressed the letter originally to Philadelphia, as that was where he thought Johns Hopkins University was located.

33. (Count) Géza Teleki to Isaiah Bowman, 21 May 1946, IB-P Ms. 58, Series II, Box 43.

34. On the ramifications of this Treaty of Paris for Hungary, Ignác Romsics, *Az 1947-es párizsi békeszerzodés* (Budapest: Osiris, 2006).

35. (Count) Géza Teleki to Isaiah Bowman, 21 May 1946, IB-P Ms. 58, Series II, Box 43.

36. Philip M. Hayden to (Count) Géza Teleki (copied to Isaiah Bowman), 13 June 1946, IB-P Ms. 58, Series II, Box 43.

37. Isaiah Bowman to Charles H. Behre, Jr., 26 June 1946, IB-P Ms. 58, Series II, Box 43.

38. Charles H. Behre, Jr., to Isaiah Bowman, 24 July 1946 and 21 August 1946, IB-P Ms. 58, Series II, Box 43.

39. Isaiah Bowman to (Count) Géza Teleki (copied to Charles H. Behre, Jr.), 26 June 1946, IB-P Ms. 58, Series II, Box 43.

40. (Count) Géza Teleki to Isaiah Bowman, 6 September 1946, IB-P Ms. 58, Series II, Box 43.

41. Ibid.

42. Dr. Joseph H. Willits, like William Morris Davis, was the son of a devout Pennsylvania Quaker farmer. He became director of the Social Science Division. Isaiah Bowman to Count Géza Teleki, 19 October 1946, c/o the Honorable Victor Csornoky, first counselor of Hungarian legation, Legation of Hungary, Washington, D.C.; Bowman to Willits, 19 October 1946, IB-P Ms. 58, Series II, Box 43.

43. Charles H. Behre, Jr., Columbia University in the City of New York, Dept. of Geology, to President Isaiah Bowman, Johns Hopkins University, 19 November 1946, IB-P Ms. 58, Series II, Box 43.

44. Isaiah Bowman to Miss Marion Elderton, the Rockefeller Foundation, 21 November 1946, IB-P Ms. 58, Series II, Box 43.

45. Isaiah Bowman to Dr. Charles H. Behre, Jr., 21 November 1946, Columbia University (copied to Miss Marion Elderton), IB-P Ms. 58, Series II, Box 43.

46. Isaiah Bowman to Dr. Paul F. Kerr, Columbia University, Dept. of Geology, 3 December 1946, IB-P Ms. 58, Series II, Box 43.

47. Lattimore taught as a professor of history at Hopkins from 1938 to 1963. Despite Bowman's inaction and the accusations of Marxism, espionage, and communism directed his way in the late 1940s, he survived McCarthy's witch hunt.

48. Quoted in Smith, *American Empire*, 444. Little has been resolved: the notorious Harvard episode, as much about the personality factor and petty grievances of these men as their ideas, has been interpreted differently, and mostly as an intradisciplinary skirmish. See Geoffrey J. Martin, "On Whittlesey, Bowman, and Harvard," *Annals of the Association of American Geographers* 78:1 (Mar. 1988): 152–58; and Smith, "'Academic War over the Field of Geography': The Elimination of Geography at Harvard, 1947–1951," *Annals of the Association of American Geographers* 77:2 (June 1987): 155–72. What Smith sees as Bowman's entrenched homophobia, gross miscalculations with Conant, even his self-delusion, Martin reads as Bowman being ill-disposed to both Rice and Whittlesey, due to his long competency as an administrator.

49. Isaiah Bowman to Robert G. Bowman, 14 May 1948, IB-P Ms. 58, Series II, Box 6.

50. Isaiah Bowman to Robert G. Bowman, IB-P Ms. 58, 12 February 1948, Series II, Box 6.

51. Robert G. Bowman to Isaiah Bowman, IB-P Ms. 58, 30 May 1946, Series II, Box 6.

52. Isaiah Bowman to Robert G. Bowman, IB-P Ms. 58, 4 November 1947, Series II, Box 6.

53. Bob started his new position as professor of geography at the University of Nebraska-Lincoln in fall 1949, also had a last exchange after Christmas. In 26 December 1949 letter, Bob from the frontier wrote to his father, "Many happy returns of the day! That the Christ Child and His Major Prophet should have been born on successive dates is one of those happy circumstances of history, like the creation of Nebraska." Robert G. to Isaiah Bowman, 26 December 1949; Isaiah to Robert G. Bowman, 30 December 1949, IB-P Ms. 58, Series II, Box 6. Isaiah sent his last letter, just a few sentences, to Bob on 30 December 1949. He offered his well-wishes to the family for Christmas.

54. Isaiah Bowman to Robert G. Bowman, 14 April 1948, IB-P Ms. 58, Series II, Box 6.

55. Isaiah Bowman to Robert G. and Jean Bowman, 6 May 1948, IB-P Ms. 58, Series II, Box 6.

56. Isaiah Bowman to Eugeniusz Romer, 3 June 1949, IB-P Ms. 58, Series II, Box 38.

57. Isaiah Bowman to Robert G. Bowman, 30 December 1949, IB-P Ms. 58, Series II, Box 6.

58. On victimhood narratives and Soviet issues facing returnees, Stephen F. Cohen, *The Victims Return: Survivors of the Gulag after Stalin* (Exeter, N.H.: Publishing Works, 2010); Steven A. Barnes, *Death and Redemption: The Gulag and the Shaping of Soviet Society* (Princeton, Princeton University Press, 2011).

59. Shablii, *Akademik Stepan Rudnyts'kyi*, 151–64.

60. Lawrence E. Gelfand, *The Inquiry: American Preparations for Peace, 1917–1919* (New Haven, Conn.: Yale University Press, 1963).

61. Edmund Romer, "Życiorys," *Wspomnienia o ojcu* (Warszawa: Czytelnik, 1985), 335–42.

62. [No author], "Wrocław: biograficzna plenerowa wystawa poświęcona Witoldowi Romerowi," 2 August 2015, http://wiadomosci.onet.pl/wroclaw/wroclaw-biograficzna-plenerowa-wystawa-poswiecona-witoldowi-romerowi/4gz34s (accessed 28 August 2015); and Sylvia Jurgiel, "W obiektywie Witolda Romer," *Reportaż*, 31 July 2014, http://www.radiowroclaw.pl/articles/view/36785/Ocalic-od-zapomnienia-Witold-Romer (accessed 23 August 2016).

63. For the exhibit's opening and context on 28 January 1989, Elżbieta Nowak-Ferdhus, "Wystawa 'Profesor Eugeniusz Romer—kartograf niepodległej Polski' w Bibliotece Jagiellońskiej w Krakowie," *Polski Przegląd Kartograficzny* 21:3–4 (1989), 173. See also Nowak-Ferdhus and Ewa Szynkiewicz, "Ewolucja 'Atlasu Geograficznego' Eugeniusza Romera z 1908 roku" and "Wystawa

'Profesor Eugeniusz Romer—kartograf niepodległej Polski' w Bibliotece Jagiellońskiej w Krakowie," *Polski Przegląd Kartograficzny* 24:1–2 (1992): 26–42.

64. "Eugeniusz Romer—sesja naukowa i wystawa w rocznicę śmierci," *Biuletyn Informacjyny Biblioteki Narodowej* 169:2 (2004): 1–19.

65. András Rónai, "Teleki Géza (1911–1983)," *Földrajzi Muzeúmi Tanulmányok* 7 (1989): 65–66.

66. Vivian Yee, "The Death of a Countess in Exile," *New York Times*, 26 April 2013.

67. [No author], *Newsletter of the Association of American Geographers*, May 1939, copy saved by Albrecht Penck and located in AP-P, K. 878, S. 5.

68. Laudations in 1943 to Penck were many, and they have never since been matched. See the AP-P, K. 871, S. 5, and K. 877, S. 5/17, for Norbert Krebs, "Volksthumsfragen der Geographie: Albrecht Penck zum 85. Geburtsgage," *Berliner Lokal-Anzeiger*, 25 September 1943; Paul Wittko, "Ein Meister der Erdkunde: Zu Albrecht Pencks 85. Geburtsage am 25. September"; [no author], *Deutsche Zeitung*, 27 September 1943; [no author], "Geograph von Weltruf: Prof. Albrecht Penck 85 Jahre alt," *Leipziger Neueste Nachrichten*, 24 September 1943; Krebs, "Albrecht Penck 85 Jahre alt," *Deutsche Allgemeine Zeitung*, 25 September 1943; and R[aimund] v[on] Klebelsberg, "Albrecht Penck und seine Beziehungen zur Heimat: Zum 85. Gebürtstag des Ehrendoktors der Deutschen Alpenuniversität," *Innsbrucher Nachrichten*, 25 November 1943.

69. My thanks to Norman Henniges for sharing knowledge on this fuzzy issue. Whereabouts of Penck's estate after 1945 remain unclear, due partly to Germany's postwar division and the family's dispersal. Around or before 1945, Penck's materials went to the Imperial Exchange Department (Reichstauschstelle) in Wannsee, others to Hamburg. Penck's boxes of books at the Geographical Institute at the University of Berlin were moved to a different building at Schönebeck on the Elbe. These were probably confiscated or sold by the Red Army; some maps and records of the institute were taken to Moscow. Penck's estate in 1949, at least in part, was returned to Berlin. Remaining boxes of correspondence as well as Penck's copious rock collection were stored in a so-called "Penck Cabinet" (Penckschrank) at the Geographic Institute of Humboldt University. When Penck's acolyte, the geographer Gerhard Engelmann, assembled a bibliography and tried to prepare for the hundredth anniversary of Penck's birth in 1858, he wanted to make a full inventory. It proved impossible. A testament to the Mitteleuropa mingling of religion and science, a so-called "cult of relics" seems to have been distributed as a legacy among Penck's disciples and colleagues such as Joachim Marcinek, Fritz Haefke, Fritz Machatschek, and Hanno Beck. Emil Meynen, his student in Berlin and colleague in Leipzig, carried on Penck's training into West German regional studies (*Landeskunde*), topographic survey (*Landesaufnahme*), and spatial planning (*Raumforschung*) projects.

70. Article by "mew," "Erstbeschreiber der glazialen Serie: Geograf und Geologe Albrecht Penck stammte aus Leipzig Strasse in Sellerhausen nach ihm benannt," *Leipziger Volkszeitung*, 11 November 2011, 6.

71. Memory follows very odd paths: Bowman Street was on the campus behind Daniel Coit Gilman Hall, the administrative building. The Bowman portrait, painted by the American artist John Howard Sanden (b. 1948), was not commissioned by the university until the 1980s, more than thirty years after Bowman's death. It appears to have been installed in Garland Hall on the Homewood Campus, where it was inventoried in 1992. A 2006 inventory locates the portrait in the Hutzler Undergraduate Reading Room of Gilman Hall (named "the Hut"). Gilman Hall was closed for renovation in 2008, and reopened in 2010. In advance of the building's reopening, Adam F. Falk, former dean of the School of Arts and Sciences at Johns Hopkins and current

president of Williams College, approved the installation of the portraits of all of Johns Hopkins's former presidents, including Bowman, in the Hut.

72. [No author], "Bowman, Erna H" [obituary], http://www.legacy.com/obituaries /democratandchronicle/obituary.aspx?page=lifestory&pid=152187110 (accessed 26 January 2014).

73. Isaiah Bowman to Robert G. Bowman, 9 April 1949; 18 June 1949; 11 October 1949, IB-P Ms. 58, Series II, Box 6.

74. Janice Brown, " 'The Leakeys of White Mountain Geology': Hanover New Hampshire's James Walter Goldthwait (1880–1947), Richard Parker Goldthwait (1911–1992) and Lawrence Goldthwait (1914–2001)," and "Cow Hampshire: New Hampshire's History Blog," 29 December 2007, http://www.cowhampshireblog.com/2007/12/29/the-leakeys-of-white-mountain -geology-hanover-new-hampshires-james-walter-goldthwait-1880-1947-richard-parker -goldthwait-1911-1992-and-lawrence-goldthwait-1914-2001/ (accessed 26 January 2014).

CONCLUSION

1. Isaiah Bowman, "Address of the President of the International Geographical Congress," *Science* 81, no. 2164 (26 April 1935): 389.

2. Volker R. Berghahn and Simone Lässig, eds., *Biography between Structure and Agency: Central European Lives in International Historiography* (New York: Berghahn Books, 2008).

3. Stoler, *Along the Archival Grain*; Stoler, ed., *Haunted by Empire: Geographies of Intimacy in North American History* (Durham: Duke University Press, 2006); and Stoler, *Carnal Knowledge and Imperial Power.*

4. On roles and alliances in epistolary networks, Amanda Herbert, *Female Friendship: Gender, Alliances, and Friendship in Early Modern Britain* (New Haven: Yale University Press, 2014).

5. Malte Rolf, "Imperiale Biographien: Lebenswege imperialer Akteure in Gross- und Kolonialreichen, 1850–1918," *Geschichte und Gesellschaft* 40:1 (2014): 5–21.

6. It bears here to keep in mind that gender as a category of analysis need not be reduced to social-constructionist studies of masculinity or femininity. See Agatha Schwartz, ed., *Gender and Modernity in Central Europe: The Austro-Hungarian Monarchy and Its Legacy* (Ottawa: University of Ottawa Press, 2010). On fraternal homosociality and homophobia, Nicholas L. Syrett, *The Company He Keeps: A History of White College Fraternities* (Chapel Hill: University of North Carolina Press, 2009), 1–11, esp. 308n5, 309n8; see also Mary Ann Clawson, *Constructing Brotherhood: Class, Gender, and Fraternalism* (Princeton: Princeton University Press, 1989); Michel Foucault, *The History of Sexuality: An Introduction*, vol. 1 (New York: Vintage, 1980); Gail Bederman, *Manliness and Civilization: A Cultural History of Race and Gender in the United States, 1880–1917* (Chicago: University of Chicago Press, 1995); Timothy Beneke, *Proving Manhood: Reflections on Men and Sexism* (Berkeley: University of California Press, 1997); J. A. Mangan and James Walvin, eds., *Manliness and Morality: Middle-class Masculinity in Britain and America, 1800–1940* (New York: St. Martin's Press, 1987); and Daphne Spain, "The Spatial Foundations of Men's Friendships and Men's Power," in Peter Mardi, ed., *Men's Friendships* (Newbury Park, Calif.: Sage, 1992), 59–73.

7. Geoffrey J. Martin, *All Possible Worlds: A History of Geographical Ideas*, 4th ed. (Oxford: Oxford University Press, 2005).

8. Yanko Tsvetkov, "Alphadesigner Identity Report" (2013), http://alphadesigner.com/about/ (accessed 9 August 2016).

9. On affect and science, Frank Biess and Daniel M. Gross, eds., *Science and Emotions after 1945: A Transatlantic Perspective* (Chicago: University of Chicago Press, 2014); in the Russian and East European field, Mark D. Steinberg and Valeria Sobol, eds., *Interpreting Emotions in Russia and Eastern Europe* (DeKalb: Northern Illinois University Press, 2011); in critical geography, Joyce Davidson, Liz Bondi, and Mick Smith, eds., *Emotional Geographies* (London: Ashgate, 2007).

10. Philippe Rekacewicz, "Cartographie radicale," *Le monde diplomatique* (February 2013), 15.

11. Denis Wood and John Fels, *The Power of Maps* (New York: Guilford Press, 1992); John Krygier and Wood, eds., *Making Maps: A Visual Guide to Map Design for GIS*, 3rd ed. (New York Guilford Press, 2016); Wood with Fels and Krygier, *Rethinking the Power of Maps* (New York: Guilford Press, 2010); and Wood and Fels, *The Natures of Maps: Cartographic Constructions of the Natural World* (Chicago: University of Chicago Press, 2008).

12. The topic is covered in David Engerman, *Know Your Enemy: The Rise and Fall of America's Soviet Experts* (Oxford: Oxford University Press, 2009).

13. As in Theodor Adorno et al., *The Authoritarian Personality* (New York: Harper Brothers, 1950), which explores the fuzzy concept of "ambiguity tolerance" and develops personality classifications according to type; and Hannah Arendt's classic *The Origins of Totalitarianism* (New York: Schocken Books, 1951). Biographies of figures involved in conceptual modeling are in Abbott Gleason, *Totalitarianism: The Inner History of the Cold War* (Oxford: Oxford University Press, 1995). On Arendt and on Jan T. Gross as "Arendtian," Snyder, *Bloodlands*, 196, 485–86n21.

14. For Smolich's life between Ukraine and Belarus, see Seegel, "Remapping the Geo-Body," 205–29; and Seegel, *Mapping Europe's Borderlands*, 277.

15. Timothy Brook, *Mr. Selden's Map of China: Decoding the Secrets of a Vanished Cartographer* (New Bloomsbury Press, 2013), xxvi.

BIBLIOGRAPHY

I. PRIMARY SOURCES

1. ALBRECHT PENCK (1858–1945)

1877. "Nordische Basalte im Diluvium von Leipzig," *Neues Jahrbuch für Mineralogie, Geologie und Paläontologie*: 243–50.

1878. *Geognostiche Karte von Mitteleuropa. Geologische Karte von Central-Europa.* Leipzig: Spamer.

1879. "Die Geschiebeformation Norddeutschlands," *Zeitschrift der Deutschen Geologischen Gesellschaft* 31: 117–203.

1887. ———, and Alfred Kirchoff. *Das Deutsche Reich.* Leipzig: Freytag.

1894. *Morphologie der Erdoberfläche.* Stuttgart: J. Engelhorn.

1896. "Die Geomorphologie als genetische Wissenschaft: Eine Einleitung über geomorphologische Nomenklatur," *International Geographical Congress* 6: 735–47.

1901–9. ———, and Eduard Brückner, *Die Alpen im Eiszeitalter*, 3 vols. Leipzig.

283

1915. "Sven Hedin über England und Deutschland," *Zeitschrift der Gesellschaft für Erdkunde zu Berlin* 50: 243–45.

1915. *Von England festgehalten: Meine Erlebnisse während des Krieges im britischen Reich.* Stuttgart: J. Engelhorn.

1915. *Was wir im Kriege gewonnen und was wir verloren haben.* Berlin, Heymann.

1916. "Der Krieg und das Studium der Geographie," *Zeitschrift der Gesellschaft für Erdkunde zu Berlin* 51: 158–76, 222–48.

1916. "Die Ukraina," *Zeitschrift der Gesellschaft für Erdkunde zu Berlin* 51: 345–61, 458–47.

1917. *Die natürlichen Grenzen Russlands: Ein Beitrag zur politischen Geographie des europäischen Ostens.* Berlin: Sammlung Meereskunde.

1917. "Polnisches," *Posener Tagblatt,* 27 March.

1917. *Rede zum Antritt des Rektorats der Universitaet Berlin, gehalten 25. September 1917.* Berlin.

1917. *U.S.-Amerika: Gedanken und Erinnerungen eines Austauchprofessors.* Stuttgart: J. Engelhorn.

1917. *Zeit- und Streitfragen. Korrespondenz des Bundes und deutscher Gelehter Künstler,* no. 10, 16 March.

1918. "Die erdkundlichen Wissenschaften an der Universität Berlin," Rede zur Gedächtnisfeier des Stifters der Berliner Universität König Friedrich Wilhelms III in der Aula am 3. Aug. 1918. Berlin.

1918. "Polen: Eine Anzeige," *Zeitschrift der Gesellschaft für Erdkunde zu Berlin* 53: 97–131.

1919. "Begrüssung der heimgekehrten Ostafrikaner," "Protest der Gesellschaft für Erdkunde gegen die Ausstossung Deutschlands aus der Reihe der kolonisierenden Mächte." *Zeitschrift der Gesellschaft für Erdkunde zu Berlin* 54: 18–29.

1919. "Die deutsch-polnische Sprachgrenze," *Zeitschrift der Gesellschaft für Erdkunde zu Berlin* 54: 108–109.

1919. "Deutsche und Polen in Westpreussen und Posen." *Deutsche Allgemeine Zeitung,* no. 27, 9 February: 1–4.

1919. ———, and H[erbert] Heyde. *Karte der Verbreitung von Deutschen und Poles längs der Warthe-Netze-Linie und der unteren Weichsel, sowie an der Westgrenze in Posen.* Berlin.

1921. "Die Deutschen im Polnischen Korridor," *Zeitschrift der Gesellschaft für Erdkunde zu Berlin* 56: 169–85.

1921. *Der Grossgau im Herzen Deutschlands.* Leipzig: Metzger & Wittig.

1921. ———, and Herbert Heyde. "Die Deutschen im Polnischen Korridor: Karte der Verbreitung der Deutsch- und Polnisch-sprechenden Bevölkerung auf Grund der Volkszählung vom 1 Dezember 1910." 1:300,000. Berlin: Preussische Landesaufnahme. [Copy stored in AGSL, 159-a.]

1924. "Mittenwald: Ein Grenzmarkt in den deutschen Alpen," in *Geographie der deutschen Alpen: Festschrift für Prof. Dr. Sieger zum 60. Geburtstage* (Wien: Seidel), 103–32.

1924. "Walther Penck 1888–1923," preface to Walther Penck, *Die morphologische Analyse* (Stuttgart), vii-xvii.

1925. "Deutscher Volks- und Kulturboden," in K[arl] C[hristian] von Loesch, ed., *Volk unter Völkern: Bücher des Deutschthums*, vol. 1 (Breslau: Ferdinand Hirt), 62–73.

1925. "Das Hauptproblem der physischen Anthropogeographie," *Zeitschrift für Geopolitik* 2: 330–48.

1925. "Sven Hedin als Forschungsreisender: Zu seinem 60. Geburstage," *Deutsche Rundschau* 51:6: 276–82.

1926. "Deutschland als geographische Gestalt," *Leopoldina: Berichte der Kaiserlichen Deutschen Akademie der Naturforscher zu Halle* 1: 72–81. Reprinted in 1928.

1926. "Die Tschechoslowakei nach Hugo Hassinger," *Zeitschrift der Gesellschaft für Erdkunde zu Berlin* 61: 425–34.

1926. "Die Zukunft Südamerikas," *Forschungen und Fortschritte* 2: 201.

1926. Letters to Bowman, *Geographical Review* 16:1 (January): 122–32; and *Geographical Review*, 16:2 (April): 350–52.

1927. "Geography among the Earth Sciences," *Proceedings of the American Philosophical Society* 67: 627–44. [Lecture read on 30 April 1927.]

1928. "Neuere Geographie," *Zeitschrift der Gesellschaft für Erdkunde zu Berlin* 63: 30–56.

1928. "Wann kamen die Indianer nach Nordamerika?" Reprinted from Proceedings of the 23rd International Congress of Americanists, September. AP-P, AfG-L-IfL, K. 871, S. 14.

1930. "Central Asia," *Geographical Journal* 76:6: 477–87.

1933. *Griechische Landschaften*. Bielefeld-Leipzig: Velhagen & Klasing (2nd ed. 1939).

1933. "Herkunft und Ablagerung von Löss," *Forschungen und Fortschritte* 9: 205–6.

1933. "Die Kartographie Preussens unter Friedrich dem Grossen," *Preussiche Akademie der Wissenschaften, Physikalisch-Matematische Klasse*: 36–62.

1933. "Nationale Erdkunde," *Zeitschrift der Gesellschaft für Erdkunde zu Berlin* 68: 321–35.

1936. *Besinnliche Rheinreise*. Bielefeld-Leipzig: Velhagen & Klasing. 2nd ed., 1937.

1937. "Die Ausbreitung des menschengeschlechtes," *Mittelungen der Gesellschaft für Erdkunde zu Leipzig* 54: 5–25.

1937. "Zur deutschen Kolonialfrage," *Zeitschrift der Gesellschaft für Erdkunde zu Berlin* 72: 43–48.

1941. "U.S.A. und Europa: Geographische Vergleiche," *Das Reich: Deutsche Wochenzeitung* no. 13, 30 March, 17–18.

1943–45. "Erinnerungen von Albrecht Penck," [Memoirs of Albrecht Penck], AP-P, L-If L, AfG, K. 871, S. 3.

2. EUGENIUSZ ROMER (1871-1954)

1892. "Studia nad rozkładem ciepła na kuli ziemskiej," *Kosmos* 17, nos. 11/12: 493–525.

1893. "Rozkład ciepła na kuli ziemskiej," *Wszechświat* 12: 193–98.

1894. "Geographische Verteilung der Niederschlagsmengen in den Karpathenländern," *Bulletin International de l'Académie des Sciences de Cracovie* 8: 257–58.

1899. "Wpływ klimatu na formy powierzchni ziemi," *Kosmos* 24: 243–71.

1903. "Wystawa środków naukowych w Wiedniu 1903 r. Geografia," *Muzeum* 19: 605–15, 738–45.

1907. *Mała geografia handlowa.* Lwów.

1909. "Próba statystyki literatury fizjograficznej ziem polskich za lata 1892–1905," *Kosmos* 32: 77–83.

1911. "O potrzebie pracowni geograficznej na naszych uniwersytetach," *Kosmos* 36: 525–36.

1912. "Geograficzne położenie ziem polskich," and "Klimat ziem polskich," *Encyklopedia polska*, vol. 1 (Warszawa), 1–8, 171–248.

1912. *Przyrodzone podstawy Polski historycznej.* Lwów: Zarzewia.

1916. *Atlas geograficzno-statystyczny Polski* [Geographical-Statistical Atlas of Poland]. Polska Akademia Umiejętności w Krakowie. Sprawozdania z czynności i posiedzeń. Rok 1915, Wydział Historyczno-Filozoficzny, Posiedzenie nadzwyczajne dnia 22 listopada 1915, no. 9. Kraków.

1916. *Wojenno-polityczna mapa Polski.* Z powodu Manifestu z 5 listopada 1916 r. Lwów.

1917. "Albrecht Penck o atlasie Polski," *Kurier Lwówski*, no. 219.

1917. "Albrecht Penck über den 'Atlas von Polen,'" *Polen*, 4 May.

1917. *Ilu nas jest?* Kraków.

1917. *Polska: ziemia i państwo.* Lwów.

1917. "Über die kriegspolitische Karte Polens: Aus Anlass der unzufriedenheit, welche sie in den 'ukrainischen Sphären' hervorgerufen hat," *Polen*
116, 15 March.

1919. *Polacy na kresach pomorskich i pojeziernych. Z siedmioma mapami*. Lwów,
Wydane zasiłkiem polskich spółek oszczędności i pożyczek pod patro-
natem Wydziału Krajowego Książnica Polska Towarz. Naucz. Szkół
Wyższych. Lwów.

1919. ——, and Witold Lutosławski, *The Ruthenian Question in Galicia*.
Paris: Impr. Levé.

1919. ——, Stanisław Zakrzewski, and Stanisław Pawłowski. *W obronie
Galicyi wschodniej*. Lwów: Książnica Polska.

1920. *Mapa Polski*. Lwów: Książnica Polska. [AGSL, 182-H.]

1920. *Poland* (*Ms. map by E. Romer, showing German population in western Poland*).
Lwów. [AGSL 186-C.]

1920. *Spis ludności na terenach administrowanych przez Zarząd cywilny ziem
wschodnich, grudzień 1919* (in Polish, French, and English). Lwów: Książnica
Polska.

1920. ——, and Henry[k] Arctowski. *Poland: A Sketch Map. Showing
Boundaries Established by Peace Conference, Plebiscite Areas, Boundaries Pro-
posed, Historical Boundaries, Etc.* New York: General Drafting Co., Inc.
[AGSL, 186-C.]

1920. ——, and T. Golachowski. *Śląsk, Tenczynskie Górne, Orawa i Spisz na
postawie spisu ludności w r. 1910*. Kraków: S. A. Krzyżanowskiego. [AGSL,
670-C.]

1920. ——, and Teofil Szumański. *Mapa etnograficzna Ziem Wschodnich* ....
Ethnographic Map of the Eastern Provinces (*Polish Lithuania and White Russia*)
According to the Official Census of December 1919. Warszawa: Książnica Polska.

1921. *Geograficzno-statystczny atlas Polski* [Geographical-Statistical Atlas of
Poland], 2nd ed. Lwów: Książnica Polska.

1921. *Polska, mapa administracyja*. Lwów: Książnica Polska. [AGSL, 186-C.]

1921. *Polska, mapa polityczno-administracyjna*, 2nd ed. Lwów: Książnica Polska.
[AGSL 186-C.]

1921. *Polski atlas kongresowy. Atlas des Problèmes territoriaux de la Pologne*.
Lwów-Warszawa.

1921. ——, and Teofil Szumanski. *Mapa Polski*. Lwów: Książnica Polska.
[AGSL 186-C.]

1922. *Rzeczpospolita Polska*. Lwów-Warszawa: Atlas. [AGSL, 186-C.]

1923. "W obronie kartografii polskiej: list do redakcji Czasopisma Geograficznego," *Czasopismo Geograficzne* 1: 283–88.

1924. "List do redakcji Czasopsimo Geograficznego w sprawie Unii Geograficznej," *Czasopismo Geograficzne* 2: 347–50.

1924. ———, and Teofil Szumański. *Atlas scienny Polski*. Lwów: Książnica-
Atlas. [AGSL, 182-H.]

1925. ———, and Józef Wasowicz. *Mały atlas scienny III. Rzeczpospolita Polska: mapa polityczna*. Lwów-Warszawa: Książnica-Atlas. [AGSL, 182-H.]

1926. "Geografia na usługach państwa. Odczyt wygłoszony na inaugurację
Towarzystwa Geograficznego we Lwowie," 23 November, k. 42–51.

1926. "Przyczynki do dziejów Międzynarodowej Mapy Świata 1/M w
Polsce," *Polski Przegląd Kartograficzny* 2:15–16: 169–80.

1927. "Pogląd na Tatry," in Ludomir Sawicki, ed., *Przewodnik Kongresowy II
Zjazdu Słowianskich Geografów i Etnografów w Polsce, 229–43*. Kraków.

1927. ———, and Teofil Szumański. *Mapa Polski* [Hypsometric map], 2nd ed.
Warszawa: Książnica-Atlas. [AGSL, 186-C.]

1928. *Atlas Polski współczesnej*, 3rd ed. Lwów: Książnica-Atlas.

1928. "Dziesąte wydanie Małego Atlasu geograficznego," *Polski Przegląd Kartograficzny* 3:23–24: 161–73.

1928. ———, and Teofil Szumański. *Atlas scienny Polski II. Polska: mapa
fizyczna*. Warszawa: Książnica-Atlas. [AGSL, 182-H.]

1928. ———, and Józef Wasowicz. *Atlas scienny Polski III. Polska: mapa polityczna*. Warszawa: Książnica-Atlas. [AGSL, 182-H.]

1929. *Polska: Mapa topograficzna, komunikacyjna i administracyjna*. Warszawa:
Książnica-Atlas. [AGSL, 186-C.]

1929. *Tatrzańska epoka lodowa*. Lwów.

1930. "Zarys moich poglądów na tatrzańską epokę lodową," *Czasopismo Geograficzne* 8: 114–40.

1931. Witold Romer, "Nowa Metoda Reprodukcji Kartograficznej (A New
Method of the Cartographical Reproduction)," *Polski Przegląd Kartograficzny* 9: 1–6.

1931. *Polska*. Warszawa: Książnica-Atlas. [AGSL, 186-C.]

1932. ———, and Teofil Szumański. *Polska fizyczna*. Warszawa: Książnica-
Atlas. [AGSL, 186-C.]

1934. *Zbiór prac: poświęcony przez Towarzystwo geograficzne we Lwowie Eugen-
juszowi Romerowi w 40-lecie jego twórzosci naukowej*, ed. Henryk Arctowski.
Lwów.

1936. "*Polski Przegląd Kartograficzny*, 1923–1934," *Czasopismo Geograficzne* 14:1:
1–15.

1936. *Polski stan posiadania na Południowym Wschodzie Rzeczypospolitej*. Lwów-
Warszawa: Książnica-Atlas.

1937. "O nowej mapie Polski," *Czasopismo Geograficzne* 15:1: 1–4.

1937. "Z biosociologii Rzeczypospolitej Polskiej." Lwów-Warszawa: Książnica-Atlas.

1938. *Rady i przestrogi, 1918–1938*. Lwów: Zarzewia.

1939. *Ziemia i państwo: kilka zagadnień geopolitycznych*. Lwów: Książnica-Atlas.

1950/1952. "Mapa jako dokument dziejów rzeźby powierzchni ziemi," *Czasopismo Geograficzne* 21/22 (1950/52): 5–57.

1955. "Fragmenty wspomnień: na drodze ku kartografii," *Czasopismo Geograficzne* 26: 201–9.

1960–64. *Wybór prac* [Collected Works], 4 vols., ed. August Zierhoffer. Polish Geographical Society. Warszawa: PWN.

1988. *Pamiętniki: problemy sumienia i wiary*. Kraków: Znak.

1989. *Pamiętnik paryski: 1918–1919*. Wrocław: Ossolineum.

1995. *Dziennik*, vols. 1–2, ed. Ryszard Jodzis. Warszawa: Interlibro.

3. STEPAN RUDNYTS'KYI (1877-1937)

1897. "Kozats'ko-pol'ska viina v 1625 r.: Istorychna rozvidka," *Zapysky NTSh* 17: 1–14.

1898. "Nove-dzherelo do istorii Khmel'nychyny," *Zapysky NTSh* 23–24: 1–22.

1899. "Rus'ki zemli pol'skoi-korony pry kintsi XV. v.: Vorozhi napady i organizatsiia pohranychnoi oborony," *Zapysky NTSh* 31–32 (Lviv): 1–54.

1903. "Fizychna heohrafiia pry kintsi XIX stolittia: naukova khronika za 1898, 1899 i 1900 rr.," *Zb[ornyk] matemat.-pryrodopys-likar. sektsii NTSh* 9: 1–116.

1905. "Znadoby do morfolohii karpats'koho stochyshcha Dnistra," *Zb. matemat.-pryrodopys-likar. sektsii NTSh* 10: 1–85.

1907. "Znadoby do morfolohii pidkarpats'koho stochyshcha Dnistra," *Zb. matemat.-pryrodopys-likar. sektsii NTSh* 12:2: 1–310; for a summary, *Kosmos* 32: 462–63.

1907. "W sprawie historii doliny Dniestra," "W sprawie dyluwialnego wypietrzenia Karpat: Odpowiedź E. Romerowi," "Oświadczenie wobec odpowiedzu prof.–dra E. Romera," "Opowiedź na recenzie D-ra E. Romera," *Kosmos* 32. [Romer-Rudnyts'kyi exchange.]

1908. *Nacherk geografichnoi terminologii*. Lviv: Shevchenko Scientific Society.

1910. *Korotka geografiia Ukrainy*, vol. 1. Kyiv-Lviv: Lan.

1914. *Korotka geografiia Ukrainy*, vol. 2. Lviv: Nakladom Ukrains'koho Pedahohichnoho Tovarystva.

1914. *Ukraina und die Ukrainer*. Wien: Verlag des Allgemeinen Ukrainischen Nationalrates.

1915. Sh. Levenko [pseudonym], *Chomu my khochemo samostiinoii Ukrainy*. Wien, reprinted as *Chomu my khochemo samostiinoi Ukrainy*, ed. Oleh Shablii. Lviv: Svit, 1994.

1915. *The Ukraine and the Ukrainians*, trans. Jacob Wittmer Hartmann. Jersey City, N.J.: Ukrainian National Council.

1916. *Ukraina: Land und Volk*. Wien: Bund zur Befreiung der Ukraina.

1917. *Ukraïna: nash ridnyi krai*. Lviv: Nakladom zahal'noi ukrains'koi kul'turnoi rady. Reprinted by *Prosvita* in 1921.

1918. *Physical Wall Map of Ukraine*. Lviv: Shevchenko Scientific Society. Wien: Freytag & Berndt. [AGSL, 288.]

1918. "Die Podolische Platte in Galizien," in *Festband Albrecht Penck, 198–211*. Stuttgart: J. Engelhorn.

1918. *Stinna fizychna karta Ukrainy*. Wien: Freytag & Berndt.

1918. *Ukraine, the Land and Its People: An Introduction to Its Geography*. New York: Rand McNally.

1918/1919. *L'Ukraine—un aperçu su son territoire, son peuple, ses conditions culturelles, ethnographiques et economiques*. Berne: Imprimerie R. Suter et cie.

1919. *Pochatkova geografiia dlia narodnikh shkil z bahat'oma maliunkamy i kartoiu*. Kyiv: Tovarystvo "Vernyhora."

1919. "Problemy heohrafii Ukrainy," *Zb. matemat.-pryrodopys.-likar. sektsii NTSh* 18–19: 1–90.

1920. ———, and Georg von Gasenko, *Ukraina u svoikh etnohrafichnykh mezhakh/ Die Ukraina in ihren ethnographischen Grenzen*. Kyiv: "Po svitu," 1920. Wien: Christopher Reissers Söhne, 1920.

1920–21. "Do osnov ukrains'koho natsionalizmu," *Volia* (Wien) vol. 2, no. 5 (1920): 206–8; vol. 2, no. 6 (1920): 238–41; vol. 2: nos. 7–8 (1920): 295–98; vol. 2, no. 9 (1920): 354–59; vol. 2, no. 10 (1920): 395–400; vol. 3, no. 2 (1920): 59–63; vol. 3, no. 4 (1920): 123–26; vol. 3, no. 5 (1920): 205–10; vol. 3, no. 9 (1920): 340–43; vol. 4, no. 7–8 (1920): 339–42; vol. 4, no. 12 (1920): 590–94; vol. 1, no. 2 (1921): 80–86; vol. 1, no. 3 (1921): 106–12. [Printed piecemeal in *Volia* in 1920–21, reprinted in full in the 163-page edition (2nd ed.) in Prague in 1923.]

1921. "Halychyna i soborna Ukraina," *Ukrains'kyi prapor*, no. 32 (Vienna), 13 August.

1921. *Ohliad terytorii Ukrainy*. Wien: [Petro] Franko & Son.

1921. "Heohrafichne stanovyshche Halychyny ta ii politychna dolia," *Ukraïns'kyi prapor*, no. 43 (Wien), 1 December.

1923. *Do osnov ukrain'skoho nationalizmu*. Praha.

1923. *Ohliad national'noi terytorii Ukrainy*. Berlin: Ukrains'ke Slovo.

1923. *Osnovy zemleznannia Ukrainy*. Praha. [Reprinted in Lviv in 1924, Uzhhorod in 1926.]

1923. *Ukrains'ka sprava zi stanovyshcha politychnoi heohrafii*. Berlin: Ukrains'ke Slovo.

1925. "Stručna předběžna správa o geologických a morfologických výzkuméch na Podkarpatské Rusi v červenci 1924," *Vestnik Statniho geologického ústavu CSR*, vol. 2 (Praha): 45–50.

1925–27. *Osnovy morfolohii i heolohii Pidkarpats'koi Rusy ta Zakarpattia vzahali*, 2 vols. Uzhhorod. [Published dissertation thesis.]

1926. "Morfolofické oblasti Východní Evropy," in J. V. Daneš, gen. ed., *Sbornik Československé společnosti zemépisné* 32 (Praha): 81–87.

1927. "Oborona kraiu i navchannia heohrafii," *Kul'tura i pobut*, no. 32 (Kharkiv), 27 August.

1927. "Perspektivy Ukrains'koho Heohrafichnoho Instytutu," *Kul'tura i pobut*, no. 34 (Kharkiv), 10 September.

1927. "Pro stanovyshche istorychnoi heohrafii v systemi suchasnoho zemleznaniia," *Zapysky istorichno-filiolohichno viddila VUAN*, vols. 13–14 (Kyiv): 345–56.

1928. "Zavdannia Ukrains'koho Heohrafichnoho Instytutu i ioho vydavnytstv," *Zapysky Ukr.-nauk.-dosl. Insytutu heohrafii ta kartohrafii* vol. 1, no. 1 (Kharkiv): 3–33.

1929. *Geografichnyi atlias na 16 tablyts'*, ed. Iu. M. Shokal's'kyi. Ukrains'ke vydannia na 18 tablyts' pid redaktsiieiu Stepana Rudnyts'koho. Kharkiv: Vydannia Derzh. Kartohrafii H.H.K. VRBG SRSR.

1930. "Dekil'ka sliv u spravi heohrafichnoi topomastyky i transliteratsii chuzhomovnykh heohradichnykh nazv na ukrains'kikh mapakh," *Visnyk Instytutu Ukrains'koi Naukovoi Movy* 2 (Kyiv): 28–33.

1931. "Beiträge zur Morphologie des Dnieprgebietes in der Ukraina," *Zeitschrift der Gesellschaft für Erdkunde zu Berlin* 5–6: 161–75.

4. ISAIAH BOWMAN (1878-1950)

1904. "A Classification of Rivers Based on Water Supply," *Journal of Geography* 4: 202–12.

1909. "Man and Climatic Change in South America," *Geographical Journal* 33: 267–78.

1909. "Regional Population Groups of Atacama," *Bulletin of the American Geographical Society* 41: 142–54, 193–211.

1911. *Forest Physiography: Physiography of the United States and Principles of Soils in Relation to Forestry*. New York: J. Wiley & Sons.

1915. *South America: A Geography Reader*. Chicago: Rand McNally.

1916. *The Andes of Southern Peru: Geographical Reconnaissance along the Seventy-third Meridian*. New York: Henry Holt & Co.

1917. "Frontier Region of Mexico: Notes to Accompany a Map of the Frontier," *Geographical Review* 3: 16–27.

1919. "The American Geographical Society's Contribution to the Peace Conference," *Geographical Review* 7: 1–10.

1921. *The New World: Problems in Political Geography*. Yonkers-on-Hudson, N.Y.: World Book Co.

1921. "Personalities and Methods at the Peace Conference" [Speech at Beacon Society of Boston], IB-P Ms. 58, Series III, 21 February, Box 3.1., Folder 1.

1922. "Modifications in the Western Boundary of Hungary," *Geographical Review* 12: 650–51.

1922. "Overpopulation in Relation to Agriculture and Famine in Eastern Europe," *Geographical Review* 12: 489–91.

1922. "Steppe and Forest in the Settlement of Southern Russia," *Geographical Review* 12: 491–92.

1923. "An American Boundary Dispute: Decision of the Supreme Court of the United States with Respect to the Texas-Oklahoma Boundary," *Geographical Review* 13: 161–89.

1923. "Boundaries of Turkey according to the Treaty of Lausanne," *Geographical Review* 13: 627–29.

1923. "Geographical Elements in the Turkish Situation: A Note on the Political Map," *Geographical Review* 13: 122–29.

1925. "Commercial Geography as a Science: Reflections on Some Recent Books," *Geographical Review* 15: 285–94.

1925. *The Field of Geography*. New York: Columbia University.

1926. "The Analysis of Land Forms: Walther Penck on the Topographic Cycle," *Geographical Review* 16:1 (January): 122–32.

1926. "Geography and Boundaries at Mosul," *Geographical Review* 16: 143–44.

1926. "Scientific Study of Settlement," *Geographical Review* 16: 647–53.

1928. "The International Geographical Congress," *Geographical Review* 18: 661–67.

1929. *Geography in Relation to the Social Sciences*. Montpelier, Vt.: Capital City Press.

1931. [no title], Bowman's Address at the International Geographical Congress (IGC) in Paris, IB-P Ms. 58, Series III, Box 3.1, Folder 7, [no date], published as "Inaugural Address," *Comptes Rendus du Congrès International de Géographie*, vol. 1 (Paris: Librarie Armand Colin), 86–87.

1931. "The Invitation of the Earth" (Abstract of an Address Illustrating Modern Geography), Bowdoin College, 14 April, IB-P Ms. 58, Series III, Box 3.1, Folder 6.

1931. *The Pioneer Fringe*. New York: American Geographical Society.

1934. "Regional Concepts and Their Application," in Eugeniusz Romer, *Zbiór prac*, ed. Henryk Arctowski (Lwów), 25–46.

1934–35. "Opening Address by Isaiah Bowman, President," *Comptes Rendus du Congrès International de Géographie, Varsovie 1934*, vol. 1, 99–103. Warsaw: International Geographical Union.

1935. "Address of the President of the International Geographical Congress," *Science* 81, no. 2164: 389–91.

1935. "The Business of Diplomacy," 28 October, Presbyterian Social Union, Baltimore, IB-P Ms. 58, Series III, Box 3.1, Folder 18.

1935. "The Fourteenth International Geographical Congress, Warsaw, 1934," *Geographical Review* 25: 142–48.

1935. "Next Steps in American Universities," 6 December, Southern University Conference, First Annual Meeting, Brown Hotel, Louisville, Kentucky, IB-P Ms. 58, Series III, Box 3.1, Folder 21.

1936. *A Design for Scholarship*. Baltimore: Johns Hopkins University Press.

1936. "Political Geography," given at Cornell University [no date, fall 1936], IB-P Ms. 58, Series III, Box 3.1, Folder 18.

1937. "Is There a Logic in International Situations?" IB-P Ms. 58, Series XIII, Box 3.2, address given 13 May 1937 before the Institute of Public Affairs, University of Georgia.

1937. *Limits of Land Settlement: A Report on Present-Day Possibilities*. New York: Council on Foreign Relations.

1937. "What Do You Do? (A Sentimental Inquiry)," 14 May, Mt. Vernon Club in Baltimore, IB-P Ms. 58, Series III, Box 3.2, Folder 10.

1937. "Why We Believe," 13 April, Founders Day Address, University of Virginia, IB-P, Series III, Box 3.2, Folder 8.

1938. "Geography in the Creative Experiment," *Geographical Review*: 1–19.

1939. *The Graduate School in American Democracy*. Washington, D.C.: Government Printing Office.

1939. "Science and Social Pioneering," *Science* 90: 309–19.

1939. "What Do You Believe?" 12 June, Gilman Country School, Baltimore, IB-P Ms. 58, Series III, Box 3.2, Folder 4.

1940. Foreword to *International Boundaries: A Study of Boundary Functions and Problems*, by S. Wittermore Boggs. New York: Columbia University Press.

1940. "The Inquiry," 5 March 1940, in Bowman's "Notes [on U.S. Foreign Policy]," used at Women's Democratic Club Forum, 12 March 1940, Carlton Hotel, Washington, D.C, IB-P, Ms. 58, Series III, Box 3.2, Folder 16, published as "The Inquiry," in James T. Adams, ed., *Dictionary of American History* (New York: Scribner, 1940), 124.

1940. "Symposium: Walther Penck's Contribution to Geomorphology," *Annals of the Association of American Geographers* 30:4 (December): 219–80.

1940. *What Do You Know?* Baltimore: Johns Hopkins University Press.

1941. "The Future of Education and Military Defense," *Educational Record* 23: 428–35.

1942. "Geography vs. Geopolitics," *Geographical Review* 32: 646–58.

1942. "What Is an American?" 5 June, University of Cincinnati Commencement, IB-P Ms. 58, Series III, Box 4, Folder 19.

1943. *Science and Our Future*. New York: Commission to Study the Organization of Peace.

1945. "Franklin Delano Roosevelt 1882–1945," *Geographical Review* 35: 349–51.

1945. "The New Geography," *Journal of Geography* 44: 213–16.

1946. "The Millionth Map of Hispanic America," *Science* 103: 319–23.

1946. "The Social Contract of an Educated Man," *Association of American Colleges Bulletin* 32: 498–505.

1947. "Geography as an Urgent University Need," 10 January, Speech before the Johns Hopkins University Board of Trustees, IB-P Ms. 58, Series III, Box 3.6, Folder 1.

1948. "The Geographical Situation of the United States in Relation to World Policies," *Geographical Journal* 112: 129–45.

1949. "Geographical Interpretation," *Geographical Review* 39: 255–70.

1950. "Mark Jefferson," *Geographical Review* 40: 134–37.

1951. "Settlement by the Modern Pioneer," in Griffith Taylor, ed., *Geography in the Twentieth Century, 248–66.* New York: Philosophical Library.

5. PÁL TELEKI (1879-1941)

1903. *Az elsodleges allamkeletkezes kerdesehez* [Dissertation]. Budapest: Franklin-Tars.

1906. "Desceliers Mappemonde-ja 1553-ból." *Földrajzi Közlemények* 34: 185–86.

1906. *Értesítés a roueni rézföldgömb és az 1554. évi, Gastaldinak tulajdonított térkép között felfedezett hasonlóságáról*. Budapest: Fritz Ny.

1906. "Japán szerepe Amerika földfedezésében" [Japans Rolle in der Geschichte der Entdeckung Amerikas]. *Földrajzi Közlemények* 34: 1–13.

1909. *Atlas zur Geschichte der Kartographie der Japanischen Inseln. Nebst dem holländischen Journal der Reise Mathys Quasts und A. J. Tasmans zur Entdeckung der Goldinseln im Osten von Japan i.d.J. 1639 und dessen deutscher Übersetzung. Budapest, Vertrieb der deutschen Ausgabe, K. W. Hiersemann, Leipzig, 1909. Atlasz a Japani szigetek cartográphiájának történetéhez* (Atlas of the history of the cartography of the Japanese islands). In Hungarian and German eds. Budapest: Kilián Figyes Utóda Magy. Kir. Egyetemi Könyvkereskedő.

1912. *A gazdaság földrajzi czélja és jelentösége: A Magyar Földrajzi Társaság gazdaság fölrajzi száokosztályának megalapitása*. Budapest: Kertész Ny.

1912. "Főtitkári jelentes," *Földrajzi Közlemények* 40: 19–21, 26–31.

1914. "Rapport sur les voyages et les travaux géographiques exécutés par les explorateurs et les savants hongrois depuis l'année 1889." Congresso Internazionale di Geografia X. Roma—27 Marzo–3 Aprile 1913. Roma: Tip. dell'Unione Editrice.

1917. *A földrajzi gondolat története*. Budapest (self-published).

1918. "A Turán földrajzi fogalom," *Turán* 3: 44–83.

1918. "Főtitkari jelentes (Keleti torekveseink): A Magyar Földrajzi Társaság 1918 Április 25-en tartott közgyűlésen elmondotta," *Földrajzi Közlemények* 46: 251–52.

1918. *Maps of Hungary by Count Paul Teleki*. Budapest: Hungarian Geographical Society. [AGSL, 160-F.]

1919. "A földrajz—tudomany es tantárgy," *Földrajzi Közlemények* 49: 20–41.

1919. "A zsidokerdes," *Teleki Pál programbeszéde, melyet Szeged első választókerületének Keresztény Nemzeti Egyesülés Pártja alakuló és képviselő nagygyűlésén mondott el, Szegeden, a Tisza-szallo nagytermeben 1919, ev december havanak 14 napjan* (Szeged: TEVÉL), 3–5.

1919. *The Ethnographical Composition of Hungary*. Berne: Bühler & Werder.

1919. *Magyarország néprajzi térképe a népsűrűség alapján, szerkesztette Gróf Teleki Pál. Carte ethnographique de la Hongrie construite en accordance avec la densité de la population, par le comte Paul Teleki. Ethnographical Map of Hungary: Based on the Density of Population, by Paul Teleki*. Budapest: Magyar Földrajzi Intézet.

1919. *Short Notes on the Economical and Political Geography of Hungary*. Budapest: V. Hornyánszky. [AGSL, Pamphlet Box 131.]

1920. *Ethnographical Map of Hungary, Based on the Density of Population.* The Hague: W. P. van Stockum & Son. [AGSL, 160-C.]

1920. *La Hongrie du sud, par le comte Paul Teleki.* Paris: H. Le Soudier, etc., etc.

1920. *The Hungarian Peace Negotiations. Report on the Activity of the Hungarian Peace Delegation at Neuilly from January to March 1920.* Budapest: Royal Hungarian Ministry for Foreign Affairs.

1920. *Magyarország gazdaságföldrajzi térképe dr. Teleki Pál gróf és dr. Cholnoky Jenő közreműködésével hivatalos adatok alapján szerkesztette dr. Fodor Ferenc.* Carte de géographie économique de la Hongrie . . . Economic-geographical map of Hungary. Budapest: Magyar Földrajzi Intézet.

1921. *Magyarország gazdasági térképekben. Gróf Teleki Pál, Béketárgyalászt Előlészítő Iroda vezetője megbizásából. The Economies of Hungary in Maps.* Compiled by Aladár de Edvi Illés and Albert Hálasz to the Commission of Count Paul Teleki, Chief of the Office for the Preparation of Peace Negotiations, 6th rev. ed. Budapest: 1921. [AGSL, At. 642 E-1921.]

1922. *Amerika gazdasági földrajza: különös tekintettel az Észak-Amerikai Egyesült Államokra* [Economic Geography of America, with special reference to the United States]. Budapest: Centrum Kiadóvállalat részvénytársaság. [AGSL (Rare), HC 103.T43x 1922.]

1922. *Ethnographical Map of Hungary: Based on the Density of Population.* Budapest: Hungarian Geographical Institute. [AGSL, 160-C.]

1922. *Loose-leaf Atlas of 29 Sheets, by Paul Teleki, of Southeastern Europe, Showing Productions of Wheat, Etc., Forest Lands, Waterways, Minerals, Etc.* Compiled by Albert Halász. Budapest. [AGSL 90–7.]

1922. *Southern Hungary.* Budapest: Hornyánszky.

1922. "Statisztika és térkép a gazdasági földrajzban" [Statistics and Maps in Economic Geography]. *Földrajzi Közlemenyek* 50: 74–91.

1923. *The Evolution of Hungary and Its Place in European History.* New York: Macmillan Company.

1926. *Magyarország a háboru elött és után gazdaságstatisztikai térképekben* [Hungary before and after the War in Economic-Statistical Maps]. Edited by Aladár Edvi Illés and Albert Halász. Under the direction of Count Paul Teleki. Budapest: Institute of Political Sciences of the Hungarian Statistical Society. [AGSL 90-F.]

1926. *Die weltpolitische und weltwirtschaftliche Lage Ungarns in Vergangenheit und Gegenwart.* Berlin-Grunewald: K. K. Vowinckel. Published also in *Zeitschrift für Geopolitik* 3, no. 6 (1926): 381–409.

1927. *Általános gazdasági földrajz, különös tekintettel az óvilági.* Budapest: Magyar Tudomanyos Akadémia.

1929. *Dékáni megnyitó*. Budapest: Élet Ny.

1929. *La Geografia dell' Ungheria*. Roma: Istituto per L' Europe Orientale.

1930. "Aktuelle Fragen internationaler Politik und die politische Geographie," *Zeitschrift für Geopolitik* 7:1: 179–81.

1930. "Lóczy Lajos, az ember és a professzor," *Földrajzi Kozlémenyek*, 58:7/8 (1930): 101–5.

1930. "Politische Geographie," *Geopolitik*: 45–57.

1930. "Some Economic Maps by Count Paul Teleki." Prepared for the *Hungarian Encyclopedia of Economics*, assisted by M. L. Gerö and F. Fodor. Budapest: Athenaeum. [AGSL 91-C.]

1930. *Ungarns Wirtschaftslage: Die Vielseitigkeitihrer Schwierigkeiten*. München: A. Dreisler.

1931. *Ungheria ed Europa*, trans. Paolo Ruzicska. Budapest: Federazione italo-ungherese. [Reprinted in 1940.]

1932. ———, and Zoltán de Magyary, "Introduction," in Gyula Hantos, *Administrative Boundaries and the Rationalisation of the Public Administration*, 3–9. Budapest: Athenaeum, 1932.

1934. *Európáról és Magyarországról*. Budapest: Athenaeum.

1936. *A gazdasági élet földrajzi alapjai* [The Geographical Basis of Economic Life]. Budapest: Centrum Kiadóvállalat.

1937. "A Magyarország házája" [The Hungarian Homeland], in Gyula Prinz, ed., *Magyar Föld, Magyar Faj*, vol. 2, 7–17. Budapest: Királyi Magyar Egyetemi Nyomda.

1937. "Concerning an Ethnic Map, Refuting Mr. Somesan's Statements on Ethnic Map of Hungary by Prof. Count Paul Teleki, with maps," *Bulletin international de la Société Hongroise de Géographie*, vol. 65, no. 1–5 (Budapest), 29–38. [AGSL, 5–31.]

1939. ———, and Cholnoky, Jenö. [no title], *Földrajzi Közlemények* 67:2: 71–80, and 67:4: 249–67.

1941. ———, and Cholnoky, Jenö. [no title], *Földrajzi Közlemenyek* 69:1: 1–11.

2000. Teleki, Pál. *Válogatott politikai írások és beszédek* [Selected Political Writings and Speeches]. Compiled and edited by Balázs Ablonczy. Budapest: Osiris.

II. SECONDARY SOURCES

Ablonczy, Balázs. *A visszatért érdély, 1940–1944*. Budapest: Jaffa Kiadó, 2011.

———. "Egy szépreményűfiatalember: száarmazáas, család, ifjúkor." *Rubicon* 15:2 (2004): 1–4.

———. *Teleki Pál*. Budapest: Osiris, 2005. Translated by Thomas J. DeKornfeld and Helen DeKornfeld as *Pál Teleki: The Life of a Controversial Hungarian Politician (1874–1941)*. Boulder: Social Science Monographs, 2006.

———. *Teleki Pál: egy politikai életrajz vázlata*. Budapest: Elektra, 2000.

———. *Trianon-Legendak*. Budapest: Jaffa Kiadó, 2010.

———. "Útkeresés: A turáni társaságban." *Rubicon* 15:2 (2004): 12–14.

Agar, John. *Science in the Twentieth Century and Beyond*. London: Polity, 2012.

Agnew, John A., and David N. Livingstone, eds. *The Sage Handbook of Geographical Knowledge*. London: Sage, 2011.

Agnew, John A., Katharyne Mitchell, and Gerard Toal, eds. *A Companion to Political Geography*. Oxford; Blackwell, 2003.

Akerman, James R., and Robert W. Karrow, Jr., eds., *Maps: Finding Our Place in the World*. Chicago: University of Chicago Press, 2007.

Aly, Götz, and Karl-Heinz Roth, eds. *The Nazi Census: Identification and Control in the Third Reich*. Philadelphia: Temple University Press, 2004.

Amar, Tarik Cyril. *The Paradox of Ukrainian Lviv: A Borderland City between Stalinists, Nazis, and Nationalists*. Ithaca: Cornell University Press, 2015.

Anderson, Benedict. *Imagined Communities: Reflections on the Origins and Spread of Nationalism*. Rev. ed. London: Verso, 1991.

Applegate, Celia. *A Nation of Provincials: The German Idea of Heimat*. Berkeley: University of California Press, 1990.

Apponyi, Albert. *Justice for Hungary*. London: Longmans, Green & Co., 1928.

Arendt, Hannah. *The Origins of Totalitarianism*. Berlin: Schocken Books, 1951.

Arnason, Johann P., Petr Hlaváček, and Stefan Troebst et al., eds. *Mitteleuropa? Zwischen Realität, Chimäre und Konzept*. Prague: Charles University Department of Philosophy, 2015.

Ash, Mitchell G., and Jan Surman, eds. *The Nationalization of Scientific Knowledge in the Habsburg Empire, 1848–1918*. New York: Palgrave Macmillan, 2012.

Baár, Monika. *Historians and Nationalism: East Central Europe in the Nineteenth Century*. Oxford: Oxford University Press, 2010.

Babak, O. I., V. M. Danylenko, and Iu. V. Plekan, eds. *Praha-Kharkiv-Solovky: arkhivno-slidcha sprava akademika Stepana Rudnyts'koho*. Kyiv: NAN Ukrainy; Instytut ukraïns'koi arkheohrafiï ta dzhereloznavstva im. M. S. Hrushevs'koho, 2007.

Babicz, Józef. "Eugeniusz Romer, 1871–1954," in T. W. Freeman et al., eds., *Geographers: Biobibliographical Studies*, vol. 1, 89–96. Salem, N.H.: Mansell, 1977.

Bachelard, Gaston. *The Poetics of Space*. Boston: Beacon Press, 1994.

Bakos, István, and István Csicsery-Rónay. *"Szobor vagyok, de fáj minden tagom!"* *Fehér könyv a Teleki szoborról* [I am a statue but all my body parts hurt! A White Book about the Teleki Statue]. Budapest: Occidental Press, 2004.

Banac, Ivo, and Frank E. Sysyn, eds. *Concepts of Nationhood in Early Modern Eastern Europe.* Cambridge: Ukrainian Research Institute, Harvard University, 1987.

Banac, Ivo, and Katherine Verdery, eds. *National Character and National Ideology in Interwar Eastern Europe.* New Haven: Yale Center for International Studies, 1995.

Barnes, Steven A. *Death and Redemption: The Gulag and the Shaping of Soviet Society.* Princeton: Princeton University Press, 2011.

Baron, Nick, and Peter Gatrell, eds. *Homelands: War, Population, and Statehood in Eastern Europe and Russia, 1918–1924.* London: Anthem Press, 2004.

———. "The Mapping of Illiberal Modernity: Spatial Science, Ideology, and the State in Early Twentieth Century Russia," in Sanna Turoma and Maxim Waldstein, eds., *Empire De/Centered: New Spatial Histories of Russia and the Soviet Union,* 105–34. London: Ashgate, 2013.

Bartov, Omer. *Erased: Vanishing Traces of Galicia in Present-Day Ukraine.* Princeton: Princeton University Press, 2007.

———. "Locating the Holocaust." *Journal of Genocide Research* 13:1 (2011): 121–29.

Bartov, Omer, and Eric D. Weitz, eds. *Shatterzone of Empires: Coexistence and Violence in the German, Habsburg, Russian, and Ottoman Borderlands.* Bloomington: Indiana University Press, 2013.

Bassa, László. "Teleki Pál és a térkép." *Tér és társadalom* 3–4 (1990): 175–83.

Bassin, Mark. "Politics from Nature," in John A. Agnew et al., eds., *A Companion to Political Geography,* 13–29 and 187–203. London: Wiley-Blackwell, 2007.

———. "Race contra Space: The Conflict between German Geopolitik and National Socialism." *Progress in Human Geography* 11 (1987): 473–95.

———. "Studying Ourselves: History and Philosophy of Geography." *Progress in Human Geography* 24 (2000): 475–87.

Bassin, Mark, Sergey Glebov, and Marlene Laruelle, eds. *Between Europe and Asia: The Origins, Theories, and Legacies of Russian Eurasianism.* Pittsburgh: University of Pittsburgh Press, 2015.

Baudrillard, Jean. *Simulacra and Simulation.* Ann Arbor: University of Michigan Press, 1994; first published in France in 1981.

Bauman, Zygmunt. *Culture in a Liquid Modern World.* Cambridge: Polity, 2011.

————. *Liquid Love: On the Frailty of Human Bonds*. Cambridge: Polity, 2003.

————. *Liquid Modernity*. Cambridge: Polity, 2000.

————. *Liquid Times: Living in an Age of Uncertainty*. London: Polity, 2006.

Baumgarten, Murray, Peter Kenez, and Bruce Thompson, eds. *Varieties of Antisemitism: History, Ideology, Discourse*. Newark, Del.: University of Delaware Press, 2009.

Beck, Hanno. "Albrecht Penck: Geograph, bahnbrechender Eiszeitforscher und Geomorphologe (1858–1945)," in Beck, ed., *Grosse Geographen: Pioniere–Aussenseiter–Gelehrte* (Berlin: Reimer, 1982), 191–212.

Bederman, Gail. *Manliness and Civilization: A Cultural History of Gender and Race in the United States, 1880–1917*. Chicago: University of Chicago Press, 1995.

Behrmann, Walther. "Albrecht Penck," *Petermanns Geographische Mitteilungen* 92:3–4 (1948): 190–93.

Bell, Morag, Robin Butlin, and Mike Heffernan, eds. *Geography and Imperialism, 1820–1940*. Manchester: Manchester University Press, 1995.

Beneke, Timothy. *Proving Manhood: Reflections on Men and Sexism*. Berkeley: University of California Press, 1997.

Benjamin, Walter. "Theses on the Philosophy of History." In *Illuminations: Essays and Reflections*, 253–64. New York: Schocken Books, 1969.

Berghahn, Volker R., and Simone Lässig, eds. *Biography between Structure and Agency: Central European Lives in International Historiography*. New York: Berghahn Books, 2008.

Berglund, Bruce R., and Brian Porter-Szűcs, eds. *Christianity and Modernity in Eastern Europe*. Budapest: Central European University Press, 2010.

Biess, Frank, and Daniel M. Gross, eds. *Science and Emotions after 1945: A Transatlantic Perspective*. Chicago: University of Chicago Press, 2014.

Bilenky, Serhiy. *Romantic Nationalism in Eastern Europe: Russian, Polish, and Ukrainian Political Imaginations*. Stanford: Stanford University Press, 2012.

Biskupski, Mieczysław B., ed. *Ideology, Politics and Diplomacy in East Central Europe*. Rochester, N.Y.: University of Rochester Press, 2003.

————. "Re-creating Central Europe: The United States 'Inquiry' into the Future of Poland in 1918." *International History Review* 12:2 (May 1990): 249–79.

Bjork, James. *Neither German Nor Pole: Catholicism and National Indifference in a Central European Borderland*. Ann Arbor: University of Michigan Press, 2008.

Black, Jeremy. *Maps and History: Constructing Images of the Past*. New Haven: Yale University Press, 1997.

Blackbourn, David. " 'The Garden of Our Hearts': Landscape, Nature, and Local Identity in the German East." In *Localism, Landscape, and the Ambiguities of Place: German-speaking Central Europe, 1860–1930*, edited by David Blackbourn and James Retallack, 149–64. Toronto: University of Toronto Press, 2007.

Blackbourn, David, and Richard J. Evans, eds. *The German Bourgeoisie: Essays on the Social History of the German Middle Class from the Late Eighteenth to the Early Twentieth Century*. London: Routledge, 1993.

Blackbourn, David, and James Retallack. *Localism, Landscape, and the Ambiguities of Place: German-speaking Central Europe, 1860–1930*. Toronto: University of Toronto Press, 2007.

Blanke, Richard. *Polish-speaking Germans? Language and National Identity among the Masurians since 1871*. Cologne: Böhlau, 2001.

Blaut, J. M. *The Colonizer's Model of the World: Geographical Diffusionism and Eurocentric History*. New York: Guilford Press, 1992.

Boemeke, Manfred F., Gerald D. Feldman, and Elisabeth Glaser, eds., *The Treaty of Versailles: A Reassessment after 75 Years*. Cambridge: Cambridge University Press, 1998.

Bohachevsky-Chomiak, Martha. *Feminists Despite Themselves: Women in Ukrainian Community Life, 1884–1939*. Edmonton: Canadian Institute of Ukrainian Studies Press, 1988.

Borzęcki, Jerzy. *The Soviet-Polish Peace of 1921 and the Creation of Interwar Europe*. New Haven: Yale University Press, 2008.

Bowler, Peter J., and John V. Pickstone, eds. *The Cambridge History of Science: The Modern Biological and Earth Sciences*, vol. 6. Cambridge: Cambridge University Press, 2009.

Bowman, Robert G. "Prospects of Land Settlement in Western Australia." *Geographical Review* 32:4 (1942): 598–621.

Bradley, Joseph. *Voluntary Associations in Tsarist Russia: Science, Patriotism, and Civil Society*. Cambridge: Harvard University Press, 2009.

Branch, Jordan. *The Cartographic State: Maps, Territory, and the Origins of Sovereignty*. Cambridge: Cambridge University Press, 2014.

Brandenberger, David, Andreas Umland, and David Marples. Debate on Stalin's Russocentrism. *Nationalities Papers* 38:5 (2010): 723–60.

Breckman, Warren, et al., eds. *The Modernist Imagination: Intellectual History and Critical Theory*. New York: Berghahn Books, 2001.

Bremer, H[anna]. "Albrecht Penck (1858–1945) and Walter Penck (1888–1923), Two German Geomorphologists." *Zeitschrift für Geomorphologie* 27 (1983): 129–38.

Briesewitz, Gernot. *Raum und Nation in der polnischen Westforschung, 1918–1948: Wissenschaftsdiskurse, Raumdeutungen und geopolitische Visionen im Kontext der deutsch-polnischen Beziehungsgeschichte.* Osnabrück: Fibre, 2014.

Brogiato, Heinz-Peter. *"Wissen ist Macht–Geographisches Wissen ist Weltmacht": Die schulgeographischen Zeitschriften im deutschsprachigen Raum (1880–1945) unter besonderer Berücksichtigung des Geographischen Anzeigers,* 2 vols. Trier: Trier Geographical Society, 1998.

Brook, Timothy. *Mr. Selden's Map of China: Decoding the Secrets of a Vanished Cartographer.* New York: Bloomsbury Press, 2013.

Brown, Kate. *A Biography of No Place: From Ethnic Borderland to Soviet Heartland.* Cambridge: Harvard University Press, 2005.

———. *Dispatches from Dystopia: Histories of Places Not Yet Forgotten.* Chicago: University of Chicago Press, 2015.

Brubaker, Rogers. *Ethnicity without Groups.* Cambridge: Harvard University Press, 2006.

Brückner, Martin, ed. *Early American Cartographies.* Chapel Hill: University of North Carolina Press, 2012.

———. *The Geographic Revolution in Early America: Maps, Literacy, and National Identity.* Chapel Hill: University of North Carolina Press, 2006.

Brunhes, Jean. "Frontier Region of Mexico: Notes to Accompany a Map of the Frontier." *Geographical Review* 3 (1917): 16–27.

———. "Non-existence of Peqary Channel." *Geographical Review* 1 (1916): 448–56.

———. "Results of an Expedition to the Central Andes." *Bulletin of the American Geographical Society* 46 (1914): 161–83.

Brzozowski, Stanisław Marian. "Eugeniusz Mikołaj Romer." *Polski Słownik Biograficzny* 31/4, no. 131 (Wrocław: Polska Akademia Umiejętności, 1989), 636–45.

Bucur, Maria. *Eugenics and Modernization in Interwar Romania.* Pittsburgh: University of Pittsburgh Press, 2002.

Bucur, Maria, Rayna Gavrilova, Wendy Goldman, Maureen Healy, and Mark Pittaway. "Forum on Everyday Life: Six Historians in Search of *Alltagsgeschichte.*" *Aspasia: The International Yearbook of Central, Eastern, and Southeastern European Women's and Gender History* 3 (2009): 189–212.

Buday, Ladislaus, *Dismembered Hungary.* London: Grant Richards, 1923.

Budde, Gunilla, Sebastian Conrad, and Oliver Janz, eds. *Transnationale Geschichte: Themen, Tendenzen und Theorien.* Göttingen: Vandenhoeck & Ruprecht, 2006.

Burleigh, Michael. *Germany Turns Eastwards: A Study of Ostforschung in the Third Reich*. London: Pan, 2002. (1st ed., Cambridge: Cambridge University Press, 1988).

Butler, Judith. *Dispossession: The Performative in the Political*. London: Polity, 2013.

———. *Gender Trouble: Feminism and the Subversion of Identity*. London: Routledge, 1989.

Butlin, Robin. *Geographies of Empire: European Empires and Colonies, c. 1880–1960*. Cambridge: Cambridge University Press, 2009.

Cadiot, Juliet. *Le laboratoire imperial: Russie-URSS, 1870–1940*. Paris: CNRS Editions, 2007.

Case, Holly. "Being European: East and West," in Jeffrey T. Checkel and Roger J. Katzenstein, eds., *European Identity*, 111–31. Cambridge: Cambridge University Press, 2009.

———. *Between States: The Transylvanian Question and the European Idea during World War II*. Stanford: Stanford University Press, 2009.

Casey, Stephen, and Jonathan Wright. *Mental Maps in the Era of Two World Wars*. New York: Palgrave Macmillan, 2008.

Cattaruzza, Marina, Stefan Dyroff, and Dieter Langewiesche, eds. *Territorial Revisionism and the Allies of Germany in the Second World War: Goals, Expectations, Practices*. New York: Berghahn Books, 2013.

Cattaruzza, Marina, and Dieter Langeweische. "Contextualizing Territorial Revisionism in East Central Europe: Goals, Expectations, and Practices," in Marina Cattaruzza, Stefan Dyroff, and Dieter Langewiesche, eds., *Territorial Revisionism and the Allies of Germany in the Second World War: Goals, Expectations, Practices*, 1–16. New York: Berghahn Books, 2013.

Certeau, Michel de. *The Practice of Everyday Life*, reprint ed. Berkeley: University of California Press, 2011.

Chałubińska, Aniela. "Kontakty Eugeniusza Romera i Albrechta Pencka, *Studia i materiały z dziejów nauki polskiej*," Seria C, vol. 24 (1980): 15–33.

Cholnoky, Jenő. "Teleki Pál gróf." *Földrajzi Közlemények* 67:4 (1939): 249–58; and 69:2 (1941): 63–71.

———. *Utazásom Amerikában Teleki Pál gróffal*. Budapest: Vajda-Wichmann, 1943.

Chorley, R. J., Robert P. Beckinsale, and Anthony J. Dunn, eds., *The History of the Study of Landforms, or the Development of Geomorphology*, vol. 2. London: Methuen, 1973.

Chu, Winson. *The German Minority in Interwar Poland*. Cambridge: Cambridge University Press, 2012.

Cienciala, Anna. "The Foreign Policy of Józef Piłsudski and Józef Beck: Misconceptions and Interpretations." *Polish Review* 56:1–2 (2011): 111–53.

Ciencała, Anna M., and Titus Komarnicki. *From Versailles to Locarno: Keys to Polish Foreign Policy, 1919–25.* Lawrence: University of Kansas Press, 1984.

Cienciala, Anna M., Natalia S. Lebedeva, and Wojciech Materski, eds. *Katyn: A Crime without Punishment.* New Haven: Yale University Press, 2008.

Clawson, Mary Ann. *Constructing Brotherhood: Class, Gender, and Fraternalism.* Princeton: Princeton University Press, 1989.

Clout, Hugh. "France, Poland, and Europe: The Experience of the XIVth International Geographical Congress, Warsaw, 1934." *Belgeo* 4 (2005): 435–44.

Cocks, Geoffrey, and Konrad H. Jarausch, eds. *German Professions, 1800–1950.* Oxford: Oxford University Press, 1990.

Coen, Deborah R. "Imperial Climatographies from Tyrol to Turkestan." *Osiris* 26 (2011): 45–65.

———. "Scaling Down: The 'Austrian' Climate between Empire and Republic," in James R. Fleming, Vladimir Jankovic, and Deborah R. Coen, eds., *Intimate Universality: Local and Global Themes in the History of Weather and Climate,* 115–40. Sagamore Beach, Mass.: Science Publications, 2006.

———. *Vienna in the Age of Uncertainty: Science, Liberalism, and Private Life.* Chicago: University of Chicago Press, 2007.

Cohen, Gary B. *Education and Middle-class Society in Imperial Austria, 1848–1918.* West Lafayette, Ind.: Purdue University Press, 1996.

Cohen, Stephen F. *The Victims Return: Survivors of the Gulag after Stalin.* Exeter, N.H.: PublishingWorks, 2010.

Connelly, John. *From Enemy to Brother: The Revolution in Catholic Teaching on the Jews, 1933–1965.* Cambridge: Harvard University Press, 2012.

Costlow, Jane T. *Heart-Pine Russia: Walking and Writing the Nineteenth-century Forest.* Ithaca: Cornell University Press, 2013.

Coverley, Merlin. *Psychogeography.* Harpenden: Oldcastle Books, 2010.

Crampton, Jeremy W. "The Cartographic Calculation of Space: Race Mapping and the Balkans at the Paris Peace Conference of 1919." *Social and Cultural Geography* 7:5 (October 2006): 731–52.

Crampton, Jeremy W., and Stuart Elden, eds. *Space, Knowledge, and Power: Foucault and Geography.* London: Ashgate, 2007.

Cseke, László. "Teleki Pál és a Moszul-bizottság." *A Földgömb* 14:4 (2012): 40–47.

Csicsery-Rónay, István, and Károly Vigh, eds. *Teleki Pal és kora: A Teleki Pál emlékév elóasásai*. Budapest: Occidental Press, 1992.

Csirpák, Lilli. *Teleki Európáról*. Budapest: Kairosz Kiadó, 2004.

Curran, Andrew S. *The Anatomy of Blackness: Science and Slavery in an Age of Enlightenment*. Baltimore: Johns Hopkins University Press, 2013.

Cvetkovski, Roland, and Alexis Hofmeister, eds. *An Empire of Others: Creating Ethnographic Knowledge in Imperial Russia and the USSR*. Budapest: Central European University Press, 2014.

Czigány, Lórant. *The Oxford History of Hungarian Literature*. Oxford: Clarendon Press, 1984.

Czyżewski, Julian. "Zycie i dzieło Eugeniusza Romera," in Eugeniusz Romer, *Wybór prac*, vol. 1, 9–115. Warszawa: PWN, 1960.

Dabrowski, Patrice M. "Borderland Encounters in the Carpathian Mountains and Their Impact on Identity Formation," in Omer Bartov and Eric D. Weitz, eds., *Shatterzone of Empires: Coexistence and Violence in the German, Habsburg, Russian, and Ottoman Borderlands*, 193–208. Bloomington: Indiana University Press, 2013.

———. " 'Discovering' the Galician Borderlands: The Case of the Eastern Carpathians," *Slavic Review* 64:2 (Summer 2005): 380–402.

Damir-Geilsdorf, Sabine, Angelika Hartmann, and Béatrice Hendrich, eds. *Mental Maps—Raum—Erinnerung: kulturwissenschaftliche Zugänge zum Verhältnis von Raum und Erinnerung*. Münster: Lit, 2005.

Danielsson, Sarah K. *The Explorer's Roadmap to National-Socialism: Sven Hedin, Geography, and the Path to Genocide*. Burlington: Ashgate, 2012.

Dashkevych, Iaroslav, et al., eds. *Kartohrafiia ta istoriia Ukrainy: zbirnyk naukovykh prats'*. Lviv, 1998. Lviv: M. P. Kots, 2000.

David-Fox, Michael, and Gyorgy Peteri, eds. *Academia in Upheaval: Origins, Transfers, and Transformations of the Communist Academic Regime in Russia and East Central Europe*. Westport, CT: Bergin & Garvey, 2000.

Davidson, Joyce, Liz Bondi, and Mick Smith, eds. *Emotional Geographies*. London: Ashgate, 2007.

Davis, W. M. "The Cycle of Erosion and the Summit Level of the Alps," *Journal of Geology* 31 (1923): 1–41.

———. *The Development of the Transcontinental Excursion of 1912*. New York: American Geographical Society, 1915.

———. "The Geographical Cycle," *Geographical Journal* 14 (1899): 481–504.

———. "Passarge's Principles of Landscape Description," *Geographical Review* 8 (1919): 266–73.

————. "The Penck Festband: A Review," *Geographical Review* 10 (1920): 249–61.

Deák, Francis. *Hungary at the Paris Peace Conference. The Diplomatic History of the Treaty of Trianon.* New York: Columbia University Press, 1942.

Dean, Carolyn J. *Sexuality and Modern Western Culture.* New York: Twayne, 1996.

Dean, Mitchell. *Governmentality: Power and Rule in Modern Society.* London: Sage, 1999.

Deloria, Philip J. *Playing Indian.* New Haven: Yale University Press, 1998.

Dibrova, Oleksii. *Heohrafiia Ukrains'koi RSR.* Kyiv: Radians'ka shkola, 1954.

Dickinson, Robert E. "Albrecht Penck," in *The Makers of Modern Geography* (New York: Prager, 1969), 100–111.

Diekmann, Irene, Peter Krüger, and Julius H. Schoeps, eds. *Geopolitik: Grenzgange im Zeitgeist. vol. 1: 1890 bis 1945.* Potsdam: Verlag für Berlin-Brandenburg, 2000.

Ditchuk, Ihor, ed. *Heohraf-akademik Stepan Rudnyts'kyi.* Ternopil': Navchal'na kniha—Bohdan, 2007.

Dnistrians'kyi, Stanislav. *Ukraina and the Peace Conference (L'Ukraine et la Conférence de la paix).* Charleston: Nabu Press, 2010; first published 1919.

Dodds, Klaus, and David Atkinson, eds. *Geopolitical Traditions: A Century of Geopolitical Thought.* London: Routledge, 2000.

Dodge, Martin, Rob Kitchin, and Chris Perkins, eds. *The Map Reader: Theories of Mapping Practice and Cartographic Representation.* Hoboken: Wiley-Blackwell, 2011.

————. *Rethinking Maps: New Frontiers in Cartographic Theory.* London: Routledge, 2009.

Drach, Ivan, ed. *Ostannia adresa: do 60-richchia solovets'koi trahedii. 3 vols.* Kyiv: Sfera, 1997–99. *For 2nd ed., see* Shapoval, Iurii.

Driver, Felix. *Geography Militant: Cultures of Exploration and Empire.* Oxford: Blackwell, 2001.

Drummond, Elizabeth A. "From 'verloren gehen' to 'verloren bleiben': Changing German Discourses on Nation and Nationalism in Poznania," in Charles W. Ingrao and Franz A.J. Szabo, eds., *The Germans and the East*, 226–40. West Lafayette, Ind.: Purdue University Press, 2009.

Dunlop, Catherine Tatiana. *Cartophilia: Maps and the Search for Identity in the French-German Borderland.* Chicago: University of Chicago Press, 2015.

Dyak, Sofia. *Home: A Century of Change.* Pittsburgh: University of Pittsburgh Press, 2013.

Dym, Jordana, and Karl Offen, eds. *Mapping Latin America: A Cartographic Reader*. Chicago: University of Chicago Press, 2011.

Eberhardt, Piotr. "Geneza i rozwój niemieckiej doktryny Lebensraumu (The Origins and Development of the German Doctrine of 'Lebensraum' [Living Space])," *Przegląd Geograficzny* 80:2 (2008): 175–98.

———. *Polska i jej granice: z historii polskiej geografii politycznej*. Lublin: Wydawnictwo Uniwersytetu Marii Curie-Skłodowskiej, 2004.

Edney, Matthew H. "The Irony of Imperial Mapping," in James R. Akerman, ed., *The Imperial Map: Cartography and the Mastery of Empire, 11–45*. Chicago: University of Chicago Press, 2008.

———. "People, Places, and Ideas in the History of Cartography," *Imago Mundi* 66 (2014): 83–106.

Edvi Illés, Aladár, ed. *The Economies of Hungary in Maps. To the Commission of Count Paul Teleki, Chief of the Office for the Preparation of Peace Negotiations*. Budapest: Hornyánsky, 1921.

Edvi Illés, Aladár, and Albert Halász, eds. *Magyarország a háboru elött és után gazdaságstatisztikai térképekben* [*Hungary before and after the War in Economic-Statistical Maps*]. Budapest: Institute of Political Sciences of the Hungarian Statistical Society, 1926.

Eile, Stanisław, *Literature and Nationalism: Literature and Nationalism in Partitioned Poland, 1795–1918*. London: MacMillan Press, 2000.

Eissmann, Lothar. "Albrecht Pencks frühes Wirken in Sachsen," *Abhandlungen und Berichte des Naturkundlichen Museums Mauritianum Altenburg* 11 (1984): 129–36.

Eley, Geoff, and James Retallack, eds. *Wilhelminism and Its Legacies: German Modernities, Imperialism, and the Meanings of Reform, 1890–1930*. New York: Berghahn Books, 2004.

Elias, Norbert. *The Germans: Power Struggles and the Development of Habitus in the Nineteenth and Twentieth Centuries*. New York: Columbia University Press, 1996.

Elshakry, Marwa. "When Science Became Western: Historiographical Reflections," *Isis* 101:1 (March 2010): 98–109.

Engerman, David. *Know Your Enemy: The Rise and Fall of America's Soviet Experts*. Oxford: Oxford University Press, 2009.

Engelmann, Gerhard. "Bibliographie Albrecht Penck," in *Wissenschaftliche Veröffentlichungen des Deutschen Instituts für Länderkunde* 17/18 (1960): 331–447.

———. *Ferdinand von Richthofen, 1833–1905; Albrecht Penck, 1858–1945: Zwei Markante Geografen Berlins*. Stuttgart: Franz Steiner, 1988.

Epple, Angelika, and Angelika Schaser, eds. *Gendering Historiography: Beyond National Canons*. Frankfurt: Campus Verlag, 2009.

Evans, Jennifer, and Matt Cook, eds. *Queer Cities, Queer Cultures: Europe since 1945*. New York: Bloomsbury, 2014.

Evans, Richard J. *The Pursuit of Power: Europe, 1815–1914*. New York: Viking, 2016.

Evtuhov, Catherine. *Portrait of a Russian Province: Economy, Society, and Civilization in Nineteenth-century Nizhnii Novgorod*. Pittsburgh: University of Pittsburgh Press, 2011.

Fábri, Mihály, ed. *Teleki Pal: becsületük előbbrevaló jólétünknél*. Gödöllő: Teleki Pál Egysület, 2004.

Fahlbusch, Michael. "Deutschthum in Ausland: Zur Volks- und Kuturbodentheorie in der Weimarer Republik," in Manfred Büttner et al., eds., *Miteinander, Nebeneinander, Gegeneinander* (Bochum: Universitätsverlag Dr. N. Brockmeyer, 1994), 213–31.

———. "The Role and Impact of German Ethnopolitical Experts in the SS Reich Security Main Office," in Ingo Haar and Michael Fahlbusch, eds., *German Scholars and Ethnic Cleansing, 1919–1945*, 28–50. New York: Berghahn Books, 2005.

———. "Volks- und Kulturbodenforschung in der Weimarer Republik: Der 'Grenzfall' Böhmen und Mähren," in Ute Wardenga and Ingrid Hönsch, eds., *Kontinuität und Diskontinuität der deutschen Geographie in Umbruchphasen: Studien zur Geschichte der Geographie* (Münster: Institut für Geographie der Westfälischen Wilhelms-Universität, 1995), 99–112.

———. *Wissenschaft im Dienste der nationalisozialistischen Politik? Die 'Volksdeutschen Forschungsgemeinschaften' von 1931–1945*. Baden-Baden: Nomos, 1999.

———. *"Wo der deutsche . . . ist, ist Deutschland!" Die Stiftung für deutsche Volks- und Kulturbodenforschung in Leipzig von 1920–1933*. Bochum: Universitätsverlag Dr. N. Brockmeyer, 1994.

Fahlbusch, Michael, and Ingo Haar, *Völkische Wissenschaften und Politikberatung im 20. Jahrhundert: "Expertise und Neuordnung" Europas*. Paderborn: Schöningh, 2010.

Faluszczak, Franciszek Paweł. *Kartografia galicji Wschodniej w latach 1772–1914*. Rzeszów: Wydawnictwo Uniwersytetu Rzeszowskiego, 2011.

Fieldhouse, D. K., ed. *Kurds, Arabs and Britons: The Memoir of Wallace Lyon in Iraq, 1918–1944*. London: I. B. Tauris, 2002.

Figes, Orlando. *The Whisperers: Private Life in Stalinist Russia*. New York: Metropolitan Books, 2007.

Fink, Carole. *Defending the Rights of Others: The Great Powers, The Jews, and International Minority Protection, 1878–1938*. Cambridge: Cambridge University Press, 2006.

Fitzpatrick, Sheila. *Everyday Stalinism: Ordinary Lives in Extraordinary Times: Soviet Russia in the 1930s*. Oxford: Oxford University Press, 1999.

Fitzpatrick, Sheila, and Yuri Slezkine, eds. *In the Shadow of Revolution: Life Stories of Russian Women from 1917 to the Second World War*. Princeton: Princeton University Press, 2000.

Flanagan, Jason C. "Woodrow Wilson's 'Rhetorical Restructuring': The Transformation of the American Self and the Construction of the German Enemy," *Rhetoric and Public Affairs* 7 (Summer 2004): 115–48.

Fleming, James R., Vladimir Jankovic, and Deborah R. Coen, eds., *Intimate Universality: Local and Global Themes in the History of Weather and Climate*. Sagamore Beach, Mass.: Science Publications, 2006.

Fodor, Ferenc. *A Magyar földrajztudomány története*. Budapest: Magyar Tudományos Akademia Földrajztudományi Kutatóintézet, 2006.

———. *A Magyar Térképírás*, 3 vols. Budapest: Honvéd Térképészeti Intézet, 1952–54.

———. "Teleki Pál geopolitikája," *Magyar Szemle* 40:6 (June 1941): 337–43.

———, and Ferenc Tibor Szávai. *Teleki Pál: egy "bujdosó könyv": megjelent Teleki Pál halálának 60. évében*. Budapest: Mike és Társa Antikvárium, 2001.

Foucault, Michel. *The History of Sexuality: An Introduction*, vol. 1. New York: Vintage, 1980.

Frank, Tibor. *Ethnicity, Propaganda, Myth-Making: Studies on Hungarian Connections to Britain and America, 1848–1945*. Budapest: Akadémiai Kiadó, 1999.

Frank, Tibor, and Frank Hadley, eds. *Disputed Territories and Shared Pasts: Overlapping National Histories in Modern Europe*. New York: Palgrave Macmillan, 2011.

Friedrichsmeyer, Sara, Sara Lennox, and Susanne Zantop, eds. *The Imperialist Imagination: German Colonialism and Its Legacy*. Ann Arbor: University of Michigan Press, 1998.

Galchen, Rivka. "Wild West Germany," *New Yorker*, 9 April 2012, 40–45.

Galloway, Colin G., Gerd Gemunden, and Suzanne Zantop, eds. *Germans and Indians: Fantasies, Encounters, Projections*. Lincoln: University of Nebraska Press, 2002.

Gantner, Eszter B., and Péter Varga, eds. *"Transfer—Interdiziplinär!": Akteure, Topographien und Praxis der Wissenstransfer*. Frankfurt: Peter Lang, 2013.

Gelfand, Lawrence E. *The Inquiry: American Preparations for Peace, 1917–1919*. New Haven: Yale University Press, 1963.

Geographers: Biobibliographical Studies, vols. 1–32. London: Mansell, 1977-present.

Getty, J. Arch, and Roberta Manning, eds. *Stalinist Terror: New Perspectives*. Cambridge: Cambridge University Press, 1993.

Glassheim, Eagle. *Noble Nationalists: The Transformation of the Bohemian Aristocracy*. Cambridge: Harvard University Press, 2005.

Gleason, Abbott. *Totalitarianism: The Inner History of the Cold War*. Oxford: Oxford University Press, 1995.

Godlewska, Anne Marie Claire. *Geography Unbound: French Geographic Science from Cassini to Humboldt*. Chicago: University of Chicago Press, 2000.

Godlewska, Anne Marie Claire, and Neil Smith, eds., *Geography and Empire*. Oxford: Blackwell, 1994.

Goffman, Erving. *The Presentation of Self in Everyday Life*. New York: Anchor Books, 1959.

Goldgar, Anne. *Impolite Learning: Conduct and Community in the Republic of Letters, 1680–1750*. New Haven: Yale University Press, 1995.

Goldgar, Anne, and Robert I. Frost. *Institutional Culture in Early Modern Society*. Boston: Brill, 2004.

Goldman, Wendy Z. *Inventing the Enemy: Denunciation and Terror in Stalin's Russia*. Cambridge: Cambridge University Press, 2011.

Gordin, Michael. *The Pseudoscience Wars: Immanuel Velikovsky and the Birth of the Modern Fringe*. Chicago: University of Chicago Press, 2012.

Górny, Maciej. "War on Paper? Physical Anthropology in the Service of States and Nations: 'Krieg der Geister,' Eastern Europe, and Physical Anthropology," in Jochen Böhler, Włodzimierz Borodziej, and Joachim von Puttkamer, eds., *Legacies of Violence: Eastern Europe's First World War*, 131–67. München: Oldenbourg, 2014.

———. *Wielka Wojna profesorów, 1912–1923*. Warszawa: Instytut PAN, 2014.

Gould, Peter, and Forrest R. Pitts, eds. *Geographical Voices: Fourteen Geographical Essays*. Syracuse: Syracuse University Press, 2002.

Greene, Mott T. "Geology," in Peter J. Bowler and John V. Pickstone, eds., *The Cambridge History of Science: The Modern Biological and Earth Sciences*, vol. 6. Cambridge: Cambridge University Press, 2009.

Gregorczyk, Joanna, ed. *Eugeniusz Romer: geograf i kartograf trzech epok*. Warszawa: Biblioteka Narodowa, 2004.

Gross, Jan T. *Fear: Anti-semitism in Poland after Auschwitz: An Essay in Historical Interpretation*. Princeton: Princeton University Press, 2006.

————. *Revolution from Abroad: The Soviet Conquest of Poland's Western Ukraine and Western Belorussia*, new ed. Princeton: Princeton University Press, 2002.

Gross, Jan T., István Deák, and Tony Judt, eds. *The Politics of Retribution in Europe*. Princeton: Princeton University Press, 2000.

Guettel, Jens-Uwe. "From the Frontier to German South-West Africa: German Colonialism, Indians, and American Westward Expansion," *Modern Intellectual History* 7:3 (2010): 523–52.

————. "The U.S. Frontier as Rationale for the Nazi East? Settler Colonialism and Genocide in Nazi-occupied Eastern Europe and the American West: A Revisionist Perspective," *Journal of Genocide Research* 15:3 (December 2013): 401–19.

Haar, Ingo. "German Ostforschung and Anti-Semitism," in Ingo Haar and Michael Fahlbusch, eds., *German Scholars and Ethnic Cleansing, 1919–1945*, 1–27. New York: Berghahn Books, 2005.

————. "Leipziger Stiftung für deutsche Volks- und Kulturbodenforschung," in Ingo Haar and Michael Fahlbusch, eds., *Handbuch der völkischen Wissenschaften: Personen—Institutionen—Forschungsprogramme—Stiftungen*, 374–82. München: K. G. Saur, 2008.

Haar, Ingo, and Michael Fahlbusch, eds. *German Scholars and Ethnic Cleansing, 1919–1945*. New York: Berghahn Books, 2005.

————. *Handbuch der völkischen Wissenschaften: Personen–Institutionen–Forschungsprogramme–Stiftungen*. München: K. G. Saur, 2008.

Häberle, D[ietrich]. "Der Anteil der Deutschen und Polen an der Bevölkerung von Westpreussen und Posen (nach A. Penck): Mit einer Kartenskizze," in *Geographische Zeitschrift* 25 (1919): 124–27.

Habermas, Jürgen. *Technik und Wissenschaft als Ideologie*. Frankfurt: Suhrkamp, 1968.

Hagen, William W. "The Moral Economy of Popular Violence: The Pogrom in Lwow, November 1918," in Robert Blobaum, ed., *Antisemitism and Its Opponents in Modern Poland*, 124–47. Ithaca: Cornell University Press, 2005.

Hajdú, Zoltán. "A Magyar Allamtér változásainak történeti politikai földrajzi szemlélete a magyar földrajztudományban 1948-ig," *Tér és társadalom* 9:3–4 (1995): 111–32.

————. "A magyar földrajztudomány szerepvállalása a trianoni békeszerződésre való tudományos felkészülésben," *Debreceni Egyetem Természetföldrajzi és geoinformatikai tanszék* (2010): 125–32.

————. *Carpathian Basin and the Development of the Hungarian Landscape Theory until 1948*. Pécs: Centre for Regional Studies of Hungarian Academy of Sciences, 2004.

————. "Friedrich Ratzel hatása a Magyar földrajztudományban," *Tér és társadalom* 12:3 (1998): 93–104.

————. "Geográfus politikus avagy politikus geográfus? A tudomány és a politika kölcsönhatása Teleki Pál életművében," *Földrajzi Közlemények* 39:1–2 (1991): 1–9.

————. "Teleki Pál a földdrajztudós," in István Csicsery-Rónay and Károly Vigh, eds., *Teleki Pal és kora: A Teleki Pál emlékév előadásai*, 44–52. Budapest: Occidental Press, 1992.

————. "Teleki Pál tájelméleti munkássága," *Földrajzi Közlemények* 49:1–2 (2001): 51–64.

Hajnal, Zsolt. *A Magyar Földrajzi Társasag története 1945-ig*. Jászbereny: Jászberényi Tanitóképzőfőiskola Közművelódési és Felnöttvelési Tanszék, 1999.

Hanebrink, Paul A. *In Defense of Christian Hungary: Religion, Nationalism, and Antisemitism, 1890–1944*. Ithaca: Cornell University Press, 2006.

Hannah, Matthew G. *Governmentality and the Mastery of Territory in Nineteenth-century America*. Cambridge: Cambridge University Press, 2000.

Hansen, Jason D. *Mapping the Germans: Statistical Science, Cartography, and the Visualization of the German Nation, 1848–1914*. Oxford: Oxford University Press, 2015.

Hänsgen, Dirk. "Heinrich Kiepert: ein Handwerker unter den Geographie-Ordinarien der ersten Stunde," in Bernhard Nitz, Hans-Dietrich Schultz, and Marlies Schulz, eds., *1810–2010: 200 Jahre Geographie in Berlin*, 51–57. Berlin: Geographisches Institut der Humboldt-Universität zu Berlin, 2011.

Happel, Jörn, Christophe von Werdt, and Mira Jovanović, eds. *Osteuropa kartiert—Mapping Eastern Europe*. Münster: LIT, 2010.

Harley, J. Brian. *The New Nature of Maps: Essays in the History of Cartography*. Baltimore: Johns Hopkins University Press, 2001.

Hartshorne, Richard. "The Polish Corridor," *Journal of Geography* 36:5 (1937): 161–76.

Haslinger, Peter. *Nation und Territorium im tschechischen politischen Diskurs, 1880–1938*. München: R. Oldenbourg, 2010.

Haslinger, Peter, and Vadim Oswalt, eds. *Kampf der Karten: Propaganda- und Geschichtskarten als politische Instrumente und Identitätstexte*. Marburg: Herder-Institut, 2012.

Hassinger, Hugo. "Albrecht Penck," *Petermanns Geographische Mitteilungen* 92 (1948): 190–93.

Haupt, Heinz-Gerhard, and Jürgen Kocka, eds. *Comparative and Transnational History: Central European Approaches and New Perspectives*. New York: Berghahn Books, 2009.

Hausmann, Guido. "Die Kultur die Niederlage: Der erste Weltkrieg in der ukrainischen Erinnerung," *Osteuropa* 64, no. 2–4 (2014): 127–40.

———. "Maps of the Borderlands: Russia and Ukraine," in Róisín Healy and Enrico Dal Lago, eds., *The Shadow of Colonialism on Europe's Modern Past*, 194–210. Houndmills: Palgrave Macmillan, 2014.

———. "Das Territorium der Ukraine: Stepan Rudnyc'kyjs Beitrag zur Geschichte räumlich-territorialen Denkens über die Ukraine," in *Die Ukraine: Prozesse der Nationsbildung*, ed. Andreas Kappeler, 145–57. Köln: Böhlau, 2011.

———. "The Ukrainian Moment of World War I," in Gearóid Barry, Enrico Dal Lago and Róisín Healy, eds., *Small Nations and Colonial Peripheries in World War I*, 177–91. Leiden: Brill, 2016.

Healy, Maureen. *Vienna and the Fall of the Habsburg Empire: Total War and Everyday Life in World War I*. Cambridge: Cambridge University Press, 2004.

Healy, Róisín, and Enrico Dal Lago, eds. "Investigating Colonialism within Europe," in Healy and Dal Lago, *The Shadow of Colonialism on Europe's Modern Past*, 3–22. Houndmills: Palgrave Macmillan, 2014.

Hedin, Sven. *Amerika im Kampf der Kontinente*. Leipzig: F. A. Brockhaus, 1942; reprinted in 1943.

Heffernan, Michael. *The Meaning of Europe: Geography and Geopolitics*. London: Arnold, 1998.

———. "The Politics of the Map in the Early Twentieth Century," *Cartography and Geographic Information Science* 29:3 (July 2002): 207–26.

———. "Professor Penck's Bluff: Geography, Espionage, and Hysteria in World War I," *Scottish Geographical Journal* 116:4 (2000): 267–82.

Hellbeck, Jochen. *Revolution on My Mind: Writing a Diary under Stalin*. Cambridge: Harvard University Press, 2006.

Herb, Guntram Henrik. *Under the Map of Germany: Nationalism and Propaganda, 1918–1945*. London: Routledge, 1997.

Herb, Guntram Henrik, and David H. Kaplan, eds. *Nested Identities: Nationalism, Territory, and Scale*. Washington, D.C.: Rowman and Littlefield, 1999.

Herbert, Amanda. *Female Friendship: Gender, Alliances, and Friendship in Early Modern Britain*. New Haven: Yale University Press, 2014.

Herf, Jeffrey. *Reactionary Modernism: Technology, Culture, and Politics in Weimar Germany and the Third Reich*. Cambridge: Cambridge University Press, 1984.

Hine, Robert V., and John Mack Faragher. *The American West: A New Interpretive History*. New Haven: Yale University Press, 2000.

Hirsch, Francine. *Empire of Nations: Ethnographic Knowledge and the Making of the Soviet Union*. Ithaca: Cornell University Press, 2005.

Hochschild, Adam. *King Leopold's Ghost: A Story of Greed, Terror, and Heroism in Colonial Africa*. Boston: Houghton Mifflin, 1998.

Holiat, Roman S. *Short History of the Ukrainian Free University*. New York: Shevchenko Scientific Society, 1964.

Holmgren, Beth, ed. *The Russian Memoir: History and Literature*. Evanston: Northwestern University Press, 2003.

Holquist, Peter J. "'Information Is the Alpha and Omega of Our Work': Bolshevik Surveillance in Its Pan-European Context," *Journal of Modern History* 69:3 (September 1997): 415–50.

———. "To Count, to Extract, and to Exterminate: Population Statistics and Population Politics in Late Imperial and Soviet Russia," in Ronald Suny and Terry Martin, eds., *A State of Nations: Empire and Nation-making in the Age of Lenin and Stalin*, 11–44. Oxford: Oxford University Press, 2001.

Hostetler, Laura. *Qing Colonial Enterprise: Ethnography and Cartography in Modern China*. Chicago: University of Chicago Press, 2005.

Howard, John. *Men Like That: A Southern Queer History*. Chicago: University of Chicago Press, 1999.

Hubbard, Phil, and Rob Kitchin, eds. *Key Thinkers on Space and Place*, 2nd ed. London: Sage, 2010.

Hurtado, Alberto. *Intimate Frontiers: Sex, Gender, and Culture in Old California*. Albuquerque: University of New Mexico Press, 1999.

Ingrao, Charles W., and Franz A. J. Szabo, eds. *The Germans and the East*. West Lafayette, Ind.: Purdue University Press, 2008.

Jackowski, Antoni, Stanisław Liszewski, and Andrzej Richling, eds. *Historia geografii polskiej*. Warszawa: PWN, 2008.

Jackowski, Antoni, and Izabela Sołjan. *Z dziejów geografii na Uniwersytecie Jagiellońskim (XV–XXI wiek)*. Kraków: Instytut Geografii i Gospodarki Przestrzennej Uniwersytetu Jagiellonskiego, 2009. Jagiellonian University Institute of Geography and Spatial Management, 2009.

Jacob, Christian. *The Sovereign Map: Theoretical Approaches to Cartography throughout History*. Chicago: University of Chicago Press, 2006.

Jacobson, Matthew Frye. *Barbarian Virtues: The United States Encounters Foreign Peoples at Home and Abroad*. New York: Hill and Wang, 2000.

Jarausch, Konrad H. *After Hitler: Recivilizing the Germans, 1945–1995*. Oxford: Oxford University Press, 2008.

————. *Students, Society, and Politics in Imperial Germany: The Rise of Academic Illiberalism.* Princeton: Princeton University Press, 1982.

————. *The Unfree Professions: German Lawyers, Teachers, and Engineers, 1900–1950.* New York: Oxford University Press, 1990.

Jarausch, Konrad H., and Geoffrey Cocks, eds. *German Professions, 1800–1950.* Oxford: Oxford University Press, 1990.

Jedlicki, Jerzy. *A Suburb of Europe: Nineteenth-Century Polish Approaches to Western Civilization.* Budapest: Central European University Press, 1999.

Jędrzejczyk, Dobiesław. "Geopolitical Essence of Central Europe in Writings of Eugeniusz Romer," *Miscellanea Geographica* 11 (2004): 199–206.

Jędrzejczyk, Dobiesław, and Waldemar Wilk. *Eugeniusz Romer jako geograf społeczno-gospodarczy.* Warszawa: Uniwersytet Warszawski, Wydział Geografii i Studiów Regionalnych, 1999.

Jobbitt, Steven. "A Geographer's Tale: Nation, Modernity, and the Negotiation of Self in 'Trianon' Hungary, 1900–1960," unpubl. Ph.D. diss., University of Toronto, 2008.

————. "Memory and Modernity in Fodor's Geographical Work on Hungary," in Steven Tötösy de Zepetnek and Louise Olga Vasvári, eds., *Comparative Hungarian Cultural Studies*, 59–71. West Lafayette, Ind.: Purdue University Press, 2011.

Johnson, Emily D. *How St. Petersburg Learned to Study Itself: The Russian Idea of Kraevedenie.* University Park: Pennsylvania State University Press, 2009.

Jones, Larry Eugene, and James Retallack, eds. *Between Reform, Reaction, and Resistance: Studies in the History of German Conservatism from 1789 to 1945.* Providence: Berg, 1993.

Joó, István, and Frigyes Raum. *A magyar földmérés és térképészet története*, 4 vols. Budapest: Földmérési es Távérzékelési Intézet, 1992–93.

Judson, Pieter M. "Frontiers, Islands, Forests, Stones: Mapping the Geography of a German Identity in the Habsburg Monarchy, 1848–1900," in Patricia Yeager, ed., *The Geography of Identity*, 382–406. Ann Arbor: University of Michigan Press, 2005.

————. *Guardians of the Nation: Activists on the Language Frontiers of Imperial Austria.* Cambridge: Harvard University Press, 2006.

————. *Inventing Germanness: Class, Ethnicity, and Colonial Fantasy at the Margins of the Habsburg Monarchy.* Minneapolis: University of Minnesota Press, 1993.

Judson, Pieter M., and Marsha L. Rozenblit, eds. *Constructing Nationalities in East Central Europe.* New York: Berghahn Books, 2005.

Kafka, Ben. *The Demon of Writing: Powers and Failures of Paperwork*. New York: Zone Books, 2012.

Kafka, Franz. *The Castle*, trans. Mark Harman. New York: Schocken Books, 1998.

Kamusella, Tomasz. *The Politics of Language and Nationalism in Modern Central Europe*. New York: Palgrave Macmillan, 2009.

Kaps, Clemens, and Jan Surman, eds. "Post-colonial Perspectives on Habsburg Galicia," *Historyka: Studia metodologiczne* 42 (2012): 7–35.

Karch, Brendan. "Regionalism, Democracy, and National Self-determination in Central Europe," *Contemporary European History* 21:4 (November 2012): 635–51.

Kasianov, Georgiy, and Philipp Ther, eds. *A Laboratory of Transnational History: Ukraine and Recent Ukrainian Historiography*. Budapest: Central European University Press, 2008.

Kasianov, Heorhii [Georgiy]. *Ukrains'ka intelihentsiia 1920-kh-30-kh rokiv: sotsial'nyi portret ta istorychna dolia*. Kyiv: Hlobus, 1992.

Kasianov, Heorhii [Georgiy], and V. M. Danylenko. *Stalinizm i ukrains'ka intelihentsiia: 20-30-i roky*. Kyiv: Naukova Dumka, 1991.

Kearns, Gerry. *Geopolitics and Empire: The Legacy of Halford Mackinder*. Oxford: Oxford University Press, 2009.

Keményfi, Róbert. "Grenzen—Karten—Ethnien: Kartenartige Konstituierungsmittel im Dienst des ungarischen nationalen Raums," in Jörn Happel and Christophe von Werdt, eds., *Osteuropa kartiert—Mapping Eastern Europe*, 201–14. Münster: LIT Verlag, 2010.

Kennedy, James, and Liliana Riga. "Mitteleuropa as Middle America? The 'Inquiry' and the Mapping of East Central Europe in 1919," *Ab Imperio* 4 (2006): 271–300.

Kern, Stephen. *The Culture of Time and Space 1880–1918*, rev. ed. Cambridge: Harvard University Press, 2000.

Kertzer, David I., and Dominique Arel, eds. *Census and Identity: The Politics of Race, Ethnicity, and Language in National Censuses*. Cambridge: Cambridge University Press, 2002.

Khagram, Sanjeev, and Peggy Levitt, eds. *The Transnational Studies Reader: Intersections and Innovations*. London: Routledge, 2008.

Kimmel, Michael S. *Manhood in America: A Cultural History*, 3rd ed. New York: Oxford University Press, 2012.

Kimmel, Michael S., and Amy Aronson, eds. *The Gendered Society Reader*, 4th ed. New York: Oxford University Press, 2011.

Kimmel, Michael S., Jeff Hearn, and R. W. Connell, eds. *Handbook of Studies on Men and Masculinities*. Thousand Oaks, Calif.: Sage Publications, 2005.

King, Jeremy. *Budweisers into Czechs and Germans: A Local History of Bohemian Politics, 1848–1948*. Princeton: Princeton University Press, 2005.

Kish, George. "Paul Teleki," *Geographical Review* 31 (1941): 514–15.

———. "Paul Teleki, 1879–1941," in *Geographers: Biobibliographical Studies*, vol. 11, ed. T. W. Freeman, 139–43. London: Mansell, 1987.

Klay, Andor. "Hungarian Counterfeit Francs: A Case of Post–World War I Political Sabotage," *Slavic Review* 33:1 (March 1974): 107–13.

Klebelsberg, R[aimund] v[on]. "Albrecht Penck und seine Beziehungen zur Heimat: Zum 85. Gebürstag des Ehrendoktors der Deutschen Alpenuniversität," *Innsbrucker Nachrichten*, 25 November 1943.

Klinghammer, István, and Gábor Gercsák. "The Hungarian Geographer Pál Teleki: Member of the Mosul Commission," *Cartographica Helvetica* 19 (1999): 17–25.

Klinghammer, István, Gyula Pápay, and Zsolt Török. *Kartográfiatörténet*. Budapest: Eötvos Kiadó, 1995.

Knight, David. *The Making of Modern Science: Science: Technology, Medicine, and Modernity: 1789–1914*. Cambridge: Polity, 2009.

Kocsis, Károly, and Ferenc Schweitzer, eds. *Hungary in Maps*. Budapest: Geographical Research Institute, Hungarian Academy of Sciences, 2009.

Koenen, Gerd. *Der Russland-Komplex: die Deutschen und der Osten, 1900–1945*. München: Beck, 2005.

Kohlrausch, Martin, Katrin Steffen, and Stefan Wiederkehr, eds. *Expert Cultures in Central Eastern Europe: The Internationalization of Knowledge and the Transformation of Nation States since World War I*. Osnabrück: Fibre, 2010.

Kónya, Sándor, and Zsigmond Pál Pach. *A Magyar Tudományos Akadémia másfél évszázada, 1825–1975*. Budapest: Akadémiai Kiadó, 1975.

Kopp, Kristin. "Cartographic Claims: Colonial Mappings of Poland in German Territorial Revisionism," in Gail Finney, ed., *Visual Culture in Twentieth-Century Germany*, 199–213. Bloomington: Indiana University Press, 2006.

———. *Germany's Wild East: Constructing Poland as Colonial Space*. Ann Arbor: University of Michigan Press, 2012.

———. "Gray Zones: On the Inclusion of 'Poland' in the Study of German Colonialism," in Michael Perraudin and Jürgen Zimmerer, eds., *German Colonialism and National Identity*, 33–42. New York: Routledge, 2011.

———. "Reinventing Poland as German Colonial Territory in the Nineteenth Century: Gustav Freytag's *Soll und Haben* as Colonial Novel," in

Robert L. Nelson, ed., *Germans, Poland, and Colonial Expansion to the East: 1850 through the Present*, 11–37. New York: Palgrave Macmillan, 2009.

Kossert, Andres. "'Grenzlandpolitik' und Ostforschung an der Peripherie des Reiches. Das ostpreussische Masuren 1919–1945," *Vierteljahrshefte für Zeitgeschichte* 51 (2001): 117–46.

Kotkin, Stephen. *Stalin: Paradoxes of Power, 1878–1928*, vol. 1. New York: Penguin, 2014.

Krebs, Norbert. "Albrecht Penck 85 Jahre alt," *Deutsche Allgemeine Zeitung*, 25 September 1943.

———. "Volksthumsfragen der Geographie: Albrecht Penck zum 85. Geburgstage," *Berliner Lokal-Anzeiger*, 25 September 1943.

Kroh, Antoni. *Tatry i Podhale*. Wrocław: Wydawnictwo Dolnośląskie, 2002.

Krueger, Rita. *Czech, German, and Noble: Status and National Identity in Habsburg Bohemia*. New York: Oxford University Press, 2009.

Krygier, John, and Denis Wood, eds. *Making Maps: A Visual Guide to Map Design for GIS*, 3rd ed. New York Guilford Press, 2016.

Kubassek, János. "A vörös térkép," *Rubicon* (2010): 35–38.

———. *Érdi kronika, Érd földrajza*. Érd: Magyar Földrajzi Múzeum, 2010.

———. "Földrajztudós és államférfi," *Földrajzi Közlemények* 49:1–2 (2001): 91–101.

———. "Gróf Teleki Pál," *A Földgömb* 25:5 (2007): 74–76.

Kubijovyč, Volodymyr. "Kharkiv," in Kubijovyč, ed., *Encyclopedia of Ukraine*, vol. 2. Toronto: University of Toronto Press, 1988.

———. "Stepan Rudnytsky," in Danylo Husar Struk, ed., *Encyclopedia of Ukraine*, vol. 4. Toronto: University of Toronto Press, 1993.

Kühn, Arthur. "Albrecht Penck," in *Westermann-Lexikon der Geographie*, vol. 3 (Braunschweig: Westermann, 1970), 79–80.

Kupchyns'kyi, Oleh, ed. *Lystuvannia Stepana Rudnyts'koho*. Lviv: Shevchenko Scientific Society, 2006.

Kuromiya, Hiroaki. *The Voices of the Dead: Stalin's Great Terror in the 1930s*. New Haven: Yale University Press, 2007.

Kuzeliia, Zenon. "Professor D-r A. Penk (Iuvileuna zamitka)," *Dilo*, 11 December 1933, 6.

Lafferton, Emese. "The Magyar Moustache: The Faces of Hungarian State Formation, 1867–1918," *Studies in History and Philosophy of Biological and the Medical Sciences* 38 (2007): 706–32.

Lam, Tong. *A Passion for Facts: Social Surveys and the Construction of the Chinese Nation-state, 1900–1949*. Berkeley: University of California Press, 2011.

Lederer, Ivo J. *Yugoslavia at the Paris Peace Conference: A Study in Frontiermaking.* New Haven: Yale University Press, 1963.

Lehmann, Edgar. "Albrecht Penck: Eine Gedächtnisrede (A Memorial Study)," *Vorträge und Schriften: Deutsche Akademie der Wissenschaften zu Berlin* 64 (1959): 1–24.

Lendvai, Paul. *The Hungarians: A Thousand Years of Victory in Defeat*, trans. Ann Major. Princeton: Princeton University Press, 2003.

Leslie, Camilo Arturo. "Territoriality, Map-mindedness, and the Politics of Place," *Theory and Society* 45 (2016): 169–201.

Levenko, Sh. [pseud. Stepan Rudnyts'kyi]. *Chomu my khochemo samostiinoi Ukrainy.* Vienna, 1915.

Lewis, G. Malcolm, ed. *Cartographic Encounters: Perspectives on Native American Mapmaking and Map Use.* Chicago: University of Chicago Press, 1998.

Litván, György. *A Twentieth-century Prophet: Oscar Jászi 1875–1957.* Budapest: Central European University Press, 2006.

Liulevicius, Vejas Gabriel. *The German Myth of the East, 1800 to the Present.* Oxford: Oxford University Press, 2009.

———. *War Land on the Eastern Front: Culture, National Identity, and German occupation in World War I.* Cambridge: Cambridge University Press, 2000.

Livingstone, David N. *The Geographical Tradition: Episodes in the History of a Contested Enterprise.* Oxford: Blackwell, 1993.

Livingstone, David N., and Charles W. J. Withers, eds. *Geographies of Nineteenth-century Science.* Chicago: University of Chicago Press, 2011.

———, eds. *Geography and Enlightenment.* Chicago: University of Chicago Press, 1999.

———, eds. *Geography and Revolution.* Chicago: The University of Chicago Press, 2005.

Lohr, Eric. *Nationalizing the Russian Empire: The Campaign against Enemy Aliens during World War I.* Cambridge: Harvard University Press, 2003.

Lohr, Eric, Vera Tolz, Alexander Semyonov, and Mark von Hagen. *The Empire and Nationalism at War.* Bloomington: Slavica, 2014.

Lord, Robert Howard. "Review of Isaiah Bowman, *The New World: Problems in Political Geography*," *American Historical Review* 27:3 (April 1922): 568–70.

Louis, Herbert. "Albrecht Penck und sein Einfluss auf Geographie und Eiszeitforschung," *Zeitschrift der Geschichte für Erdkunde zu Berlin* 89: 3–4 (1958): 161–82.

Lovell, Stephen. *Summerfolk: A History of the Dacha, 1710–2000.* Ithaca: Cornell University Press, 2003.

Łukomski, Grzegorz. *Problem "korytarza" w stosunkach polsko-niemieckie i na arenie międzynarodowej 1919–1939: studium polityczne.* Warszawa: Adiutor, 2000.

Lundgreen-Nielsen, Kay. "Aspects of American Policy towards Poland at the Paris Peace Conference and the Role of Isaiah Bowman," in Paul Latawski, ed., *The Reconstruction of Poland, 1914–23*, 95–116. New York: St. Martin's Press, 1992.

———. *The Polish Problem at the Paris Peace Conference: A Study of the Policies of the Great Powers and the Poles, 1918–1919*, translated by Alison Borch-Johansen. Odense: Odense University Press, 1979.

Mace, James E. *Communism and the Dilemmas of National Liberation: National Communism in Soviet Ukraine, 1918–1933.* Cambridge: Harvard University Press, 1983.

Mackenzie, John. *Museums and Empire: Natural History, Human Cultures, and Colonial Identities.* Manchester: Manchester University Press, 2009.

Macmillan, Margaret. *Paris 1919: Six Months That Changed the World.* New York: Random House, 2001.

———. *The War that Ended Peace: The Road to 1914.* New York: Random House, 2013.

Magocsi, Paul Robert. *A History of Ukraine: The Land and Its Peoples*, 2nd ed. Toronto: University of Toronto Press, 2011.

Maier, Charles S. "Transformations of Territoriality, 1600–2000," in Gunilla Budde, Sebastian Conrad, and Oliver Janz, eds., *Transnationale Geschichte: Themen, Tendenzen und Theorien*, 32–55. Göttingen: Vandenhoeck & Ruprecht, 2010.

Manela, Erez. *The Wilsonian Moment: Self-Determination and the International Origins of Anticolonial Nationalism.* Oxford: Oxford University Press, 2007.

Mangan, J. A., and James Walvin, eds. *Manliness and Morality: Middle-class Masculinity in Britain and America, 1800–1940.* New York: St. Martin's Press, 1987.

Marcinek, Joachim. "Die Bedeutung von Albrecht Penck für die Eiszeitforschung: zu seinem 125. Geburtstag," *Geographische Berichte* 28, no. 3 (1983): 153–64.

Marochko, V. I., and H'otts Khillih. *Represovani pedahohy Ukrainy: zhertvy politychnoho teroru, 1929–1941.* Kyiv: Naukovyi Svit, 2003.

Marples, David R. *Heroes and Villains: Creating National History in Contemporary Ukraine.* Budapest: Central European University Press, 2007.

Massey, Doreen B. *For Space.* London: Sage, 2005.

————. *Space, Place, and Gender*. Minneapolis: University of Minnesota Press, 1994.

Martin, Geoffrey J. *All Possible Worlds: A History of Geographical Ideas*, 4th ed. Oxford: Oxford University Press, 2005.

————. *American Geography and Geographers: Toward Geographic Science*. Oxford: Oxford University Press, 2015.

————. "A Fragment on the Penck-Davis Conflict," *Geography and Map Division: Special Libraries Association Bulletin* 98 (1974): 11–27.

————. *The Life and Thought of Isaiah Bowman*. Hamden: Archon, 1980.

————. "On Whittlesey, Bowman, and Harvard," *Annals of the Association of American Geographers* 78:1 (March 1988): 152–58.

Martin, Terry. *The Affirmative Action Empire: Nations and Nationalism in the Soviet Union, 1923–1939*. Ithaca: Cornell University Press, 2001.

Matelski, Dariusz. "Rewindikacja i repatriacja z Rosji Radzieckiej i ZSRR," in *Grabież i restytucja polskich dóbr kultury od czasów nowożytnych do współczesnych*, 266–88. Kraków: Pałac Sztuki, 2006.

Mayer, Arno J. *Politics and Diplomacy of Peacemaking: Containment and Counterrevolution at Versailles, 1918–1919*. New York: Knopf, 1967.

Mazower, Mark. *Governing the World: The History of an Idea*. New York: Penguin, 2012.

————. *Hitler's Empire: How the Nazis Ruled Europe* (New York: Penguin, 2009)

————. *No Enchanted Palace: The End of Empire and the Ideological Origins of the United Nations*. Princeton: Princeton University Press, 2009.

Mazurkiewicz-Herzowa, Łucja. *Eugeniusz Romer*. Warszawa: Wiedza Powszechna, 1966.

McDowell, Linda. *Gender, Identity, and Place: Understanding Feminist Geographies*. Cambridge: Polity, 1999.

McElligott, Anthony, ed. *Weimar Germany*. Oxford University Press, 2009.

Mehmel, Astrid. "Deutsche Revisionspolitik in der Geographie nach dem ersten Weltkrieg," *Geographische Rundschau* 47 (1995): 498–505.

Mendelsohn, Ezra. *The Jews of East Central Europe between the Wars*. Bloomington: Indiana University Press, 1987.

Meyer, Henry Cord. *Mitteleuropa in German Thought and Action, 1815–1945*. The Hague: Martinus Nijhoff, 1955.

Meynen, Emil. "Albrecht Penck, 1858–1945," in T. W. Freeman, ed., *Geographers: Biobibliographical Studies*, vol. 7 (London: Mansell, 1983), 101–108.

Mick, Christoph. *Lemberg, Lwów, L'viv: Violence and Ethnicity in a Contested City*. West Lafayette, Ind.: Purdue University Press, 2015.

Migacz, Władysław. "'Atlas geograficzny' Eugeniusza Romera z 1908 r.," *Czasopismo Geograficzne* 31:2 (1960): 149–62.

———. "Działalność naukowa Józefa Wąsowicza w geografii i kartografii," *Polski Przegląd Kartograficzny* 6:4 (1974): 145–48.

———. "Eugeniusz Romer," in Jan Szeliga and Wiesława Wernerowa, eds., *Materiały do słownika kartografów i geodetów polskich*, 114–18. Warszawa: Instytut Historii Nauki PAN; Zespół Historii Kartografii, Polskie Towarzystwo Geograficzne, 1999.

Miller, Michael L., and Scott Ury, eds. *Cosmopolitanism, Nationalism, and the Jews of East Central Europe*. London: Routledge, 2015.

Miłosz, Czesław, *The Captive Mind* (New York: Knopf, 1953)

Molik, Witold. *Polskie peregrynacje uniwersyteckie do Niemiec 1871–1914*. Poznań: Wydawnictwo Naukowe Uniwersytetu Adama Mickiewicza, 1989.

Monmonier, Mark S. *How to Lie with Maps*. Chicago: University of Chicago Press, 1996.

Moss, Pamela J., ed. *Placing Autobiography in Geography*. Syracuse: Syracuse University Press, 2000.

Moss, Pamela J., and Karen Falconer Al-Hindi, eds. *Feminisms in Geography: Rethinking Space, Place, and Knowledges*. Lanham, Md.: Rowman & Littlefield, 2008.

Mosse, George L. *The Crisis of German Ideology: Intellectual Origins of the Third Reich*. New York: Grosset & Dunlap, 1964.

———. *The Image of Man: The Creation of Modern Masculinity*. New York: Oxford University Press, 1996.

———. *Nationalism and Sexuality: Middle-class Morality and Sexual Norms in Modern Europe*. Madison: University of Wisconsin Press, 1988.

Mroczko, Marian. *Eugeniusz Romer: biografia polityczna*. Słupsk: Wydawnictwo Naukowe Akademii Pomorskiej, 2008.

———. *Polska myśl zachodnia 1918–1939: kształtowanie i upowszechnianie*. Poznań: Instytut Zachodni, 1989.

Murphy, David Thomas. *The Heroic Earth: Geopolitical Thought in Weimar Germany 1918–1933*. Kent: Kent State University Press, 1997.

———. "Hitler's Geostrategist? The Myth of Karl Haushofer and the Institut für Geopolitik," *Historian* 76:1 (Spring 2014): 1–25.

Mushnyk, Mykola, ed. *Lysty Stepana Rudnyts'koho do Sofii ta Stanislava Dnistrians'kykh: 1926–1932*. Edmonton: Canadian Institute of Ukrainian Studies, 1991.

Nail, Thomas. *Theory of the Border*. Oxford: Oxford University Press, 2016.

Natter, Wolfgang. "Geopolitics in Germany, 1919–45: Karl Haushofer and the *Zeitschrift für Geopolitik*," in John Agnew, Katharyne Mitchell, and Gerard Toal, eds., *A Companion to Political Geography, 187–203*. London: Blackwell, 2003.

Nelson, Lise, and Joni Seager, eds. *A Companion to Feminist Geography*. Malden, Mass.: Blackwell, 2005.

Nemes, Robert. *Another Hungary: The Nineteenth-century Provinces in Eight Lives*. Stanford: Stanford University Press, 2016.

———. "Mapping the Hungarian Borderlands," in Omer Bartov and Eric D. Weitz, eds., *Shatterzone of Empires: Coexistence and Violence in the German, Habsburg, Russian, and Ottoman Borderlands*, 209–27. Bloomington: Indiana University Press, 2013.

Nietzel, Benno. "Im Bann des Raums: Der 'Osten' im deutschen Blick vom 19. Jahrhundert bis 1945," in Guenter Gebhard, Oliver Geisler, and Steffen Schröter, eds., *Das Prinzip "Osten": Geschichte und Gegenwart eines symbolischen Raums*, 21–49. Bielefeld: Transcript, 2010.

Novick, Peter. *That Noble Dream: The "Objectivity Question" and the American Historical Profession*. Cambridge: Cambridge University Press, 1988.

Ó Tuathail, Gearóid [Gerard Toal]. *Critical Geopolitics: The Politics of Writing Global Space*. Minneapolis: University of Minnesota Press, 1996.

O'Donnell, Krista, Renate Bridenthal, and Nancy Reagin, eds. *The Heimat Abroad: The Boundaries of Germanness*. Ann Arbor: University of Michigan Press, 2005.

Oglivie, A. G. "Albrecht Penck, 1858–1945," *Royal Society Year Book* (1951): 46–47.

———. "Isaiah Bowman: An Appreciation," *Geographical Journal* 115, no. 4/6 (April–June 1950): 226–30.

Olsson, Gunnar. *Abysmal: A Critique of Cartographic Reason*. Chicago: University of Chicago Press, 2007.

Orlowski, Hubert. *"Polnische Wirtschaft": Zum deutschen Polendiskurs der Neuzeit*. Wiesbaden: Harrassowitz, 1996.

Orzoff, Andrea. *Battle for the Castle: The Myth of Czechoslovakia in Europe, 1914–1948*. New York: Oxford University Press, 2009.

Osterhammel, Jürgen. *Die Verwandlung der Welt: Eine Geschichte des 19. Jahrhunderts*. München: C. H. Beck, 2009. English translation: *The Transformation of the World: A Global History of the Nineteenth Century*, trans. Patrick Camiller. Princeton: Princeton University Press, 2014.

Ostrowski, Donald G., and Marshall Poe, eds. *Portraits of Old Russia: Imagined Lives of Ordinary People, 1300–1725*. Armonk, N.Y.: M. E. Sharpe, 2011.

Ostrowski, Jerzy, Jacek Pasławski, and Lucyna Szaniawska, eds. *Eugeniusz Romer: geograf i kartograf trzech epok*. Warszawa: Biblioteka Narodowa, 2004.

Oushakine, Serguei Alex. *The Patriotism of Despair: Nation, War, and Loss in Russia*. Ithaca: Cornell University Press, 2009.

Outram, Dorinda. *The Enlightenment*, 3rd ed. Cambridge: Cambridge University Press, 2013.

Padrón, Ricardo. *The Spacious Word: Cartography, Literature, and Empire in Early Modern Spain*. Chicago: University of Chicago Press, 2004.

Panayi, Panikos, ed. *Weimar and Nazi Germany: Continuities and Discontinuities*, new ed. New York: Longman Press, 2014.

Papp-Váry, Árpad. *Magyarország története térképeken*. Budapest: Kossuth Kiadó, 2002.

———. "Teleki Pál a kartográfus," in István Csicsery-Rónay and Károly Vigh, eds., *Teleki Pál és kora: A Teleki Pál emlékév előadásai*, 79–101. Budapest: Occidental Press, 1992.

Parenti, Christian. "The World Was Not Enough," *In These Times*, 28 July 2003.

Partsch, Joseph. "Die Festgaben zu Albrecht Pencks sechzigsten Geburtstage," *Zeitschrift der Gesellschaft für Erdkunde zu Berlin* 53:7–8 (1918): 326–35.

Pasierb, Bronisław. "Profesor Eugeniusz Romer jako konsultant na rokowania pokojowe w Rydze," in Mieczysław Wojciechowski, ed., *Traktat ryski 1921 roku po 75 latach: studia*, 87–107. Toruń: Wydawnictwo Uniwersytetu Mikołaja Kopernika, 1998.

Paul, Allen. *Katyn: Stalin's Massacre and the Triumph of Truth*. DeKalb: Northern Illinois University Press, 2010.

Pawlak, Władysław. "Eugeniusz Romer jako geograf i kartograf," in Lucyna Szaniawska et al., eds., *Eugeniusz Romer: geograf i kartograf trzech epok*, 11–63. Warszawa: Biblioteka Narodowa, 2004.

Pease, Neal. "'This Troublesome Question': The United States and the 'Polish Pogroms' of 1918–9," in Mieczysław Biskupski, ed., *Ideology, Politics, and Diplomacy in East Central Europe*, 58–79. Rochester: University of Rochester Press, 2003.

Pedersen, Susan. *The Guardians: The League of Nations and the Crisis of Empire*. Oxford: Oxford University Press, 2015.

Penck, Walther. *Die morphologische Analyse: Ein Kapitel der physikalischen Geologie*. Stuttgart: Geographische Abhandlungen, 1924. Translated as *Morphological Analysis of Land Forms: A Contribution to Physical Geology* (London: Macmillan, 1953).

Penny, H. Glenn. *Kindred by Choice: Germans and American Indians since 1800.* Chapel Hill: University of North Carolina Press, 2013.

Penrose, Jan. "Nations, States, and Homelands: Territory and Territoriality in Nationalist Thought," *Nations and Nationalism* 8:3 (2002): 277–97.

Perraudin, Michael, and Jürgen Zimmerer, eds. *German Colonialism and National Identity.* London: Routledge, 2011.

Petrovsky-Shtern, Yohanan. *The Anti-imperial Choice: The Making of the Ukrainian Jew.* New Haven: Yale University Press, 2009.

Peukert, Detlev. *The Weimar Republic: The Crisis of Classical Modernity.* New York: Hill and Wang, 1992.

Pickles, John. *A History of Spaces: Cartographic Reason, Mapping, and the Geocoded World.* London: Routledge, 2004.

Pidhainyi, Semyon. "Akademik Stepan Rudnyts'kyi," *Ukrains'kyi visnyk* 152, no. 26 (11 November 1944).

———. *Ukrains'ka intelihentsia na Solovkakh: Spomnyky 1933–34* (Novyi Ul'm: Prometei, 1947), 60–62.

Piskorski, Jan, and Jörg Hackmann. "Polish myśl zachodnia and German Ostforschung: An Attempt at Comparison," in Ingo Haar and Michael Fahlbusch, eds. *German Scholars and Ethnic Cleansing, 1919–1945,* 260–71. New York: Berghahn Books, 2005.

Piskorski, Jan M., Jörg Hackmann, and Rudolf Jaworski, eds. *Deutsche Ostforschung und polnische Westforschung im Spannungsfeld von Wissenschaft und Politik: Disziplinen im Vergleich.* Poznań: Poznańskie Towarzystwo Przyjaciół Nauk, 2002.

Pittaway, Mark. "National Socialism and the Production of German-Hungarian Borderland Space on the Eve of the Second World War," *Past and Present* 216:1 (2012): 143–80.

Piványi, Eugene. *Some Facts about the Proposed Dismemberment of Hungary, with a Map, Statistical Table and Two Appendices.* Cleveland: Hungarian American Federation, 1919.

Plokhii, Serhii. *Unmaking Imperial Russia: Mykhailo Hrushevsky and the Writing of Ukrainian History.* Toronto: University of Toronto Press, 2005.

Porter, Brian A. *When Nationalism Began to Hate: Imagining Modern Politics in Nineteenth-century Poland.* Oxford: Oxford University Press, 2002.

Porter-Szűcs, Brian A. *Faith and Fatherland: Catholicism, Modernity, and Poland.* Oxford: Oxford University Press, 2011.

Pratt, Mary Louise. *Imperial Eyes: Travel Writing and Transculturation.* London: Routledge, 1992.

Potul'nyts'kyi, Volodymyr. "Galician Identity in Ukrainian Historical and

Political Thought," in Chris Hann and Paul Robert Magocsi, eds., *Galicia: A Multicultured Land*, 82–102. Toronto: University of Toronto Press, 2005.

Präsent, Hans von. "Beiträge zur polnischen Landeskunde: Das Quellenmaterial zur Bevölkerungsstatistik Polens," *Zeitschrift der Gesellschaft für Erdkunde zu Berlin*, no. 3 (1917): 245–49.

Promitzer, Christian, Sevasti Trubeta, and Marius Turda, eds. *Health, Hygiene, and Eugenics in Southeastern Europe to 1945*. Budapest: Central European University Press, 2010.

Przyłuska, Barbara. *Atlasy, mapy i globusy Eugeniusza Romera*. Warszawa: Biblioteka Narodowa, 2004.

Przyłuska, Barbara, and Zbigniew Kolek. *Atlasy i mapy Książnicy-Atlas z wyłączeniem prac Eugeniusza Romera*. Warszawa: Biblioteka Narodowa, 2007.

Raj, Kapil. *Relocating Modern Science: Circulation and the Construction of Knowledge in South Asia and Europe, 1650–1900*. New York: Palgrave Macmillan, 2007.

Ramaswamy, Sumathi. *The Goddess and the Nation: Mapping Mother India*. Durham: Duke University Press, 2010.

Randolph, John, and Eugene M. Avrutin, eds. *Russia in Motion: Cultures of Human Mobility since 1850*. DeKalb: University of Illinois Press, 2012.

Rekacewicz, Philippe. "Cartographie radicale," *Le monde diplomatique* (February 2013), 15.

Retallack, James, ed. *Saxony in German History: Culture, Society, and Politics, 1830–1933*. Ann Arbor: University of Michigan Press, 2000.

Riga, Liliana, and James Kennedy, "*Mitteleuropa* as Middle America? 'The Inquiry' and the Mapping of East Central Europe in 1919," *Ab Imperio* 4 (2006): 271–300.

Ringer, Fritz. *The Decline of the German Mandarins: The German Academic Community, 1890–1933*. Cambridge: Harvard University Press, 1969.

———. *Fields of Knowledge: French Academic Culture in Comparative Perspective, 1890–1920*. Oxford: Oxford University Press, 1992.

Risch, William Jay. *The Ukrainian West: Culture and the Fate of Empire in Soviet Lviv*. Cambridge: Harvard University Press, 2011.

Rodgers, Daniel T. *Atlantic Crossings: Social Politics in a Progressive Age*. Cambridge: Harvard University Press, 2000.

Rolf, Malte. "Imperiale Biographien: Lebenswege imperialer Akteure in Gross- und Kolonialreichen, 1850–1918," *Geschichte und Gesellschaft* 40:1 (2014): 5–21.

———. "Importing the 'Spatial Turn' to Russia: Recent Studies on the Spatialization of Russian History," *Kritika* 11:2 (Spring 2010): 359–80.

Romer, Edmund. *Geograf trzech epok: wspomnienia o ojcu.* Warszawa: Czytelnik, 1985.

———. "Życiorys," in *Wspomnienia o ojcu,* 335–42. Warszawa: Czytelnik, 1985.

Romer, Jan Edward. *Pamiętniki.* Warszawa: Muzeum Historii Polski, 2011.

Romer, Tomasz E. "Eugeniusz Romer we wspomnieniach wnuka," in Jerzy Ostrowski, Jacek Pasławski, and Lucyna Szaniawska, eds., *Eugeniusz Romer: Geograf i kartograf trzech epok,* 7–10. Warsaw: Biblioteka Narodowa, 2004.

Romsics, Ignác, ed. *A magyar jobboldal hagyomány 1900–1948.* Budapest: Osiris, 2009.

———. *A trianoni békeszerződés.* Budapest: Osiris, 2007.

———. *Az 1947-es párizsi békeszerzodés.* Budapest: Osiris, 2006.

———. *The Dismantling of Historic Hungary: The Peace Treaty of Trianon, 1920.* New York: Columbia University Press, 2002.

———. *Magyarország története a XX. században.* Budapest: Osiris, 2010.

———. *Mítoszok, legendák, tévhitek a 20. századi magyar történelemrol.* Budapest: Osiris, 2002.

Rónai, András. "Teleki Géza (1911–1983)," *Földrajzi Múzeumi Tanulmányok* 7 (1989): 65–66.

———. "Teleki Pál, a geográfus," *Földrajzi Múzeumi Tanulmányok* 6 (1989): 3–8.

———. "Teleki Pál és a korabeli földtani tudomány." *Magyarkoni Földtani Társulat Tudománytörténeti Szakosztály* 8 (1981): 185–193.

———. *Térképezett történelem,* 2nd ed. Budapest: Püski, 1993.

Rose, Gillian. *Feminism and Geography: The Limits of Geographical Knowledge.* Minneapolis: University of Minnesota Press, 1993.

Rosenbaum, Julia B., and Sven Beckert, eds. *The American Bourgeoisie: Distinction and Identity in the Nineteenth Century.* New York: Palgrave Macmillan, 2010.

Rössler, Mechtild. *"Wissenschaft und Lebensraum": geographische Ostforschung im Nationalsozialismus: ein Beitrag zur Disziplingeschichte der Geographie.* Berlin: D. Reimer, 1990.

Rothschild, Joseph. *East Central Europe between the World Wars,* rev. ed. Seattle: University of Washington Press, 1998.

Rotundo, E. Anthony. *American Manhood: Transformations in Masculinity from the Revolution to the Modern Era.* New York: Basic Books, 1993.

Rudling, Per Anders. *The OUN, the UPA, and the Holocaust: A Study in the Manufacturing of Historical Myths*. Pittsburgh: University of Pittsburgh, 2011.

Sack, Robert David. *Homo Geographicus: A Framework for Action, Awareness, and Moral Concern*. Baltimore: Johns Hopkins University Press, 1997.

Sakmyster, Thomas. *Hungary, the Great Powers, and the Danubian Crisis, 1936–1939*. Athens: University of Georgia Press, 1980.

———. *Hungary's Admiral on Horseback: Miklós Horthy, 1918–1944*. Boulder: East European Monographs, 1984.

Sammartino, Annemarie H. *The Impossible Border: Germany and the East, 1914–1922*. Ithaca: Cornell University Press, 2010.

Sammonds, Jeffrey L. *Ideology, Mimesis, Fantasy: Charles Sealsfield, Friedrich Gerstäcker, Karl May, and Other German Novelists of America*. Chapel Hill: University of North Carolina Press, 1998.

Sandage, Scott A. *Born Losers: A History of Failure in America*. Cambridge: Harvard University Press, 2005.

Schäfer, Ingo. "Der Weg Albrecht Pencks nach München, zur Geographie und zur alpinen Eiszeitforschung," *Mitteilungen des Geographischen Gesellschaft in München* 74 (1989): 5–25.

Schenk, Frithjof Benjamin. "Mental Maps: Die Konstruktion von geographischen Räumen in Europa seit der Aufklärung," *Geschichte und Gesellschaft* 28 (2002): 493–514.

Schlögel, Karl. *Im Raume lese wir die Zeit: Über Zivilisationsgeschichte und Geopolitik*. München: Carl Hanser, 2003.

Schmiedt, Helmut. *Karl May oder die Macht der Phantasie: Eine Biographie*. München: C. H. Beck, 2011.

Schneider, Ute. *Die Macht der Karten: Eine Geschichte der Kartographie vom Mittelalter bis heute*. Darmstadt: Primus, 2004.

Schöpflin, George, and Nancy Wood, eds. *In Search of Central Europe*. Cambridge: Polity Press, 1989.

Schulten, Susan. *The Geographical Imagination in America, 1880–1950*. Chicago: University of Chicago Press, 2001.

———. *Mapping the Nation: History and Cartography in Nineteenth-century America*. Chicago University of Chicago Press, 2012.

Schultz, Hans-Dietrich. "Albrecht Penck—herausagender Forscher und Mitbegründer einer neuzeitlichen naturwissenschaftlichen Geographie," *Wissenschaftliche Zeitschrift der Humboldt-Universität zu Berlin; Mathem.-Naturwiss. Reihe* 39:3 (1990): 257–63.

———. "Deutschlands 'natuerliche' Grenzen," in Reimer Hansen and Alexander Demandt, eds., *Deutschlands Grenzen in der Geschichte* (München: Beck, 1993), 32–93.

———. "'Ein wachsendes Volk braucht Raum': Albrecht Penck als politischer Geograph," in Bernhard Nitz, Hans-Dietrich Schultz, and Marlies Schultz, eds., *1810–2010: 200 Jahre Geographie in Berlin*, 99–153. Berlin: Geographisches Institut der Humboldt-Universität zu Berlin, 2011.

Schwartz, Agatha, ed., *Gender and Modernity in Central Europe: The Austro-Hungarian Monarchy and Its Legacy*. Ottawa: University of Ottawa Press, 2010.

Schweiger, Alexandra. *Polens Zukunft liegt im Osten: Polnische Ostkonzepte der späten Teilungszeit, 1890–1918*. Marburg: Herder-Institut, 2014.

Seegel, Steven. *Mapping Europe's Borderlands: Russian Cartography in the Age of Empire*. Chicago: University of Chicago Press, 2012.

———. "Remapping the Geo-Body: Transnational Dimensions of Stepan Rudnyts'kyi and His Contemporaries," in Serhii Plokhy, ed., *The Future of the Past: New Perspectives on Ukrainian History* (Cambridge: Harvard University Press, 2016), 205–29.

Self, Will. *Psychogeography: Disentangling the Modern Conundrum of Psyche and Place*. New York: Bloomsbury USA, 2007.

Seton-Watson, Hugh, and Christopher Seton-Watson. *The Making of a New Europe: R. W. Seton-Watson and the Last Years of Austria-Hungary*. Seattle: University of Washington Press, 1981.

Shablii, Oleh. *Akademik Stepan Rudnyts'kyi: fundator Ukrains'koi heohrafii, heohrafii*, 2nd ed. Lviv: Vydavnychyi Tsentr LNU im. Ivana Franka, 2007.

———. *Ukraina prostovora v kontseptsiynomu Stepan Rudnyts'koho: monografiia*. Kyiv: Ukrains'ka Vydavnycha Spilka, 2003.

———, ed. *Doktor heohrafii Hryhorii Velychko*. Lviv: Vydavnychyi Tsentr LNU im. Ivana Franka, 2012.

Shallcross, Bożena. *Framing the Polish Home: Postwar Cultural Constructions of Heart, Nation, and Self*. Athens: Ohio University Press, 2002.

Shapoval, Iurii, and Serhii Bohunov, eds. *Ostannia adresa: rozstrily solovets'kykh v'iazniv z Ukrainy u 1937–1938 rokakh*, 2 vols. Kyiv: Sfera, 2003.

Shapoval, Iurii, and Hiroaki Kuromiya, eds. *Ukraina v dobu "Velykoho teroru" 1936–1938 roky*. Kyiv: Lybid', 2009.

Sharp, Alan. *The Versailles Settlement: Peacemaking After the First World War, 1919–1923*, 2nd ed. London: Palgrave Macmillan, 2008.

Shields, Sarah. "Mosul, the Ottoman Legacy, and the League of Nations," *International Journal of Contemporary Iraqi Studies* 3:2 (2009): 217–30.

Shore, Marci. *Caviar and Ashes: A Warsaw Generation's Life and Death in Marxism, 1918–1968*. New Haven: Yale University Press, 2006.

———. *The Taste of Ashes: The Afterlife of Totalitarianism in Eastern Europe*. New York: Crown, 2013.

Shtoiko, Pavlo. *Stepan Rudnyts'kyi, 1887–1937: Zhyttepysno-bibliohrafichnyi narys*. Lviv: NTSh, 1997.

———. *Stepan Rudnyts'kyi: Vyzhachni diiachi NTSh*. Lviv: NTSh, 1997.

Silk, Leonard, and Mark Silk. *The American Establishment*. New York: Basic Books, 1980.

Skelton, R. A. *Maps: A Historical Survey of Their Study and Collecting*. Chicago: University of Chicago Press, 1975.

Slotkin, Richard. *The Fatal Environment: The Myth of the Frontier in the Age of Industrialization, 1800–1890*. New York: Harper Perennial, 1985.

———. *Gunfighter Nation: The Myth of the Frontier in Twentieth-century America*. New York: Harper Perennial, 1992.

Smith, Bonnie G. *The Gender of History: Men, Women, and Historical Practice*. Cambridge: Harvard University Press, 2000.

———, ed. *Women's History in Global Perspective*, 3 vols. Urbana: University of Illinois Press, 2004–5.

Smith, Neil. "'Academic War over the Field of Geography': The Elimination of Geography at Harvard, 1947–1951," *Annals of the Association of American Geographers* 77:2 (June 1987): 155–72.

———. *American Empire: Roosevelt's Geographer and the Prelude to Globalization*. Berkeley: University of California Press, 2003.

———. "Bowman's New World and the Council on Foreign Relations," *Geographical Review* 76:4 (October 1986): 438–60.

Smith, Richard J. *Mapping China and Managing the World: Culture, Cartography and Cosmology in Late Imperial Times*. London: Routledge, 2013.

Smith, Sidonie, and Julia Watson. *Reading Autobiography: A Guide for Interpreting Life Narratives*, 2nd ed. Minneapolis: University of Minnesota Press, 2010.

Smith, Woodruff D. "Friedrich Ratzel and the Origins of Lebensraum," *German Studies Review* 3 (1980): 51–68.

Snyder, Timothy. *Black Earth: The Holocaust as History and Warning*. New York: Tim Duggan Books, 2015.

———. *Bloodlands: Europe between Hitler and Stalin*. New York: Basic Books, 2010.

————. *The Reconstruction of Nations: Poland, Ukraine, Lithuania, Belarus, 1569–1999*. New Haven: Yale University Press, 2003.

————. *The Red Prince: The Secret Lives of a Habsburg Archduke*. New York: Basic Books, 2008.

Sobczak, Janusz. *Propaganda zagraniczna niemiec weimarskich wobec Polski*. Poznań: Instytut Zachodni, 1973.

Solnit, Rebecca. *Wanderlust: A History of Walking*. New York: Penguin Books, 2001.

Sölch, Johann. "Albrecht Penck," *Mitteilungen der Geographische Gesellschaft zu Wien* 89 (1946): 88–122.

Somogyi, Sándor. "Cholnoky Jenő és Teleki Pál amerikai tanulmányútja," *Földrajzi Múzeumi Tanulmányok*, vol. 7 (1989): 43–48.

Sonevytsky, Leonid C. "The Ukrainian Question in R. H. Lord's Writings on the Paris Peace Conference in 1919," *Annals of the Ukrainian Academy of Arts and Sciences in the U.S.*, 10:1–2 (1962): 68–84.

Sossa, Rostyslav. *Istoriia kartohrafyvannia terytorii Ukrainy*. Kyiv: Lybid', 2007.

Spain, Daphne. "The Spatial Foundations of Men's Friendships and Men's Power," in Peter Mardi, ed., *Men's Friendships*, 59–73. Newbury Park, Calif.: Sage, 1992.

Spector, Scott. *Prague Territories: National Conflict and Cultural Innovation in Franz Kafka's Fin de Siècle*. Berkeley: University of California Press, 2002.

Spreitzer, Hans. "Albrecht Pencks letztes Lebensjahr: Erinnerung an einen grossen Forscher und Lehrer," *Zeitschrift für Gletscherkunde und Glazialgeologie* 1 (1950): 187–92.

————. "Albrecht Pencks Wirk in Wien," *Mitteilungen Österreichen Geographischen Gesellschaft* 101 (1959): 375–80.

Staeheli, Lynn A. "Place," in John Agnew, Katharyne Mitchell, and Gerard Toal, eds., *A Companion to Political Geography*, 2nd ed., 158–70. London: Blackwell, 2008.

Staeheli, Lynn A., Eleonore Kofman, and Linda J. Peake, eds. *Mapping Women, Making Politics: Feminist Perspectives on Political Geography*. Routledge: London, 2004.

Statiev, Alexander. *The Soviet Counterinsurgency in the Western Borderlands*. Cambridge: Cambridge University Press, 2010.

Stebelsky, Ihor. *Placing Ukraine on the Map: Stepan Rudnyts'kyi's Nation-Building Geography*. Self-published by the author, 2014.

————. "Putting Ukraine on the Map: The Contribution of Stepan Rudnytsky to Ukrainian Nation-building." *Nationalities Papers* 39:4 (July 2011): 587–613.

Stefanov, Nenad. *Wissenschaft als nationaler Beruf: Die Serbische Akademie der Wissenschaften 1944–1992: Tradierung und Modifizierung nationaler Ideologie.* Wiesbaden: Harrasowitz, 2011.

Steinberg, Mark D., and Valeria Sobol, eds. *Interpreting Emotions in Russia and Eastern Europe.* DeKalb: Northern Illinois University Press, 2011.

Steinweis, Alan E. "Eastern Europe and the Notion of the 'Frontier' in Germany to 1945," *Yearbook of European Studies* 13 (1999): 56–69.

Stern, Fritz. *Dreams and Delusions: The Drama of German History.* New York: Knopf, 1987.

———. *The Politics of Cultural Despair: A Study in the Rise of the Germanic Ideology.* Berkeley: University of California Press, 1961.

Stoler, Ann Laura. *Along the Archival Grain: Epistemic Anxieties and Colonial Common Sense.* Princeton: Princeton University Press, 2009.

———. *Carnal Knowledge and Imperial Power: Race and the Intimate in Colonial Rule.* Berkeley: University of California Press, 2002.

———, ed. *Haunted by Empire: Geographies of Intimacy in North American History.* Durham: Duke University Press, 2006.

Stoler, Ann Laura, and Frederick Cooper, eds. *Tensions of Empire: Colonial Cultures in a Bourgeois World.* Berkeley: University of California Press, 1997.

Sugar, Peter F., Péter Hanák, and Tibor Frank. *A History of Hungary.* Bloomington: Indiana University Press, 1994.

Sunderland, Willard. *The Baron's Cloak: A History of the Russian Empire in War and Revolution.* Ithaca: Cornell University Press, 2014.

———. "The Emperor's Men at the Emperor's Edges," *Kritika: Explorations in Russian and Eurasian History* 5:3 (2004): 515–25.

Sunderland, Willard, and Stephen Norris, eds. *Russia's People of Empire: Life Stories from Eurasia, 1500 to the Present.* Bloomington: Indiana University Press, 2012.

Syrett, Nicholas L. *The Company He Keeps: A History of White College Fraternities.* Chapel Hill: University of North Carolina Press, 2009.

Sziklay, Andor. *Magyar lábnyomok: körkep Washingtonból.* [U.S.A.]: Liberty Media, 1988.

Szporluk, Roman. *The Political Thought of Thomas G. Masaryk.* Boulder, Colo.: East European Monographs, 1981.

Taylor, Charles. *Sources of the Self: The Making of the Modern Identity.* Cambridge: Harvard University Press, 1989.

Tazbir, Janusz. *Kultura szlachecka w Polsce: rozkwit, upadek, relikty.* Poznań: Wydawnictwo Poznańskie, 1998.

Theweleit, Klaus. *Male Fantasies*, 2 vols. Minneapolis: University of Minnesota Press, 1987–89.

Thrower, Norman J. W. *Maps and Civilization: Cartography in Culture and Society*. Chicago: University of Chicago Press, 1999.

Thum, Gregor. "Megalomania and Angst: The Nineteenth-century Mythicization of Germany's Eastern Borderlands," in Omer Bartov and Eric D. Weitz, eds., *Shatterzone of Empires: Coexistence and Violence in the German, Habsburg, Russian, and Ottoman Borderlands*, 42–60. Bloomington: Indiana University Press, 2013.

———, ed. *Traumland Osten: Deutsche Bilder vom östlichen Europa im 20. Jahrhundert*. Göttingen: Vandenhoeck & Ruprecht, 2006.

———. *Uprooted: How Breslau Became Wrocław during the Century of Expulsions*. Princeton: Princeton University Press, 2011.

Tilkovszky, Lóránt. *Pál Teleki, 1879–1941: A Biographical Sketch*. Budapest: Akadémiai Kiadó, 1974.

Tilley, Helen. *Africa as a Living Laboratory: Empire, Development, and the Problem of Scientific Knowledge, 1870–1950*. Chicago: University of Chicago Press, 2011.

Tisseyre, Charles. *An Error in Diplomacy: Dismembered Hungary*. Paris: Mercure, 1924.

Todorova, Maria. "The Trap of Backwardness: Modernity, Temporality, and the Study of Eastern European Nationalism," *Slavic Review* 64:1 (Spring 2005): 140–64.

Tolz, Vera. *Russian Academicians and the Revolution: Combining Professionalism and Politics*, 2nd ed. New York Palgrave Macmillan, 2014.

Török, Zsolt. *Teleki Pál: A kísérő tanulmányt írta, a szövegeket vál*. Budapest: Országos Pedagógiai Könyvtár és Múzeum, 2001.

Trautmann, Györgyné. [Bibliography], in István Csicsery-Rónay, and Károly Vigh, eds., *Teleki Pal és kora: A Teleki Pál emlékév előadásai* (Budapest: Occidental Press, 1992), 187–222.

Tripp, Charles. *A History of Iraq*. Cambridge: Cambridge University Press, 2007.

Tuan, Yi-Fu. *Space and Place: The Perspective of Experience*. Minneapolis: University of Minnesota Press, 2001.

Tuathail, Gearóid Ó. *See* Ó Tuathail.

Tumblety, Joan. *Remaking the Male Body: Masculinity and the Uses of Physical Culture in Interwar and Vichy France*. Oxford: Oxford University Press, 2012.

Turda, Matthias, and Paul J. Weindling, eds. *"Blood and Homeland": Eugenics and Racial Nationalism in Central and Southeast Europe, 1900–1940*. Budapest: Central European University Press, 2007.

Tymowski, Stanisław Janusz. "Częściowy wykaz strat wojennych środowiska mierniczego w okresie 1939–1945," in *Zarys historii organizacji społecznych geodetów polskich*, 152–76. Warszawa: Państwowe Przedsiębiorstwa Wydawnictw Kartograficznych; Stowarzyszenie Geodetów Polskich, 1970.

Uhorczak, Franciszek, and Józef Wąsowicz. "Bibliografia adnotowana dzieł kartograficznych E. Romera," in Eugeniusz Romer, *Wybór prac*, vol. 1 (Warszawa, 1960), 137–54.

Urry, John *Mobilities*. London: Polity, 2007.

Vick, Brian. *Defining Germany: The 1848 Frankfurt Parliamentarians and National Identity*. Cambridge: Harvard University Press, 2002.

Vigh, Károly, ed. *Ismeretlen fejzetek Teleki Pál életéből*. Budapest: Szazadveg Kiadó, 2001.

Vushko, Iryna. *The Politics of Cultural Retreat: Imperial Bureaucracy in Austrian Galicia, 1772–1867*. New Haven: Yale University Press, 2015.

Wandycz, Piotr S. *France and Her Eastern Allies, 1919–1925: French-Czechoslovak-Polish relations from the Paris Peace Conference to Locarno*. Minneapolis: University of Minnesota Press, 1962.

———. "The Polish Question," in Manfred F. Boemeke, Gerald D. Feldman, and Elisabeth Glaser, eds., *The Treaty of Versailles: A Reassessment after 75 Years*, 313–35. Cambridge: Cambridge University Press, 1998.

———. *The United States and Poland*. Cambridge: Harvard University Press, 1980.

Wardenga, Ute. "Vor 125 Jahren—Albrecht Penck weist eine dreimalige Vereisung Norddeutschlands nach," *Petermanns Geographische Mitteilungen* 148:3 (2004): 94–95.

Wardenga, Ute, Norman Henniges, Heinz-Peter Brogiato, and Bruno Schelhaas, eds. *Der Verband deutscher Berufsgeographen 1950–1979: Eine sozialgeschichtliche Studie zur Frühphase des DVAG*. Leipzig: Leibniz-Institut für Landeskunde, 2011.

Wardenga, Ute, and Ingrid Hönsch, eds., *Kontinuität und Diskontinuität der deutschen Geographie in Umbruchphasen: Studien zur Geschichte der Geographie*. Münster: Institut für Geographie der Westfälischen Wilhelms-Universität, 1995.

Warf, Barney, and Santa Arias, eds., *The Spatial Turn: Interdisciplinary Perspectives*. New York: Routledge, 2009.

Wąsowicz, Józef. "Kartografia romerowska," *Czasopismo Geograficzne* 26:1–2 (1955): 167–79.

Weber, Eugen. *Peasants into Frenchmen: The Modernization of Rural France, 1870–1914*. Stanford: Stanford University Press, 1976.

Weiner, Amir, ed. *Landscaping the Human Garden: Twentieth-century Population Management in a Comparative Framework*. Stanford: Stanford University Press, 2003.

Weitz, Eric D. *Weimar Germany: Promise and Tragedy*. Princeton: Princeton University Press, 2007.

Wendland, Anna Veronika. "Ikonografen des Raumbilds Ukraine: Eine europäische Wissenstransfergeschichte," in Peter Haslinger and Vadim Oswalt, eds., *Kampf der Karten: Propaganda- und Geschichtskarten als politische Instrumente und Identitätstexte*, 85–120. Marburg: Herder-Institut, 2012.

Werner, Michael, and Bénédicte Zimmermann. "Beyond Comparison: Histoire Croisée and the Challenge of Reflexivity," *History and Theory* 45 (2006): 30–50.

———. "Vergleich, Transfer, Verflechtung: Der Ansatz der Histoire croisée und die Herausforderungdes Transnationalen," *Geschichte und Gesellschaft* (October–December 2002) 28:4: 607–36.

Whisnant, Clayton J. *Queering Identities and Politics in Germany: A History, 1880–1945*. New York: Harrington Park Press, 2016.

Wigen, Kären. *A Malleable Map: Geographies of Restoration in Central Japan, 1600–1912*. Berkeley: University of California Press, 2010.

Wigen, Kären, Sugimoto Fumiko, and Cary Karacas, eds. *Cartographic Japan: A History in Maps*. Chicago: University of Chicago Press, 2016.

Winid, Bohdan. "Profesor Eugeniusz Romer–jakiego znałem," in Dobiesław Jędrzejczyk and Waldemar Wilk, eds., *Eugeniusz Romer jako geograf społeczno-gospodarczy*, 16–18. Warszawa: Uniwersytet Warszawski, Wydział Geografii i Studiów Regionalnych, 1999.

Wingfield, Nancy M. *Creating the Other: Ethnic Conflict and Nationalism in Habsburg Central Europe*. New York: Berghahn Books, 2003.

———. *Flag Wars and Stone Saints: How the Bohemian Lands Became Czech*. Cambridge: Harvard University Press, 2007.

Winichakul, Thongchai. *Siam Mapped: A History of the Geo-body of a Nation*. Honolulu: University of Hawai'i Press, 1994.

Wippermann, Wolfgang. *Die Deutschen und der Osten: Feinbild und Traumland*. Darmstadt: Primus, 2007.

Withers, Charles W. J. *Geography and Science in Britain, 1831–1939: A Study of the British Association for the Advancement of Science*. Manchester: Manchester University Press, 2010.

———. "Scale and the Geographies of Civic Science: Practice and Experience in the Meetings of the British Association for the Advancement of Science in Britain and in Ireland, c. 1845–1900," in David N. Livingstone and Charles W. J. Withers, eds., *Geographies of Nineteenth-century Science*, 99–122. Chicago: University of Chicago Press, 2011.

Wolff, Larry. *The Idea of Galicia: History and Fantasy in Habsburg Political Culture*. Stanford: Stanford University Press, 2010.

———. *Inventing Eastern Europe: The Map of Civilization on the Mind of the Enlightenment*. Stanford: Stanford University Press, 1994.

———. "The Traveler's View of Central Europe: Gradual Transitions and Degrees of Difference in European Borderlands," in Omer Bartov and Eric D. Weitz, eds., *Shatterzone of Empires: Coexistence and Violence in the German, Habsburg, Russian, and Ottoman Borderlands*, 23–71. Bloomington: Indiana University Press, 2013.

Wolff, Larry, and Marco Cipolloni, eds. *The Anthropology of the Enlightenment*. Stanford: Stanford University Press, 2007.

Wolff, Stefan. *Disputed Territories: The Transnational Dynamics of Ethnic Conflict Settlement*. New York: Berghahn Books, 2004.

Wood, Denis, and John Fels. *The Natures of Maps: Cartographic Constructions of the Natural World*. Chicago: University of Chicago Press, 2008.

———. *The Power of Maps*. New York: Guilford Press, 1992.

———. *Rethinking the Power of Maps*. New York: Guilford Press, 2010.

Worobec, Christine. *The Human Tradition in Imperial Russia*. Lanham, Md.: Rowman & Littlefield, 2009.

Wright, John Kirtland. *Geography in the Making: The American Geographical Society, 1851–1951*. New York: American Geographical Society, 1952.

Wrigley, Gladys. "Isaiah Bowman," *Geographical Review* 41 (1951): 7–65.

Wroniak, Zdzisław. "Rola delegacji polskiej na Konferencję Paryską w ustaleniu polskiej granicy zachodniej," in Janusz Pajewski, ed., *Problem polsko-niemiecki w Traktacie Wersalskim: praca zbiorowa*, 219–70. Poznań, Instytut Zachodni, 1963.

Wulf, Andrea. *The Invention of Nature: Alexander von Humboldt's New World*. New York: Knopf, 2015.

Wunderlich, Dr. E[rich], ed. *Handbuch von Polen (Kongress-Polen): Beiträge zu einer allgemeinen Landeskunde, Veröffentlichungen der Landeskundlichen Kommission beim Kaiserl. Deutschen Generalgouvernement Warschau*. Berlin: D. Reimer, 1918.

Yee, Vivian. "The Death of a Countess in Exile," *New York Times*, 26 April 2013.

Yekelchyk, Serhy. *Stalin's Empire of Memory: Russian-Ukrainian Relations in the Soviet Historical Imagination*, 2nd ed. Toronto: University of Toronto Press, 2004.

————. *Ukraine: Birth of a Modern Nation*. Oxford: Oxford University Press, 2007.

Zahra, Tara. *The Great Departure: Mass Migration from Eastern Europe and the Making of the Free World*. New York: W.W. Norton, 2016.

————. "Imagined Non-communities: National Indifference as a Category of Analysis," *Slavic Review* 69 (Spring 2010): 93–119.

————. "Looking East: East Central European Borderlands in German History and Historiography," *History Compass* 3 (2005): 1–23.

Zakar, András. *Teleki Pal halála: Mitosz és rejtély*. Budapest: Kairosz Kiadó, 2009.

Zaremba, Marcin. *Im nationalen Gewande Strategien kommunistischer Herrschaftslegitimation in Polen 1944–1980*. Osnabrück: Fibre, 2011.

Zeidler, Miklós. *A revíziós gondolat*. Pozsony: Kalligram, 2009.

————. *Ideas on Territorial Revision in Hungary, 1920–45*, trans. Thomas and Helen DeKornfeld. Boulder: Social Science Monographs, 2008.

Zieliński, Władysław. *Polska i niemiecka propaganda plebiscytowa na Górnym Śląsku*. Wrocław: Zakład Narodowy im. Ossolińskich, 1972.

Zierhoffer, August. "Eugeniusz Romer," *Przegląd Geograficzny* 26:2 (1954): 178–82.

Zögner, Lothar. "Albrecht Penck und die Kartographie," *Kartographische Nachrichten* 59:4 (August 2009): 206–8.

INDEX

The letter *f* following a page number denotes a figure.

Edvi Illés, Aladár, 87
Enabling Act (Ermächtigungsgesetz) of 1933, 143
Engelhorn (publisher), 246n2
erosion cycle. *See* Davis, William Morris: erosion
cycle
Evans, Richard J., 3

Fodor, Ferenc, 88, 115, 118, 130, 209, 263n30
Foerster, Friedrich Wilhelm, 199
Földrajzi Közlemények. See *Geographical Bulletin*
Franc Affair (1925), 135–37
Francis I, Emperor, 18
Francis Joseph I, Emperor, 17, 49
Franko, Ivan, 109
Friedrich Wilhelm III, 59

Galicia: Habsburg, 23, 110; Polish-Ukrainian ten-
sion in, 25, 109; Romer's views on, 90, 168;
as a zone of conflict, 54. *See also* Poland: bor-
ders after World War I
Gallois, Lucien, 12
Ganghofer, August von, 16
Ganghofer, Gastel, 202
Ganghofer, Ludwig von, 16
Garrigue, Charlotte, 67
Geisler, Walther, 147–48
Gelfand, Lawrence, 218
geo-body, 239n30
Geographical Bulletin, 36, 51
geography: Bolsheviks and, 109; epistolary, 6–7;
in Poland in the 1920s, 99–101; as a science,
3, 139–40, 227; World War I and, 6
Geography Department, Harvard University, 213
Geography Department, Jagiellonian University,
243n31
geopolitics, 227
German expansionism, post–World War I, 126–27
Gerő, László, Major, 136–37
Gerwig, Walter H., Jr., 222
Gilbert, Zsuzsanna (Suzanne), 220
Gilman, Daniel Coit, 154
Glatzel, Jan Karol, 195
Goetz, Walter, 135
Goffman, Erving, 8
Gömbös, Gyula, 180
Górny, Maciej, 9
Gottmann, Jean, 193, 213
Grabski, Stanisław, 195, 208–9
Gross, Jan T., 257n59

Haar, Ingo, 121
Häberle, Dietrich, 63
Haits, Lajos, General, 136–37
Halász, Albert, 87

Halász, Gyula, 263n30
Harding, Warren G., 113
Hartshorne, Richard, 157, 194
Haslinger, Peter, 7
Haushofer, Karl, 97, 108, 126, 128, 132, 230
Hausmann, Guido, 7
Hayden, Philip M., 211
Hedin, Sven, 14, 42, 176–77, 179–80, 203–4
Heinrich, Francis de, 87
Hennig, Richard, 132
Henniges, Norman, 9
Herb, Guntram Henrik, 7, 61
Herder, Johann Gottfried von, 29, 51
Herynovych, Volodymyr, 111, 141
Hessler, John, 274n45
Hettner, Alfred, 93, 103, 247n22
Heyde, Herbert, 62, 80
Hirsch, Francine, 109
Hochschild, Adam, 5
Holodomor (1932–33), 143
Holquist, Peter, 8
Horn, Ernest, 156, 161, 190
Horthy, Miklós, Admiral, 88, 113, 180–81
House, Edward M., 49, 74, 174
Hrushevs'kyi, Mykhailo, 23–24, 109, 110
Hübl, Alfred, General, 68
Hughes, Charles Evans, 49, 113
Hunfalvy, Janos, 36
Hungarian Academy of Sciences (Magyar Tu-
dományos Akadémia, or MTA), 37, 128
Hungarian Geographical Museum (Magyar Földra-
jzi Muzeúm), 220–21
Hungarian Geographical Society (Magyar Földra-
jzi Társaság, or MFT), 36–37, 51, 119, 248n31
Hungarian Revisionist League, 130
Hungarian Soviet Republic, 67, 85
Hungary: annexation of part of Romania (Second
Vienna Award, 1940), 186; annexation of South-
ern Slovakia (First Vienna Award, 1938), 181;
autonomy, 37, 65–66 (*see also* Trianon, Treaty
of); border disputes in 1918–19, 66–67 (*see also*
Trianon, Treaty of); borders after World War II,
210–11; Carpathian Rus' and, 181–82; neutrality in World War II, 182, 187; treaty with Yugoslavia (1940), 186–87

Imrédy, Béla, 180–81
Inquiry, The (1917), 11, 49, 74, 77–78, 91, 133, 157,
173–74
Institute for Geographical Exploration, Harvard
University, 213
Institute of Geography, Kyiv University, 163
Institute of Political Sciences, Hungarian Academy
of Sciences, 128